T0099972

WATER POLLUTION ISSUES AND DEVELOPMENTS

WATER POLLUTION ISSUES AND DEVELOPMENTS

SARAH V. THOMAS
EDITOR

Nova Science Publishers, Inc.
New York

LIBRARY OF CONGRESS CATALOGING-IN-PUBLICATION DATA

Water pollution issues and developments / Sarah V. Thomas, editor.
  p. cm.
Includes bibliographical references and index.
ISBN 978-1-60456-208-8 (hardcover)
1. Water--Pollution. I. Thomas, Sarah V. II. Copeland, Claudia.
TD420.W36   2007
363.739'4--dc22   2007047041

Published by Nova Science Publishers, Inc. ✣ New York

CONTENTS

PREFACE

Pollution is undesirable state of the natural environment being contaminated with harmful substances as a consequence of human activities so that the environment becomes harmful or unfit for living things; especially applicable to the contamination of soil, water, or the atmosphere by the discharge of harmful substances. In addition to the harm to living beings, both present or future and known or unknown, pollution cleanup and surveillance are enormous financial drains of the economies of the world. This book focuses on issues and developments critical for the field.

Chapter 1 - The cruise industry is a significant and growing contributor to the U.S. economy, providing more than $32 billion in benefits annually and generating more than 330,000 U.S. jobs, but also making the environmental impacts of its activities an issue to many. Although cruise ships represent a small fraction of the entire shipping industry worldwide, public attention to their environmental impacts comes in part from the fact that cruise ships are highly visible and in part because of the industry's desire to promote a positive image.

Cruise ships carrying several thousand passengers and crew have been compared to "floating cities," and the volume of wastes that they produce is comparably large, consisting of sewage; wastewater from sinks, showers, and galleys (graywater); hazardous wastes; solid waste; oily bilge water; ballast water; and air pollution. The waste streams generated by cruise ships are governed by a number of international protocols (especially MARPOL) and U.S. domestic laws (including the Clean Water Act and the Act to Prevent Pollution from Ships), regulations, and standards, but there is no single law or rule. Some cruise ship waste streams appear to be well regulated, such as solid wastes (garbage and plastics) and bilge water. But there is overlap of some areas, and there are gaps in others. Some, such as graywater and ballast water, are not regulated (except in the Great Lakes), and concern is increasing about the impacts of these discharges on public health and the environment. In other areas, regulations apply, but critics argue that they are not stringent enough to address the problem — for example, with respect to standards for sewage discharges. Environmental advocates have raised concerns about the adequacy of existing laws for managing these wastes, and they contend that enforcement is weak.

In 2000, Congress enacted legislation restricting cruise ship discharges in U.S. navigable waters within the state of Alaska. California, Alaska, and Maine have enacted state-specific laws concerning cruise ship pollution, and a few other states have entered into voluntary agreements with industry to address management of cruise ship discharges. Meanwhile, the cruise industry has voluntarily undertaken initiatives to improve pollution prevention, by adopting waste management guidelines and procedures and researching new technologies. Concerns about cruise ship pollution raise issues for Congress in three broad areas: adequacy of laws and regulations, research needs, and oversight and enforcement of existing requirements. Legislation to regulate cruise ship discharges of sewage, graywater, and bilge water nationally was introduced in the 109th Congress, but there was no further congressional action.

This report describes the several types of waste streams that cruise ships may discharge and emit. It identifies the complex body of international and domestic laws that address pollution from cruise ships. It then describes federal and state legislative activity concerning cruise ships in Alaskan waters and recent activities in a few other states, as well as current industry initiatives to manage cruise ship pollution. Issues for Congress are discussed.

Chapter 2 - As gasoline prices have risen in March and April 2006, renewed attention has been given to methyl tertiary butyl ether (MTBE), a gasoline additive being phased out of the nation's fuel supply. Many argue that the phaseout of MTBE and its replacement by ethanol have been a major factor in driving up prices.

MTBE has been used by refiners since the late 1970s. It came into widespread use when leaded gasoline was phased out — providing an octane boost similar to that of lead, but without fouling the catalytic converters used to reduce auto emissions since the mid-1970s. MTBE has also been used to produce cleaner-burning Reformulated Gasoline (RFG), which the Clean Air Act has required in the nation's most polluted areas since 1995. The act didn't mandate the use of MTBE (ethanol or other substances could have been used to meet the act's oxygenate requirement), but price and handling characteristics of the additive led to its widespread use.

Under the Energy Policy Act of 2005 (P.L. 109-58), the RFG program's oxygen mandate terminates on May 6, 2006, and refiners are scrambling to remove MTBE from the nation's gasoline supply by that date. The phaseout of MTBE (like its use) is not required by federal law, but gasoline refiners have focused on the May 6 date because of concerns over their potential liability for its continued presence. MTBE has contaminated drinking water in a number of states, and about half have passed legislation to ban or restrict its use. Hundreds of suits have been filed to require petroleum refiners and marketers to pay for cleanup of contaminated water supplies, the cost of which has been estimated to be in the billions of dollars. The petroleum industry has maintained that it used MTBE to meet the RFG program's oxygen mandate and therefore should not be held liable. That position could become more difficult to maintain once the oxygen mandate is removed.

To replace MTBE, refiners are switching to ethanol as swiftly as they can, leading temporarily to supply shortages and higher prices. The ethanol industry maintains that there will be sufficient ethanol to meet demand but concedes that temporary shortages exist in some parts of the country that could affect prices until the end of June. These shortages and higher prices have led to renewed discussion by some of exempting gasoline refiners from liability for MTBE cleanup (a so-called "safe harbor" provision). Others have renewed their call for federal legislation to stimulate the construction of new refining capacity.

Besides removing the RFG program's oxygen requirement, Congress provided a major incentive to the production of ethanol in the Energy Policy Act of 2005. Under a Renewable Fuels Standard, an increasing amount of the nation's motor fuels must consist of renewable fuel, such as ethanol. The law requires 4.0 billion gallons in 2006 (a level already being achieved) and an increase of 700 million gallons each year through 2011, before reaching 7.5 billion gallons in 2012.

This report provides background regarding MTBE and summarizes the actions taken by states and Congress to address problems raised by MTBE contamination of the nation's water supplies. It will be updated if future developments warrant.

Chapter 3 - According to the Environmental Protection Agency, the release of waste from animal feedlots to surface water, groundwater, soil, and air is associated with a range of human health and ecological impacts and contributes to degradation of the nation's surface waters. The most dramatic ecological impacts are massive fish kills. A variety of pollutants in animal waste can affect human health, including causing infections of the skin, eye, ear, nose, and throat. Contaminants from manure can also affect human health by polluting drinking water sources.

Although agricultural activities are generally not subject to requirements of environmental law, discharges of waste from large concentrated animal feeding operations (CAFO) into the nation's waters are regulated under the Clean Water Act. In the late 1990s, the Environmental Protection Agency (EPA) initiated a review of the Clean Water Act rules that govern these discharges, which had not been revised since the 1970s, despite structural and technological changes in some components of the animal agriculture industry that have occurred during the last two decades. A proposal to revise the existing rules was released by the Clinton Administration in December 2000. The Bush Administration promulgated final revised regulations in December 2002; the rules took effect in February 2003.

The final rules were generally viewed as less stringent than the proposal, a fact that strongly influenced how interest groups have responded to them. Agriculture groups said that the final rules were workable, and they were pleased that some of the proposed requirements were scaled back, such as changes that would have made thousands more CAFOs subject to regulation. However, some continue to question EPA's authority to issue portions of the rules. Many states had been seeking more flexible approaches than EPA had proposed and welcomed the fact that the final rules retain the status quo to a large extent. Environmentalists contended that the rules relied too heavily on voluntary measures and fail to require improved technology.

This report provides background on the revised environmental rules, the previous Clean Water Act rules and the Clinton Administration proposal, and perspectives of key interest groups on the proposal and final regulations. It also identifies several issues that could be of congressional interest as implementation of the revised rules proceeds. Issues include adequacy of funding for implementing the rules, research needs, oversight of implementation of the rules, and possible need for legislation.

The revised CAFO rules were challenged by multiple parties, and in February 2005, a federal court issued a ruling that upheld major parts of the rules, vacated other parts, and remanded still other parts to EPA for clarification. In June 2006, EPA proposed revisions to the rules in response to the 2005 court decision; for information on the status of this proposal, see CRS Report RL33656, *Animal Waste and Water Quality: EPA's Response to the "Waterkeeper Alliance" Court Decision on Regulation of CAFOs*, which will be updated as warranted by developments.

Chapter 4 - On June 30, 2006, the Environmental Protection Agency (EPA) proposed regulations that would revise a 2003 Clean Water Act rule governing waste discharges from large confined animal feeding operations CAFO. This proposal was necessitated by a 2005 federal court decision (*Waterkeeper Alliance et al. v. EPA*, 399 F.3d 486 (2nd Cir. 2005)), resulting from challenges brought by agriculture industry groups and environmental advocacy groups, that vacated parts of the 2003 rule and remanded other parts.

The Clean Water Act prohibits the discharge of pollutants from any "point source" to waters of the United States unless authorized under a permit that is issued by EPA or a qualified state, and the act expressly defines CAFOs as point sources. Permits limiting the type and quantity of pollutants that can be discharged are derived from effluent limitation guidelines promulgated by EPA. The 2003 rule, updating rules that had been in place since the 1970s, revised the way in which discharges of manure, wastewater, and other process wastes from CAFOs are regulated, and it modified both the permitting requirements and applicable effluent limitation guidelines. It contained important first-time requirements: all CAFOs must apply for a discharge permit, and all CAFOs that apply such waste on land must develop and implement a nutrient management plan.

EPA's proposal for revisions addresses those parts of the 2003 rule that were affected by the federal court's ruling: (1) it would eliminate the "duty to apply" requirement that all CAFOs either apply for discharge permits or demonstrate that they have no potential to discharge, which was challenged by industry plaintiffs, (2) it would add procedures regarding review of and public access to nutrient management plans, challenged by environmental groups, and (3) it would modify aspects of the effluent limitation guidelines, also challenged by environmental groups. EPA's proposal also considers modifying a provision of the rule that the court upheld, concerning the treatment of a regulatory exemption for agricultural stormwater discharges.

Public comments addressed a number of points, with particular focus on the "duty to apply" for a permit and agricultural stormwater exemption provisions of the proposal. Industry's comments were generally supportive of the proposal, approving deletion of the previous "duty to apply" provision and also EPA's efforts to provide flexibility regarding nutrient management plan modifications. Environmental groups strongly criticized the proposal, arguing that the *Waterkeeper Alliance* court left in place several means for the agency to accomplish much of its original permitting approach, but instead EPA chose not to do so. State permitting authorities also have a number of criticisms, focusing on key parts that they argue will greatly increase the administrative and resource burden on states. In July 2007 EPA extended compliance deadlines for permitted CAFOs from July 2007, subject to revised rules, to February 2009. Congress has shown some interest in CAFO issues, primarily through oversight hearings in 1999 and 2001.

Chapter 5 - Congress established the statutory formula governing distribution of financial aid for municipal wastewater treatment in the Clean Water Act (CWA) in 1972. Since then, Congress has modified the formula and incorporated other eligibility changes five times. Federal funds are provided to states through annual appropriations according to the statutory formula to assist local governments in constructing wastewater treatment projects in compliance with federal standards. The most recent formula change, enacted in 1987, continues to apply to distribution of federal grants to capitalize state revolving loan funds (SRFs) for similar activities

The current state-by-state allotment is a complex formulation consisting basically of two elements, state population and "need." The latter refers to states' estimates of capital costs for wastewater projects necessary for compliance with the act. Funding needs surveys have been done since the 1960s and became an element of distributing CWA funds in 1972. The

Environmental Protection Agency (EPA) in consultation with states has prepared 13 clean water needs surveys since then to provide information to policymakers on the nation's total funding needs, as well as needs for certain types of projects.

This report describes the formula and eligibility changes adopted by Congress since 1972, revealing the interplay and decisionmaking by Congress on factors to include in the formula. Two types of trends and institutional preferences can be discerned in these actions. First, there are differences over the use of "need" and population factors in the allocation formula itself. Over time, the weighting and preference given to certain factors in the allocation formula have become increasingly complex and difficult to discern. Second, there is a gradual increase in restrictions on types of wastewater treatment projects eligible for federal assistance.

Crafting an allotment formula has been one of the most controversial issues debated during past reauthorizations of the Clean Water Act. The dollars involved are significant, and considerations of "winner" and "loser" states bear heavily on discussions of policy choices reflected in alternative formulations. This is likely to be the case again, when Congress moves to reauthorize the act. In the 109th Congress, legislation to extend CWA infrastructure financing was approved by the Senate Environment and Public Works Committee (S. 1400, S.Rept. 109-186). It included changes to the allotment process. However, the bill did not receive further action, partly because of controversies over the proposed allocation formulas. Because the current allocation formula is now 20 years old, and because needs and population have changed, the issue of how to allocate state-by-state distribution of federal funds is likely to be an important topic in debate over water infrastructure legislation. This report will be updated as developments warrant.

Chapter 6 - In 2001, the Environmental Protection Agency (EPA) promulgated a new regulation for arsenic in drinking water, as required by 1996 Safe Drinking Water Act Amendments. The rule set the legal limit for arsenic in tap water at 10 parts per billion (ppb), replacing a 50 ppb standard set in 1975, before arsenic was classified as a carcinogen. When issuing the rule, the EPA projected that compliance could be costly for some small systems, but many water utilities and communities expressed concern that the EPA had underestimated the rule's costs significantly. The arsenic rule was to enter into effect on March 23, 2001, and public water systems were given until January 23, 2006, to comply. Subsequently, the EPA postponed the rule's effective date to February 22, 2002, to review the science and cost and benefit analyses supporting the rule. After completing the review in October 2001, the EPA affirmed the 10 ppb standard. The new standard became enforceable for water systems in January 2006.

Since the rule was completed, Congress and the EPA have focused on how to help communities comply with the new standard. In the past several Congresses, numerous bills have been offered to provide more financial and technical assistance and/or compliance flexibility to small systems; however, none of the bills has been enacted.

Chapter 7 - Perchlorate is the explosive component of solid rocket fuel, fireworks, road flares, and other products. Used mainly by the Department of Defense (DOD) and related industries, perchlorate occurs naturally and is present in some organic fertilizer. This soluble, persistent compound has been detected in sources of drinking water for more than 11 million people. It also has been found in milk, fruits, and vegetables. Concern over the potential health risks of perchlorate exposure has increased, and some states and Members of Congress have urged the Environmental Protection Agency (EPA) to set a drinking water standard for perchlorate. The EPA has not determined whether to regulate perchlorate and has cited the need for more research on health effects, water treatment techniques, and occurrence. Related issues have involved environmental cleanup and water treatment costs, which will be driven by federal and state standards. Interagency disagreements over the risks of perchlorate exposure led several federal agencies to ask the National Research Council (NRC) to evaluate perchlorate's health effects and EPA's risk analyses. In 2005, the NRC issued its report, and the EPA adopted the NRC's recommended reference dose (i.e., the expected safe dose) for perchlorate exposure. However, new studies raise more concerns about potential health risks of low-level exposures, particularly for infants. Perchlorate bills in the 110th Congress include S. 150 and H.R. 1747, which direct the EPA to set a standard. This report reviews perchlorate water contamination issues and developments.

Chapter 8 - To address a nationwide water pollution problem caused by leaking underground storage tanks (USTs), Congress created a leak prevention, detection, and cleanup program in 1984. In 1986, Congress established the Leaking Underground Storage Tank Trust Fund (LUST) to help the Environmental Protection Agency (EPA) and states oversee LUST cleanup activities and pay the costs of remediating leaking petroleum USTs where owners fail to do so. Despite progress in the program, challenges remain. A key issue has been that state resources have not met the demands of administering the UST leak prevention program. States have long sought larger appropriations from the trust fund to support the LUST cleanup program, and some have sought flexibility to use the fund to administer and enforce the UST leak prevention program. Another issue has involved the detection of methyl tertiary butyl ether (MTBE) in groundwater at many LUST sites. This gasoline additive was used widely to reduce air pollution from auto emissions. However, MTBE is very water soluble and, once released, tends to travel farther than conventional gas leaks, making it more likely to reach water supplies and more costly to remediate. For more than a decade, Congress considered various bills to broaden the use of the trust fund and strengthen the leak prevention program.

The Energy Policy Act of 2005 (P.L. 109-58) added new leak prevention provisions to the UST program, imposed new requirements on states, and authorized EPA and states to use LUST Trust Fund appropriations for both cleanup and prevention activities. A key now issue concerns the level of resources available to states to meet the new mandates. Some stakeholders have called for increased trust fund appropriations, while EPA has asked Congress to modify UST inspection requirements. This report reviews the LUST program, legislative changes made by P.L. 109-58, and related developments.

In: Water Pollution Issues and Developments
Editor: S. V. Thomas, pp. 1-24

ISBN: 978-1-60456-208-8
© 2008 Nova Science Publishers, Inc.

Chapter 1

CRUISE SHIP POLLUTION: BACKGROUND, LAWS AND REGULATIONS, AND KEY ISSUES*

Claudia Copeland

ABSTRACT

The cruise industry is a significant and growing contributor to the U.S. economy, providing more than $32 billion in benefits annually and generating more than 330,000 U.S. jobs, but also making the environmental impacts of its activities an issue to many. Although cruise ships represent a small fraction of the entire shipping industry worldwide, public attention to their environmental impacts comes in part from the fact that cruise ships are highly visible and in part because of the industry's desire to promote a positive image.

Cruise ships carrying several thousand passengers and crew have been compared to "floating cities," and the volume of wastes that they produce is comparably large, consisting of sewage; wastewater from sinks, showers, and galleys (graywater); hazardous wastes; solid waste; oily bilge water; ballast water; and air pollution. The waste streams generated by cruise ships are governed by a number of international protocols (especially MARPOL) and U.S. domestic laws (including the Clean Water Act and the Act to Prevent Pollution from Ships), regulations, and standards, but there is no single law or rule. Some cruise ship waste streams appear to be well regulated, such as solid wastes (garbage and plastics) and bilge water. But there is overlap of some areas, and there are gaps in others. Some, such as graywater and ballast water, are not regulated (except in the Great Lakes), and concern is increasing about the impacts of these discharges on public health and the environment. In other areas, regulations apply, but critics argue that they are not stringent enough to address the problem — for example, with respect to standards for sewage discharges. Environmental advocates have raised concerns about the adequacy of existing laws for managing these wastes, and they contend that enforcement is weak.

In 2000, Congress enacted legislation restricting cruise ship discharges in U.S. navigable waters within the state of Alaska. California, Alaska, and Maine have enacted state-specific laws concerning cruise ship pollution, and a few other states have entered

---

* Excerpted from CRS Report RL32450, dated June 14, 2007.

into voluntary agreements with industry to address management of cruise ship discharges. Meanwhile, the cruise industry has voluntarily undertaken initiatives to improve pollution prevention, by adopting waste management guidelines and procedures and researching new technologies. Concerns about cruise ship pollution raise issues for Congress in three broad areas: adequacy of laws and regulations, research needs, and oversight and enforcement of existing requirements. Legislation to regulate cruise ship discharges of sewage, graywater, and bilge water nationally was introduced in the 109th Congress, but there was no further congressional action.

This report describes the several types of waste streams that cruise ships may discharge and emit. It identifies the complex body of international and domestic laws that address pollution from cruise ships. It then describes federal and state legislative activity concerning cruise ships in Alaskan waters and recent activities in a few other states, as well as current industry initiatives to manage cruise ship pollution. Issues for Congress are discussed.

INTRODUCTION

More than 46,000 commercial vessels — tankers, bulk carriers, container ships, barges, and passenger ships — travel the oceans and other waters of the world, carrying cargo and passengers for commerce, transport, and recreation. Their activities are regulated and scrutinized in a number of respects by international protocols and U.S. domestic laws, including those designed to protect against discharges of pollutants that could harm marine resources, other parts of the ambient environment, and human health. However, there are overlaps of some requirements, gaps in other areas, geographic differences in jurisdiction based on differing definitions, and questions about the adequacy of enforcement.

Public attention to the environmental impacts of the maritime industry has been especially focused on the cruise industry, in part because its ships are highly visible and in part because of the industry's desire to promote a positive image. It represents a relatively small fraction of the entire shipping industry worldwide. As of January 2005, passenger ships (which include cruise ships and ferries) composed about 12% of the world shipping fleet.[1] The cruise industry is a significant and growing contributor to the U.S. economy, providing more than $32 billion in total benefits annually and generating more than 330,000 U.S. jobs,[2] but also making the environmental impacts of its activities an issue to many. Since 1980, the average annual growth rate in the number of cruise passengers worldwide has been 8.4%, and in 2005, cruises hosted an estimated 11.5 million passengers. Cruises are especially popular in the United States. In 2005, U.S. ports handled 8.6 million cruise embarkations (75% of global passengers), 6.3% more than in 2004. The worldwide cruise ship fleet consists of more than 230 ships, and the majority are foreign-flagged, with Liberia and Panama being the most popular flag countries.[3] Foreign-flag cruise vessels owned by six companies account for nearly 95% of passenger ships operating in U.S. waters. Each year, the industry adds new ships to the total fleet (12 new cruise ships debuted in 2004 and 4 more in 2005), vessels that are bigger, more elaborate and luxurious, and that carry larger numbers of passengers and crew.

To the cruise ship industry, a key issue is demonstrating to the public that cruising is safe and healthy for passengers and the tourist communities that are visited by their ships. Cruise ships carrying several thousand passengers and crew have been compared to "floating cities," in part because the volume of wastes produced and requiring disposal is greater than that of many small cities on land. During a typical one-week voyage, a large cruise ship (with 3,000 passengers and crew) is estimated to generate 210,000 gallons of sewage; 1 million gallons of graywater (wastewater from sinks, showers, and laundries); more than 130 gallons of hazardous wastes; 8 tons of solid waste; and 25,000 gallons of oily bilge water.[4] Those wastes, if not properly treated and disposed of, can pose risks to human health, welfare, and the environment. Environmental advocates have raised concerns about the adequacy of existing laws for managing these wastes, and suggest that enforcement of existing laws is weak.

A 2000 General Accounting Office (GAO) report focused attention on problems of cruise vessel compliance with environmental requirements.[5] GAO found that between 1993 and 1998, foreign-flag cruise ships were involved in 87 confirmed illegal discharge cases in U.S. waters. A few of the cases included multiple illegal discharge incidents occurring over the six-year period. GAO reviewed three major waste streams (solids, hazardous chemicals, and oily bilge water) and concluded that 83% of the cases involved discharges of oil or oil-based products, the volumes of which ranged from a few drops to hundreds of gallons. The balance of the cases involved discharges of plastic or garbage. GAO judged that 72% of the illegal discharges were accidental, 15% were intentional, and 13% could not be determined. The 87 cruise ship cases represented 4% of the 2,400 illegal discharge cases by foreign-flag ships (including tankers, cargo ships and other commercial vessels, as well as cruise ships) confirmed during the six years studied by GAO. Although cruise ships operating in U.S. waters have been involved in a relatively small number of pollution cases, GAO said, several have been widely publicized and have led to criminal prosecutions and multimillion-dollar fines.

In 2000, a coalition of 53 environmental advocacy groups petitioned the Environmental Protection Agency (EPA) to take regulatory action to address pollution by cruise ships.[6] The petition called for an investigation of wastewater, oil, and solid waste discharges from cruise ships. In response, EPA agreed to study cruise ship discharges and waste management approaches. As part of that effort, in 2000 EPA issued a background document with preliminary information and recommendations for further assessment through data collection and public information hearings.[7] The agency reportedly is developing a cruise ship discharge assessment report that may be released in the fall of 2007, but it has not made a final decision on the environmental groups' original petition, which has been pending for seven years.[8] In May 2007, the groups sued EPA, seeking to compel the agency to act on the petition.

This report presents information on issues related to cruise ship pollution. It begins by describing the several types of waste streams and contaminants that cruise ships may discharge and emit. It identifies the complex body of international and domestic laws that address pollution from cruise ships, as there is no single law in this area. Some wastes are covered by international standards, some are subject to U.S. law, and for some there are gaps in law, regulation, or possibly both. The report then describes federal and state legislative activity concerning cruise ships in Alaskan waters and recent activities in a few other states. Cruise ship companies have taken a number of steps to prevent illegal waste discharges and have adopted waste management plans and practices to improve their environmental operations. Environmental critics acknowledge these initiatives, even as they have petitioned the federal government to strengthen existing regulation of cruise ship wastes. Environmental groups endorsed companion bills in the 109th Congress (the Clean Cruise Ship Act, S. 793/H.R. 1636) that would have required stricter standards to control wastewater discharges from cruise ships. Congress did not act on either bill.

CRUISE SHIP WASTE STREAMS

Cruise ships generate a number of waste streams that can result in discharges to the marine environment, including sewage, graywater, hazardous wastes, oily bilge water, ballast water, and solid waste. They also emit air pollutants to the air and water. These wastes, if not properly treated and disposed of, can be a significant source of pathogens, nutrients, and toxic substances with the potential to threaten human health and damage aquatic life. It is important, however, to keep these discharges in some perspective, because cruise ships represent a small — although highly visible — portion of the entire international shipping industry, and the waste streams described here are not unique to cruise ships. However, particular types of wastes, such as sewage, graywater, and solid waste, may be of greater concern for cruise ships relative to other seagoing vessels, because of the large numbers of passengers and crew that cruise ships carry and the large volumes of wastes that they produce. Further, because cruise ships tend to concentrate their activities in specific coastal areas and visit the same ports repeatedly (especially Florida, California, New York, Galveston, Seattle, and the waters of Alaska), their cumulative impact on a local scale could be significant, as can impacts of individual large-volume releases (either accidental or intentional).

Blackwater is sewage, wastewater from toilets and medical facilities, which can contain harmful bacteria, pathogens, diseases, viruses, intestinal parasites, and harmful nutrients. Discharges of untreated or inadequately treated sewage can cause bacterial and viral contamination of fisheries and shellfish beds, producing risks to public health. Nutrients in sewage, such as nitrogen and phosphorous, promote excessive algal growth, which consumes oxygen in the water and can lead to fish kills and destruction of other aquatic life. A large cruise ship (3,000 passengers and crew) generates an estimated 15,000 to 30,000 gallons per day of blackwater waste.[9]

Graywater is wastewater from the sinks, showers, galleys, laundry, and cleaning activities aboard a ship. It can contain a variety of pollutant substances, including fecal coliform bacteria, detergents, oil and grease, metals, organics, petroleum hydrocarbons, nutrients, food waste, and medical and dental waste. Graywater has potential to cause adverse environmental effects because of concentrations of nutrients and other oxygen-demanding materials, in particular. Graywater is typically the largest source of liquid waste generated by cruise ships (90%-95% of the total). Estimates of graywater range from 30 to 85 gallons per day per person, or 90,000 to 255,000 gallons per day for a 3,000-person cruise ship.[10]

Cruise ships produce hazardous wastes from a number of on-board activities and processes, including photo processing, dry-cleaning, and equipment cleaning. These materials contain a wide range of substances such as hydrocarbons, chlorinated hydrocarbons, heavy metals, paint waste, solvents, fluorescent and mercury vapor light bulbs, various types of batteries, and unused or outdated pharmaceuticals. Although the quantities of hazardous waste generated on cruise ships are small, their toxicity to sensitive marine organisms can be significant. Without careful management, these wastes can find their way into graywater, bilge water, or the solid waste stream.

Solid waste generated on a ship includes glass, paper, cardboard, aluminum and steel cans, and plastics. Much of this solid waste is incinerated on board, and the ash typically is discharged at sea, although some is landed ashore for disposal or recycling. Marine mammals, fish, sea turtles, and birds can be injured or killed from entanglement with plastics and other solid waste that may be released or disposed off of cruise ships. On average, each cruise ship passenger generates at least two pounds of non-hazardous solid waste per day and disposes of two bottles and two cans.[11] With large cruise ships carrying several thousand passengers, the amount of waste generated in a day can be massive. For a large cruise ship, about 8 tons of solid waste are generated during a one-week cruise.[12] It has been estimated that 24% of the solid waste generated by vessels worldwide (by weight) comes from cruise ships.[13] Most cruise ship garbage is treated on board (incinerated, pulped, or ground) for discharge overboard. When garbage must be off-loaded (for example, because glass and aluminum cannot be incinerated), cruise ships can put a strain on port reception facilities, which are rarely adequate to the task of serving a large passenger vessel (especially at non-North American ports).[14]

On a ship, oil often leaks from engine and machinery spaces or from engine maintenance activities and mixes with water in the bilge, the lowest part of the hull of the ship. Oil, gasoline, and byproducts from the biological breakdown of petroleum products can harm fish and wildlife and pose threats to human health if ingested. Oil in even minute concentrations can kill fish or have various sub-lethal chronic effects. Bilge water also may contain solid wastes and pollutants containing high amounts of oxygen-demanding material, oil and other chemicals. A typical large cruise ship will generate an average of 8 metric tons of oily bilge water for each 24 hours of operation.[15] To maintain ship stability and eliminate potentially hazardous conditions from oil vapors in these areas, the bilge spaces need to be flushed and periodically pumped dry. However, before a bilge can be cleared out and the water discharged, the oil that has been accumulated needs to be extracted from the bilge water, after which the extracted oil can be reused, incinerated, and/or offloaded in port. If a separator, which is normally used to extract the oil, is faulty or is deliberately bypassed, untreated oily bilge water could be discharged directly into the ocean, where it can damage marine life. A number of cruise lines have been charged with environmental violations related to this issue in recent years.

Cruise ships, large tankers, and bulk cargo carriers use a tremendous amount of ballast water to stabilize the vessel during transport. Ballast water is often taken on in the coastal waters in one region after ships discharge wastewater or unload cargo, and discharged at the next port of call, wherever more cargo is loaded, which reduces the need for compensating ballast. Thus, it is essential to the proper functioning of ships (especially cargo ships), because the water that is taken in compensates for changes in the ship's weight as cargo is loaded or unloaded, and as fuel and supplies are consumed. However, ballast water discharge typically contains a variety of biological materials, including plants, animals, viruses, and bacteria. These materials often include non-native, nuisance, exotic species that can cause extensive ecological and economic damage to aquatic ecosystems. Ballast water discharges are believed to be the leading source of invasive species in U.S. marine waters, thus posing public health and environmental risks, as well as significant economic cost to industries such as water and power utilities, commercial and recreational fisheries, agriculture, and tourism.[16] Studies suggest that the economic cost just from introduction of pest mollusks (zebra mussels, the Asian clam, and others) to U.S. aquatic ecosystems is more than $6 billion per year.[17] These problems are not limited to cruise ships, but there is little cruise-industry specific data on the issue, and further study is needed to determine cruise ships' role in the overall problem of introduction of non-native species by vessels.

Air pollution from cruise ships is generated by diesel engines that burn high sulfur content fuel, producing sulfur dioxide, nitrogen oxide and particulate matter, in addition to carbon monoxide, carbon dioxide, and hydrocarbons. EPA recognizes that these emissions from marine diesel engines contribute to ozone and carbon monoxide nonattainment, as well as adverse health effects associated with ambient concentrations of particulate matter and visibility, haze, acid deposition, and eutrophication and nitrophication of water.[18] EPA estimates that large marine diesel engines accounted for about 1.6% of mobile source nitrogen oxide emissions and 2.8% of mobile source particulate emissions in the United States in 2000. Contributions of marine diesel engines can be higher on a port-specific basis.

One source of environmental pressures on maritime vessels recently has come from states and localities, as they assess the contribution of commercial marine vessels to regional air quality problems when ships are docked in port. For instance, large marine diesel engines are believed to contribute 7% of mobile source nitrogen oxide emissions in Baton Rouge/New Orleans. Ships can also have a significant impact in areas without large commercial ports: they contribute about 37% of total area nitrogen oxide emissions in the Santa Barbara area, and that percentage is expected to increase to 61% by the year 2015.[19] Again, there is little cruise-industry specific data on this issue. They comprise only a small fraction of the world shipping fleet, but cruise ship emissions may exert significant impacts on a local scale in specific coastal areas that are visited repeatedly. Shipboard incinerators also burn large volumes of garbage, plastics, and other waste, producing ash that must be disposed of. Incinerators may release toxic emissions as well.

APPLICABLE LAWS AND REGULATIONS

The several waste streams generated by cruise ships are governed by a number of international protocols and U.S. domestic laws, regulations and standards, which are described in this section, but there is no single law or regulation. Moreover, there are overlaps in some areas of coverage, gaps in other areas, and differences in geographic jurisdiction, based on applicable terms and definitions.

International Legal Regime

The International Maritime Organization (IMO), a body of the United Nations, sets international maritime vessel safety and marine pollution standards. It consists of representatives from 152 major maritime nations, including the United States. The IMO implements the 1973 International Convention for the Prevention of Pollution from Ships, as modified by the Protocol of 1978, known as MARPOL 73/78. Cruise ships flagged under countries that are signatories to MARPOL are subject to its requirements, regardless of where they sail, and member nations are responsible for vessels registered under their respective nationalities.[20] Six Annexes of the Convention cover the various sources of pollution from ships and provide an overarching framework for international objectives, but they are not sufficient alone to protect the marine environment from waste discharges, without ratification and implementation by sovereign states.

- Annex I deals with regulations for the prevention of pollution by oil.
- Annex II details the discharge criteria and measures for the control of pollution by noxious liquid substances carried in bulk.
- Annex III contains general requirements for issuing standards on packing, marking, labeling, and notifications for preventing pollution by harmful substances.
- Annex IV contains requirements to control pollution of the sea by sewage.
- Annex V deals with different types of garbage, including plastics, and specifies the distances from land and the manner in which they may be disposed of.
- Annex VI sets limits on sulfur oxide, nitrogen oxide, and other emissions from marine vessel operations and prohibits deliberate emissions of ozone- depleting substances.

In order for IMO standards to be binding, they must first be ratified by a total number of member countries whose combined gross tonnage represents at least 50% of the world's gross tonnage, a process that can be lengthy. All six have been ratified by the requisite number of nations; the most recent is Annex VI, which took effect in May 2005. The United States has ratified all except Annex IV. The country where a ship is registered (flag state) is responsible for certifying the ship's compliance with MARPOL's pollution prevention standards. IMO also has established a large number of other conventions, addressing issues such as ballast water management, and the International Safety Management Code, with guidelines for passenger safety and pollution prevention.

Each signatory nation is responsible for enacting domestic laws to implement the convention and effectively pledges to comply with the convention, annexes, and related laws of other nations. In the United States, the Act to Prevent Pollution from Ships (APPS, 33 U.S.C. §§1905-1915) implements the provisions of MARPOL and the annexes to which the United States is a party. The most recent U.S. action concerning MARPOL occurred in April 2006, when the Senate approved Annex VI, which regulates air pollution (Treaty Doc. 108-7, Exec. Rept. 109-13). Following that approval, in March 2007, the House approved legislation to implement the standards in Annex VI (H.R. 802), through regulations to be promulgated by EPA in consultation with the U.S. Coast Guard.[21] APPS applies to all U.S.-flagged ships anywhere in the world and to all foreign-flagged vessels operating in navigable waters of the United States or while at port under U.S. jurisdiction. The Coast Guard generally has primary responsibility to prescribe and enforce regulations necessary to implement APPS in these waters. The regulatory mechanism established in APPS to implement MARPOL is separate and distinct from the Clean Water Act and other federal environmental laws.

One of the difficulties in implementing MARPOL arises from the very international nature of maritime shipping. The country that the ship visits can conduct its own examination to verify a ship's compliance with international standards and can detain the ship if it finds significant noncompliance. Under the provisions of the Convention, the United States can take direct enforcement action under U.S. laws against foreign-flagged ships when pollution discharge incidents occur within U.S. jurisdiction. When incidents occur outside U.S. jurisdiction or jurisdiction cannot be determined, the United States refers cases to flag states, in accordance with MARPOL. The 2000 GAO report documented that these procedures require substantial coordination between the Coast Guard, the State Department, and other flag states and that, even when referrals have been made, the response rate from flag states has been poor.[22]

Domestic Laws and Regulations

In the United States, several federal agencies have some jurisdiction over cruise ships in U.S. waters, but no one agency is responsible for or coordinates all of the relevant government functions. The U.S. Coast Guard and EPA have principal regulatory and standard-setting responsibilities, and the Department of Justice prosecutes violations of federal laws. In addition, the Department of State represents the United States at meetings of the IMO and in international treaty negotiations and is responsible for pursuing foreign-flag violations. Other federal agencies have limited roles and responsibilities. For example, the National Oceanic and Atmospheric Administration (NOAA, Department of Commerce) works with the Coast Guard and EPA to report on the effects of marine debris. The Animal and Plant Health Inspection Service (APHIS) is responsible for ensuring quarantine inspection and disposal of food-contaminated garbage (these APHIS responsibilities are part of the Department of Homeland Security). In some cases, states and localities have responsibilities as well. This section describes U.S. laws and regulations that apply to cruise ship discharges.

Sewage
The Federal Water Pollution Control Act, or Clean Water Act (CWA), is the principal U.S. law concerned with limiting polluting activity in the nation's streams, lakes, estuaries, and coastal waters. The act's primary mechanism for controlling pollutant discharges is the National Pollutant Discharge Elimination System (NPDES) program, authorized in Section 402. In accordance with the NPDES program, pollutant discharges from point sources — a term that includes vessels — are prohibited unless a permit has been obtained. While sewage is defined as a pollutant under the act, sewage from cruise ships and other vessels is exempt from this statutory definition and is therefore exempt from the requirement to obtain an NPDES permit. Further, EPA regulations implementing the NPDES permit program provide that "discharges incidental to the normal operation of vessels" are excluded from regulation and thus from permit requirements (40 CFR §122.3(a)). However, a 2006 federal court ruling could result in changes to these regulations that would remove the current permitting exemption (see discussion of "Ballast Water" on page 13).

Marine Sanitation Devices
Section 312 of the Clean Water Act seeks to address this gap by prohibiting the dumping of untreated or inadequately treated sewage from vessels into the navigable waters of the United States (defined in the act as within 3 miles of shore). Cruise ships are subject to this prohibition. It is implemented jointly by EPA and the Coast Guard. Under Section 312,

commercial and recreational vessels with installed toilets are required to have marine sanitation devices (MSDs), which are designed to prevent the discharge of untreated sewage. EPA is responsible for developing performance standards for MSDs, and the Coast Guard is responsible for MSD design and operation regulations and for certifying MSD compliance with the EPA rules. MSDs are designed either to hold sewage for shore-based disposal or to treat sewage prior to discharge. Beyond 3 miles, raw sewage can be discharged.

The Coast Guard regulations cover three types of MSDs (33 CFR Part 159). Large vessels, including cruise ships, use either Type II or Type III MSDs. In Type II MSDs, the waste is either chemically or biologically treated prior to discharge and must meet limits of no more than 200 fecal coliform per 100 milliliters and no more than 150 milligrams per liter of suspended solids. Type III MSDs store wastes and do not treat them; the waste is pumped out later and treated in an onshore system or discharged outside U.S. waters. Type I MSDs use chemicals to disinfect the raw sewage prior to discharge and must meet a performance standard for fecal coliform bacteria of not greater than 1,000 per 100 milliliters and no visible floating solids. Type I MSDs are generally only found on recreational vessels or others under 65 feet in length. The regulations, which have not been revised since 1976, do not require ship operators to sample, monitor, or report on their effluent discharges.

Critics point out a number of deficiencies with this regulatory structure as it affects cruise ships and other large vessels. First, the MSD regulations only cover discharges of bacterial contaminants and suspended solids, while the NPDES permit program for other point sources typically regulates many more pollutants such as chemicals, pesticides, heavy metals, oil, and grease that may be released by cruise ships as well as land-based sources. Second, sources subject to NPDES permits must comply with sampling, monitoring, recordkeeping, and reporting requirements, which do not exist in the MSD rules.

In addition, the Coast Guard, responsible for inspecting cruise ships and other vessels for compliance with the MSD rules, has been heavily criticized for poor enforcement of Section 312 requirements. In its 2000 report, the GAO said that Coast Guard inspectors "rarely have time during scheduled ship examinations to inspect sewage treatment equipment or filter systems to see if they are working properly and filtering out potentially harmful contaminants." GAO reported that a number of factors limit the ability of Coast Guard inspectors to detect violations of environmental law and rules, including the inspectors' focus on safety, the large size of a cruise ship, limited time and staff for inspections, and the lack of an element of surprise concerning inspections.[23] The Coast Guard carries out a wide range of responsibilities that encompass both homeland security (ports, waterways, and coastal security, defense readiness, drug and migrant interdiction) and non-homeland security (search and rescue, marine environmental protection, fisheries enforcement, aids to navigation). Since the September 11 terrorist attacks on the United States, the Coast Guard has focused more of its resources on homeland security activities.[24] One likely result is that less of the Coast Guard's time and attention are available for vessel inspections for MSD or other environmental compliance.

Annex IV of MARPOL was drafted to regulate sewage discharges from vessels. It has entered into force internationally and would apply to cruise ships that are flagged in ratifying countries, but because the United States has not ratified Annex IV, it is not mandatory that ships follow it when in U.S. waters. However, its requirements are minimal, even compared with U.S. rules for MSDs. Annex IV requires that vessels be equipped with a certified sewage treatment system or holding tank, but it prescribes no specific performance standards. Treated waste may be discharged in waters more than 3 nautical miles from land. Vessels are permitted to meet alternative, less stringent requirements when they are in the jurisdiction of countries where less stringent requirements apply. In U.S. waters, cruise ships and other vessels must comply with the regulations implementing Section 312 of the Clean Water Act.

No Discharge Zones

Section 312 has another means of addressing sewage discharges, through establishment of no-discharge zones (NDZs) for vessel sewage. A state may completely prohibit the discharge of both treated and untreated sewage from all vessels with installed toilets into some or all waters over which it has jurisdiction (up to 3 miles from land). To create a no-discharge zone to protect waters from sewage discharges by cruise ships and other vessels, the state must apply to EPA under one of three categories.

- NDZ based on the need for greater environmental protection, and the state demonstrates that adequate pumpout facilities for safe and sanitary removal and treatment of sewage from all vessels are reasonably available. As of 2006, this category of designation has been used for 58 areas representing part or all of the waters of 26 states, including a number of inland states.
- NDZ for special waters found to have a particular environmental importance (e.g., to protect environmentally sensitive areas such as shellfish beds or coral reefs); it is not necessary for the state to show pumpout availability. This category of designation has been used twice (state waters within the Florida Keys National Marine Sanctuary and the Boundary Waters Canoe area of Minnesota).
- NDZ to prohibit the discharge of sewage into waters that are drinking water intake zones; it is not necessary for the state to show pumpout availability. This category of designation has been used to protect part of the Hudson River in New York.

Graywater

Under current law, graywater is not defined as a pollutant, nor is it generally considered to be sewage (thus, no NPDES permit is required). There are no separate federal effluent standards for graywater discharges. The Clean Water Act only includes graywater in its definition of sewage for the express purpose of regulating commercial vessels in the Great Lakes, under the Section 312 MSD requirements. Thus, graywater can be discharged by cruise ships anywhere — except in the Great Lakes, where the Section 312 MSD rules apply, but those rules limit only bacterial contaminant content and total suspended solids of graywater.

Hazardous Waste

The Resource Conservation and Recovery Act (RCRA, 42 U.S.C. 6901-6991k) is the primary federal law that governs the generation, transport, and disposal of hazardous waste. Under this act, a waste is hazardous if it is ignitable, corrosive, reactive, or toxic, or appears on a list of about 100 industrial process waste streams and more than 500 discarded commercial products and chemicals. Treatment, storage, and disposal facilities are required to have permits and comply with operating standards and other EPA regulations.

A range of activities on board cruise ships generate hazardous wastes and toxic substances that would ordinarily be presumed to be subject to RCRA. However, it is not entirely clear what regulations apply to the management and disposal of these wastes.[25] RCRA rules that cover small-quantity generators (those that generate more than 100 kilograms but less than 1,000 kilograms of hazardous waste per month) are less stringent than those for large-quantity generators (generating more than 1,000 kilograms per month), and it is unclear whether cruise ships are classified as large or small generators of hazardous waste. Moreover, some cruise companies argue that they generate less than 100 kilograms per month and therefore should be classified in a third category, as "conditionally exempt small-quantity generators," a categorization that allows for less rigorous requirements for notification, recordkeeping, and the like.[26]

A release of hazardous substances by a cruise ship or other vessel could also theoretically trigger the Comprehensive Environmental Response, Compensation, and Liability Act (CERCLA, or Superfund, 42 U.S.C. 9601-9675), but it does not appear to have been used in response to cruise ship releases. CERCLA requires that any person in charge of a vessel shall immediately notify the National Response Center of any release of a hazardous substance (other than discharges in compliance with a federal permit under the Clean Water Act or other environmental law) into waters of the United States or the contiguous zone. Notification is required for releases in amounts determined by EPA that may present substantial danger to the public health, welfare, or the environment. EPA has identified 500 wastes as hazardous substances under these provisions and issued rules on quantities that are reportable, covering releases as small as 1 pound of some substances (40 CFR Part 302). CERCLA authorizes the President (acting through the Coast Guard in coastal waters) to remove and provide for remedial action relating to the release. The law distinguishes between short-term and long-term responses to threats posed by hazardous substances. Short-term responses, also referred to as removal actions, address immediate threats to public health and the environment and would most likely be the type of response invoked for a release from a cruise ship. Long-term responses, also called remedial actions, involve complex and highly contaminated sites that often require several years to study and clean up the hazardous waste.

Solid Waste

Cruise ship discharges of solid waste are governed by two laws. Title I of the Marine Protection, Research and Sanctuaries Act (MPRSA, 33 U.S.C. 1402-1421) applies to cruise ships and other vessels and makes it illegal to transport garbage from the United States for the purpose of dumping it into ocean waters without a permit or to dump any material transported from a location outside the United States into U.S. territorial seas or the contiguous zone (within 12 nautical miles from shore) or ocean waters. EPA is responsible for issuing permits that regulate the disposal of materials at sea (except for dredged material disposal, for which the U.S. Army Corps of Engineers is responsible). Beyond waters that are under U.S. jurisdiction, no MPRSA permit is required for a cruise ship to discharge solid waste. The routine discharge of effluent incidental to the propulsion of vessels is explicitly exempted from the definition of dumping in the MPRSA.[27]

The Act to Prevent Pollution from Ships (APPS, 33 U.S.C. 1901-1915) and its regulations, which implement U.S.-ratified provisions of MARPOL, also apply to cruise ships. APPS prohibits the discharge of all garbage within 3 nautical miles of shore, certain types of garbage within 12 nautical miles offshore, and plastic anywhere. It applies to all vessels, whether seagoing or not, regardless of flag, operating in U.S. navigable waters and the Exclusive Economic Zone (EEZ). It is administered by the Coast Guard.

Bilge Water

Section 311 of the Clean Water Act, as amended by the Oil Pollution Act of 1990 (33 U.S.C. 2701-2720), applies to cruise ships and prohibits discharge of oil or hazardous substances in harmful quantities into or upon U.S. navigable waters, or into or upon the waters of the contiguous zone, or which may affect natural resources in the U.S. EEZ (extending 200 miles offshore). Coast Guard regulations (33 CFR §151.10) prohibit discharge of oil within 12 miles from shore, unless passed through a 15-ppm oil water separator, and unless the discharge does not cause a visible sheen. Beyond 12 miles, oil or oily mixtures can be discharged while a vessel is proceeding en route and if the oil content without dilution is less than 100 ppm. Vessels are required to maintain an Oil Record Book to record disposal of oily residues and discharges overboard or disposal of bilge water.

Ballast Water

Clean Water Act regulations currently exempt ballast water discharges incidental to the normal operation of cruise ships and other vessels from NPDES permit requirements (see above discussions concerning sewage and graywater). Because of the growing problem of introduction of invasive species into U.S. waters via ballast water, in January 1999, a number of conservation organizations, fishing groups, native American tribes, and water agencies petitioned EPA to repeal its 1973 regulation exempting ballast water discharge, arguing that ballast water should be regulated as the "discharge of a pollutant" under the Clean Water Act's Section 402 permit program. EPA rejected the petition in September 2003, saying that the "normal operation" exclusion is long-standing agency policy, to which Congress has acquiesced twice (in 1979 and 1996) when it considered the issue of aquatic nuisance species in ballast water and did not alter EPA's CWA interpretation.[28] Further, EPA said that other ongoing federal activities related to control of invasive species in ballast water are likely to be more effective than changing the NPDES rules.[29] Until recently, these efforts to limit ballast water discharges by cruise ships and other vessels were primarily voluntary, except in the Great Lakes. Since 2004, all vessels equipped with ballast water tanks must have a ballast water management plan.[30]

After the denial of their administrative petition, the environmental groups filed a lawsuit seeking to force EPA to rescind the regulation that exempts ballast water discharges from CWA permitting. In March 2005, a federal district court ruled in favor of the groups, and in September 2006, the court remanded the matter to EPA with an order that the challenged regulation be set aside by September 30, 2008 (*Northwest Environmental Advocates v. EPA*, No. C 03-05760 SI (N.D.Cal, September 18, 2006)). The district court rejected EPA's contention that Congress had previously acquiesced in exempting the "normal operation" of vessels from CWA permitting and disagreed with EPA's argument that the court's two-year deadline creates practical difficulties for the agency and the affected industry. Significantly, while the focus of the environmental groups' challenge was principally to EPA's permitting exemption for ballast water discharges, the court's ruling — and its mandate to EPA to rescind the exemption in 40 CFR §122.3(a) — applies fully to other types of vessel discharges that are covered by the regulatory exemption, including sewage, gray water, and bilge water. The government has appealed the district court's ruling.

Air Pollution

The Clean Air Act (42 U.S.C. 7401 et seq.) is the principal federal law that addresses air quality concerns. It requires EPA to set health-based standards for ambient air quality, sets standards for the achievement of those standards, and sets national emission standards for large and ubiquitous sources of air pollution, including mobile sources. Cruise ships emissions were not regulated until February 2003. At that time, EPA promulgated emission standards for new marine diesel engines on large vessels (called Category 3 marine engines) such as container ships, tankers, bulk carriers, and cruise ships flagged or registered in the United States.[31] The 2003 rule resulted from settlement of litigation brought by the environmental group Bluewater Network after it had petitioned EPA to issue stringent emission standards for large vessels and cruise ships.[32] Standards in the rule are equivalent to internationally negotiated standards set in Annex VI of the MARPOL protocol for nitrogen oxides, which engine manufacturers currently meet, according to EPA.[33] Emissions from these large, primarily ocean-going vessels had not previously been subject to EPA regulation. The rule is one of several EPA regulations establishing emissions standards for nonroad engines and vehicles, under Section 213(a) of the Clean Air Act. Smaller marine diesel engines are regulated under rules issued in 1996 and 1999.

In the February 2003 rule, EPA announced that over the next two years it would continue to review issues and technology related to emissions from large marine vessel engines in order to promulgate additional, more stringent emission standards for very large marine engines and vessels by April 2007. Addressing long-term standards in a future rulemaking, EPA said, could facilitate international efforts through the IMO (since the majority of ships used in international commerce are flagged in other nations), while also permitting the United States to proceed, if international standards are not adopted in a timely manner. Environmental groups criticized EPA for excluding foreign-flagged vessels that enter U.S. ports from the marine diesel engine rules and challenged the 2003 rules in federal court. The rules were upheld in a ruling issued June 22, 2004.[34] EPA has said that it will consider including foreign vessels in the future rulemaking to consider more stringent standards. In April 2007, EPA announced an extension of the deadline that had been announced in 2003 for new Category 3 marine diesel engine standards until December 17, 2009. EPA explained that more time was needed to assess advanced emission control technologies and to coordinate with the IMO. Some groups are displeased with EPA's delay, and in response, legislation has been introduced in the 110th Congress that would set specific standards and deadlines to limit the sulfur content of fuel used by U.S. and foreign-flagged marine vessels when they enter or leave U.S. ports, and to require advanced pollution controls for other air emissions from such marine vessels (S. 1499/H.R. 2548).

Considerations of Geographic Jurisdiction

The various laws and regulations described here apply to different geographic areas, depending on the terminology used. For example, the Clean Water Act treats navigable waters, the contiguous zone, and the ocean as distinct entities. The term "navigable waters" is defined to mean the waters of the United States, including the territorial seas (33 U.S.C. §1362(7)). In turn, the territorial seas are defined in that act as extending a distance of 3 miles seaward from the baseline (33 U.S.C. §1362(8)); the baseline generally means the land or shore. In 1988, President Reagan signed a proclamation (Proc. No. 5928, December 27, 1988, 54 *Federal Register* 777) providing that the territorial sea of the United States extends to 12 nautical miles from the U.S. baseline. However, that proclamation had no effect on the geographic reach of the Clean Water Act.

The contiguous zone is defined in the CWA to mean the entire zone established by the United States under Article 24 of the Convention of the Territorial Sea and the Contiguous Zone (33 U.S.C. §1362(9)). That convention defines "contiguous zone" as extending from the baseline from which the territorial sea is measured to not beyond 12 miles. In 1999, President Clinton signed a proclamation (Proc. No. 7219 of August 2, 1999, 64 *Federal Register* 48701) giving U.S. authorities the right to enforce customs, immigration, or sanitary laws at sea within 24 nautical miles from the baseline, doubling the traditional 12-mile width of the contiguous zone. As with the 1988 presidential proclamation, this proclamation did not amend any statutory definitions (as a general matter, a presidential proclamation cannot amend a statute). Thus, for purposes of the Clean Water Act, the territorial sea remains 3 miles wide, and the contiguous zone extends from 3 to 12 miles. Under CERCLA, "navigable waters" means waters of the United States, including the territorial seas (42 U.S.C. §9601(15)), and that law incorporates the Clean Water Act's definitions of "territorial seas" and "contiguous zone" (42 U.S.C. §9601(30)).

The CWA defines the "ocean" as any portion of the high seas beyond the contiguous zone (33 U.S.C. §1362(10)). In contrast, the MPRSA defines "ocean waters" as the open seas lying seaward beyond the baseline from which the territorial sea is measured, as provided for in the Convention of the Territorial Sea and the Contiguous Zone (33 U.S.C. §1402(b)).

Limits of jurisdiction are important because they define the areas where specific laws and rules apply. For example, the Clean Water Act MSD standards apply to sewage discharges from vessels into or upon the navigable waters, and Section 402 NPDES permits are required for point source discharges (excluding vessels) into the navigable waters. Section 311 of the CWA, as amended by the Oil Pollution Act, addresses discharges of oil or hazardous substances into or upon the navigable waters of the United States or the waters of the contiguous zone. Provisions of the Act to Prevent Pollution from Ships (APPS, 33 U.S.C. §§1901-1915) concerning discharges of oil and noxious substances apply to navigable waters. Other provisions of that same act concerning garbage and plastics apply to navigable waters or the EEZ, but the term "navigable waters" is not defined in APPS. The MPRSA regulates ocean dumping within the area extending 12 nautical miles seaward from the baseline and regulates transport of material by U.S.-flagged vessels for dumping into ocean waters.

Further complicating jurisdictional considerations is the fact that the Clean Water Act refers to these distances from shore in terms of miles, without other qualification, which is generally interpreted to mean an international mile or statute mile. APPS, the MPRSA, and the two presidential proclamations refer to distances in terms of nautical miles from the baseline. These two measures are not identical: a nautical mile is a unit of distance used primarily at sea and in aviation; it equals 6,080 feet and is 15% longer than an international or statute mile.[35]

Alaskan Activities

In Alaska, where tourism and commercial fisheries are key contributors to the economy, cruise ship pollution has received significant attention. After the state experienced a three-fold increase in the number of cruise ship passengers visits during the 1990s,[36] concern by

Alaska Natives and other groups over impacts of cruise ship pollution on marine resources began to increase. In one prominent example of environmental violations, in July 1999, Royal Caribbean Cruise Lines entered a federal criminal plea agreement involving total penalties of $6.5 million for violations in Alaska, including knowingly discharging oil and hazardous substances (including dry-cleaning and photo processing chemicals). The company admitted to a fleet-wide practice of discharging oil-contaminated bilge water. The Alaska penalties were part of a larger $18 million total federal plea agreement involving environmental violations in multiple locations, including Florida, New York, and California.

Public concern about the Royal Caribbean violations led the state to initiate a program in December 1999 to identify cruise ship waste streams. Voluntary sampling of large cruise ships in 2000 indicated that waste treatment systems on most ships did not function well and discharges greatly exceeded applicable U.S. Coast Guard standards for Type II MSDs. Fecal coliform levels sampled during that period averaged 12.8 million colonies per 100 milliliters in blackwater and 1.2 million in graywater, far in excess of the Coast Guard standard of 200 fecal coliforms per 100 milliliters.

Federal Legislation

Concurrent with growing regional interest in these problems, attention to the Alaska issues led to passage of federal legislation in December 2000 (Certain Alaskan Cruise Ship Operations, Division B, Title XIV of the Miscellaneous Appropriations Bill, H.R. 5666, in the Consolidated Appropriations Act, 2001 (P.L. 106-554); 33 U.S.C. 1901 Note). This law established standards for vessels with 500 or more overnight passengers and generally prohibited discharge of untreated sewage and graywater in navigable waters of the United States within the state of Alaska. It authorized EPA to promulgate standards for sewage and graywater discharges from cruise ships in these waters.[37] Until such time as EPA issues regulations, cruise ships may discharge treated sewage wastes in Alaska waters only while traveling at least 6 knots and while at least 1 nautical mile from shore, provided that the discharge contains no more than 200 fecal coliforms per 100 ml and no more than 150 mg/l total suspended solids (the same limits prescribed in federal regulations for Type II MSDs).

The law also allows for discharges of treated sewage and graywater inside of one mile from shore and at speeds less than 6 knots (thus including stationary discharges while a ship is at anchor) for vessels with systems that can treat sewage and graywater to a much stricter standard. Such vessels must meet these minimum effluent standards: no more than 20 fecal coliforms per 100 ml, no more than 30 mg/l of total suspended solids, and total residual chlorine concentrations not to exceed 10 mg/l. The legislation requires sampling, data collection, and recordkeeping by vessel operators to facilitate Coast Guard oversight and enforcement. Regulations to implement the federal law were issued by the U.S. Coast Guard in July 2001 and became effective immediately upon publication.[38] The regulations stipulate minimum sampling and testing procedures and provide for administrative and criminal penalties for violations of the law, as provided in the legislation.

In the 109th Congress, the House approved legislation, H.R. 5681, with a provision (Section 410) directing the Coast Guard to conduct a demonstration project on the methods and best practices of the use of smokestack scrubbers on cruise ships that operate in the Alaska cruise trade. The Senate did not act on H.R. 5681 before the 109th Congress adjourned in December 2006.

Alaska State Legislation and Initiatives

Building on the federal legislation enacted in 2000, the state of Alaska enacted its own law in June 2001 (AS 46.03.460-AS 46.03.490). The state law sets standards and sampling requirements for the underway discharge of blackwater in Alaska that are identical to the blackwater/sewage standards in the federal law. However, because of the high fecal coliform counts detected in graywater in 2000, the state law also extends the effluent standards to discharges of graywater. Sampling requirements for all ships took effect in 2001, as did effluent standards for blackwater discharges by large cruise ships (defined as providing overnight accommodations to 250 or more). Effluent standards for graywater discharges by large vessels took effect in 2003. Small ships (defined as providing overnight accommodations for 50 to 249 passengers) were allowed three years to come into compliance with all effluent standards. The law also established a scientific advisory panel to evaluate the effectiveness of the law's implementation and to advise the state on scientific matters related to cruise ship impacts on the Alaskan environment and public health.

In February 2004, the state reported on compliance with the federal and state requirements for the years 2001-2003.[39] According to the state, the federal and state standards have prompted large ships to either install advanced wastewater treatment systems that meet the effluent standards or to manage wastes by holding all of their wastewater for discharge outside of Alaskan waters (beyond 3 miles from shore). As of 2003, the majority of large ships (56%) had installed advanced technology (compared with 8% that had done so in 2001), while the remaining 44% discharge outside of Alaska waters. As a result, the quality of wastewater discharged from large ships has improved dramatically, according to the state: the majority of conventional and toxic pollutants that ships must sample for were not detected, and test results indicate that wastewater from large ships with advanced wastewater treatment systems does not pose a risk to aquatic organisms or to human health, even during stationary discharge.

Small ships, however, had not installed new wastewater treatment systems, and the effluent quality has remained relatively constant, with discharge levels for several pollutants regularly exceeding state water quality standards. In particular, test results indicated that concentrations of free chlorine, fecal coliform, copper, and zinc from stationary smaller vessels pose some risk to aquatic life and also to human health in areas where aquatic life is harvested for raw consumption.

In addition to the state's 2001 action, in August 2006 Alaska voters approved a citizen initiative requiring cruise lines to pay the state a $50 head tax for each passenger and a corporate income tax, increasing fines for wastewater violations, and mandating new environmental regulations for cruise ships (such as a state permit for all discharges of treated wastewater). Revenues from the taxes will go to local communities affected by tourism and into public services and facilities used by cruise ships. Supporters of the initiative contend that the cruise industry does not pay enough in taxes to compensate for its environmental harm to the state and for the services it uses. Opponents argued that the initiative would hurt Alaska's competitiveness for tourism.

Other State Activities

Activity to regulate or prohibit cruise ship discharges also has occurred in several other states.

In April 2004, the state of Maine enacted legislation governing discharges of graywater or mixed blackwater/graywater into coastal waters of the state (Maine LD. 1158). The legislation applies to large cruise ships (with overnight accommodations for 250 or more passengers) and allows such vessels into state waters after January 1, 2006, only if the ships have advanced wastewater treatment systems, comply with discharge and recordkeeping requirements under the federal Alaska cruise ship law, and get a permit from the state Department of Environmental Protection. Under the law, prior to 2006, graywater dischargers were allowed if the ship operated a treatment system conforming to requirements for continuous discharge systems under the Alaska federal and state laws. In addition, the legislation required the state to apply to EPA for designation of up to 50 No Discharge Zones, in order that Maine may gain federal authorization to prohibit blackwater discharges into state waters. EPA approved the state's NDZ request for Casco Bay in June 2006.

California enacted three bills in 2004. One bars cruise ships from discharging treated wastewater while in the state's waters (Calif. A.B. 2672). Another prohbits vessels from releasing graywater (Calif. A.B. 2093), and the third measure prevents cruise ships from operating waste incinerators (Calif. A.B. 471). Additionally, in 2003 California enacted a law that bans passenger ships from discharging sewage sludge and oil bilge water (Calif. A.B. 121), as well as a bill that prohibits vessels from discharging hazardous wastes from photo-processing and dry cleaning operations into state waters (Calif. A.B. 906). Another measure was enacted in 2006: California S.B. 497 requires the state to adopt ballast water performance standards by January 2008 and sets specific deadlines for the removal of different types of species from ballast water, mandating that ship operators remove invasive species (including bacteria) by the year 2020.

Several states, including Florida, Washington, and Hawaii, have entered into memoranda of agreement with the industry (through the International Council of Cruise Lines and related organizations) providing that cruise ships will adhere to certain practices concerning waste minimization, waste reuse and recycling, and waste management. For example, under a 2001 agreement between industry and the state of Florida, cruise lines must eliminate wastewater discharges in state waters within 4 nautical miles off the coast of Florida, report hazardous waste off-loaded in the United States by each vessel on an annual basis, and submit to environmental inspections by the U.S. Coast Guard.

Similarly, in April 2004 the Washington Department of Ecology, Northwest Cruise Ship Association, and Port of Seattle signed a memorandum of understanding (MOU) that would allow cruise ships to discharge wastewater treated with advanced wastewater treatment systems into state waters and would prohibit the discharge of untreated wastewater and sludge. Environmental advocates are generally critical of such voluntary agreements, because they lack enforcement and penalty provisions. States respond that while the Clean Water Act limits a state's ability to control cruise ship discharges, federal law does not bar states from entering into voluntary agreements that have more rigorous requirements.[40] In January 2005 the Department of Ecology reported that cruise ships visiting the state during the 2004 sailing season mostly complied with the MOU to stop discharging untreated wastewater, leading to some improvement in management of wastes. Although enforcement of what is essentially a voluntary agreement is difficult, having something in place to protect water quality while not lessening the state's authority is beneficial.[41]

Industry Initiatives

Pressure from environmental advocates, coupled with the industry's strong desire to promote a positive image, have led the cruise ship industry to respond with several initiatives. In 2001, members of the International Council of Cruise Lines (ICCL), which represents 15 of the world's largest cruise lines, adopted a set of waste management practices and procedures for their worldwide operations building on regulations of the IMO and U.S. EPA. The guidelines generally require graywater and blackwater to be discharged only while a ship is underway and at least 4 miles from shore and require that hazardous wastes be recycled or disposed of in accordance with applicable laws and regulations.

Twelve major cruise line companies also have implemented Safety Management System (SMS) plans for developing enhanced wastewater systems and increased auditing oversight. These SMS plans are certified in accordance with the IMO's International Safety Management Code. The industry also is working with equipment manufacturers and regulators to develop and test technologies in areas such as lower emission turbine engines and ballast water management for elimination of non-native species. Environmental groups commend industry for voluntarily adopting improved management practices but also believe that enforceable standards are preferable to voluntary standards, no matter how well intentioned.[42]

The ICCL joined with the environmental group Conservation International (CI) to form the Ocean Conservation and Tourism Alliance to work on a number of issues. In December 2003 they announced conservation efforts in four areas to protect biodiversity in coastal areas: improving technology for wastewater management aboard cruise ships, working with local governments to protect the natural and cultural assets of cruise destinations, raising passenger and crew awareness and support of critical conservation issues, and educating vendors to lessen the environmental impacts of products from cruise ship suppliers. Because two-thirds of the top cruise destinations in the world are located in the Caribbean and Mediterranean, two important biodiversity regions, in March 2006 ICCL and CI announced a joint initiative to develop a map integrating sensitive marine areas into cruise line navigational charts, with the goal of protecting critical marine and coastal ecosystems.

In May 2004, Royal Caribbean Cruises Ltd. announced plans to retrofit all vessels in its 29-ship fleet with advanced wastewater treatment technology by 2008, becoming the first cruise line to commit to doing so completely. The company had been the focus of efforts by the environmental group Oceana to pledge to adopt measures that will protect the ocean environment and that could serve as a model for others in the cruise ship industry, in part because of the company's efforts to alter its practices following federal enforcement actions in the 1990s for environmental violations that resulted in RCCL paying criminal fines that totaled $27 million.

ISSUES FOR CONGRESS

Concerns about cruise ship pollution raise issues for Congress in three broad areas: adequacy of laws and regulations, research needs, and oversight and enforcement of existing programs and requirements. Attention to these issues is relatively recent, and more assessment is needed of existing conditions and whether current steps (public and private) are adequate. Bringing the issues to national priority sufficient to obtain resources that will address the problems is a challenge.

Laws and Regulations

A key issue is whether the several existing U.S. laws, international protocols and standards, state activities, and industry initiatives described in this report adequately address management of cruise ship pollution, or whether legislative changes are needed to fill in gaps, remedy exclusions, or strengthen current requirements. As noted by EPA in its 2000 white paper, certain cruise ship waste streams such as oil and solid waste are regulated under a comprehensive set of laws and regulations, but others, such as graywater, are excluded or treated in ways that appear to leave gaps in coverage.[43] Graywater is one particular area of interest, since investigations, such as sampling by state of Alaska officials, have found substantial contamination of cruise ship graywater from fecal coliform, bacteria, heavy metals, and dissolved plastics. State officials were surprised that graywater from ships' galley and sink waste streams tested higher for fecal coliform than did the ships' sewage lines.[44] One view advocating strengthened requirements came from the U.S. Commission on Ocean Policy. In its 2004 final report, the Commission advocated clear, uniform requirements for controlling the discharge of wastewater from large passenger vessels, as well as consistent interpretation and enforcement of those requirements. It recommended that Congress establish a new statutory regime that should include:

- uniform discharge standards and waste management procedures.
- thorough recordkeeping requirements to track the waste management process.
- required sampling, testing, and monitoring by vessel operators using uniform protocols ! flexibility and incentives to encourage industry investment in innovative treatment technologies.[45]

A proposal reflecting some of these concepts, the Clean Cruise Ship Act, was introduced in the 109th Congress as S. 793 (Durbin) and H.R. 1636 (Farr), but Congress did not act on either bill. The bills were free-standing legislation that would not have amended any current law, nor ratified Annex IV of MARPOL. The legislation would have prohibited cruise vessels entering a U.S. port from discharging sewage, graywater, or bilge water into waters of the United States, including the Great Lakes, except in compliance with prescribed effluent limits and management standards. It further would have directed EPA and the Coast Guard to promulgate effluent limits for sewage and graywater discharges from cruise vessels that were no less stringent than the more restrictive standards under the existing federal Alaska cruise ship law described above. It would have required cruise ships to treat wastewater wherever they operate and authorized broadened federal enforcement authority, including inspection, sampling, and testing. Environmental advocates supported this legislation. Industry groups argued that it targeted an industry that represents only a small percentage of the world's ships and that environmental standards of the industry, including voluntary practices, already meet or exceed current international and U.S. regulations.

As noted above, some states have passed legislation to regulate cruise ship discharges. If this state-level activity increases, Congress could see a need to develop federal legislation that would harmonize differences in the states' approaches.

Another issue for Congress is the status of EPA's efforts to manage or regulate cruise ship wastes. As discussed previously, in 2000 Congress authorized EPA to issue standards for sewage and graywater discharges from large cruise ships operating in Alaska. The agency has been collecting information and assessing the need for additional standards, beyond those provided in P.L. 106-554, but has not yet proposed any rules. Further, EPA is reportedly preparing a Cruise Ship Assessment Report that builds on the 2000 White Paper and is intended to respond to the 2000 petition by Bluewater Network and other groups that seek to force EPA to address cruise ship pollution (see page 2). Some of these groups and Senator Durbin (sponsor of S. 793 in the 109th Congress) have pressed EPA for more action. An agency spokesman reportedly stated in September 2006 that EPA plans to release a draft of the cruise ship discharge assessment report for public comment in the fall of 2007. However, as noted previously, in May 2007, environmental groups sued EPA, seeking to compel the agency to act on the 2000 petition.[46]

Other related issues of interest could include harmonizing the differences presented in U.S. laws for key jurisdictional terms as they apply to cruise ships and other types of vessels; providing a single definition of "cruise ship," which is defined variously in federal and state laws and rules, with respect to gross tonnage of ships, number of passengers carried, presence of overnight passenger accommodations, or primary purpose of the vessel; or requiring updating of existing regulations to reflect improved technology (such as the MSD rules that were issued in 1976).

Research

Several areas of research might help improve understanding of the quantities of waste generated by cruise ships, impacts of discharges and emissions, and the potential for new control technologies. EPA's Cruise Ship Discharge Assessment Report, when completed, may answer some of these questions. The U.S. Commission on Ocean Policy noted in its 2004 final report that research can help identify the degree of harm represented by such activities and can assist in prioritizing limited resources to address the most significant threats. The commission identified several directions for research by the Coast Guard, EPA, NOAA, and other appropriate entities on the fates and impacts of vessel pollution:[47]

- Processes that govern the transport of pollutants in the marine environment.
- Small passenger vessel practices, including the impacts of stationary discharges.
- Disposal options for concentrated sludge resulting from advanced sewage treatment on large passenger vessels.

- Cumulative impacts of commercial and recreational vessel pollution on particularly sensitive ecosystems, such as coastal areas with low tidal exchange and coral reef systems.
- Impacts of vessel air emissions, particularly in ports and inland waterways where the surrounding area is already having difficulty meeting air quality standards.

Oversight and Enforcement

The 2000 GAO report documented — and EPA's 2000 cruise ship white paper acknowledged — that existing laws and regulations may not be adequately enforced or implemented. GAO said there is need for monitoring of the discharges from cruise ships in order to evaluate the effectiveness of current standards and management. GAO also said that increased federal oversight of cruise ships by the Coast Guard and other agencies is needed concerning maintenance and operation of pollution prevention equipment, falsifying of oil record books (which are required for compliance with MARPOL), and analysis of records to verify proper off-loading of garbage and oily sludge to onshore disposal facilities.[48]

The Coast Guard has primary enforcement responsibility for many of the federal programs concerning cruise ship pollution. A key oversight and enforcement issue is the adequacy of the Coast Guard's resources to support its multiple homeland and non-homeland security missions. The resource question as it relates to vessel inspections was raised even before the September 11 terrorist attacks, in the GAO's 2000 report. The same question has been raised since then, in light of the Coast Guard's expanded responsibilities for homeland security and resulting shift in operations, again by the GAO and others.[49]

In its 2000 report, GAO also found that the process for referring cruise ship violations to other countries does not appear to be working, either within the Coast Guard or internationally, and GAO recommended that the Coast Guard work with the IMO to encourage member countries to respond when pollution cases are referred to them and that the Coast Guard make greater efforts to periodically follow up on alleged pollution cases occurring outside U.S. jurisdiction.

REFERENCES

[1]   Lloyd's Maritime Information Services, on the website of the Maritime International Secretaries Services, Shipping and World Trade Facts, at [http://www.marisec.org/shippingfacts/keyfactsnoofships.htm]

[2]   International Council of Cruise Lines, "The Cruise Industry, 2005 Economic Summary."

[3]   U.S. Environmental Protection Agency, "Cruise Ship White Paper," August 22, 2000, p. 3. Hereafter, EPA White Paper.

[4]   Bureau of Transportation Statistics, Department of Transportation, "Summary of Cruise Ship Waste Streams."

[5]   U.S. General Accounting Office, *Marine Pollution: Progress Made to Reduce Marine Pollution by Cruise Ships, but Important Issues Remain*, GAO/RCED-00-48, February 2000. 70 pp. Hereafter, 2000 GAO Report.

[6]   Bluewater Network, Petition to the Administrator, U.S. Environmental Protection Agency, March 17, 2000.

[7]   EPA White Paper.

[8]   The petition was amended in 2000 to request that EPA also examine air pollution from cruise ships; see discussion below.

[9]   The Ocean Conservancy, "Cruise Control, A Report on How Cruise Ships Affect the Marine Environment," May 2002, p. 13. Hereafter, "Cruise Control."

[10] Ibid., p. 15.

[11] The Center for Environmental Leadership in Business, "A Shifting Tide, Environmental Challenges and Cruise Industry Responses," p. 14. Hereafter, "Shifting Tide."

[12] Bluewater Network, "Cruising for Trouble: Stemming the Tide of Cruise Ship Pollution," March 2000, p. 5. Hereafter, "Cruising for Trouble." A report prepared for an industry group estimated that a 3,000-person cruise ship generates 1.1 million gallons of graywater during a seven-day cruise. Don K. Kim, "Cruise Ship Waste Dispersion Analysis Report on the Analysis of Graywater Discharge," presented to the International Council of Cruise Lines, September 14, 2000.

[13] National Research Council, Committee on Shipboard Wastes, *Clean Ships, Clean Ports, Clean Oceans: Controlling Garbage and Plastic Wastes at Sea* (National Academy Press, 1995), table 2-3, pp. 38-39.

[14] Ibid., p. 126.

[15] "Shifting Tide," p. 16.

[16] Statement of Catherine Hazelwood, The Ocean Conservancy, "Ballast Water Management: New International Standards and NISA Reauthorization," Hearing, House Transportation and Infrastructure Subcommittee on Water Resources and Environment, 108th Cong., 2nd sess., March 25, 2004.

[17] David Pimentel, Lori Lach, Rodolfo Zuniga, and Doug Morrison, "Environmental and Economic Costs Associated with Non-indigenous Species in the United States," presented at AAAS Conference, Anaheim, CA, January 24, 1999.

[18] 68 *Federal Register* 9751, 9753, February 28, 2003.

[19] Ibid., pp. 9751, 9756.

[20] The majority of cruise ships are foreign-flagged, primarily in Liberia and Panama. Both of these countries have ratified all six of the MARPOL annexes. For information, see [http://www.imo.org/].

[21] The 109th Congress also considered but did not enact legislation to implement the standards in Annex VI, as part of H.R. 5681, the Coast Guard Authorization Act of 2006. The House passed H.R. 5681 on September 28, 2006. Title V of the bill incorporated the text of H.R. 5811, the MARPOL Annex VI Implementation Act of 2006, which was separately approved by the House Transportation and Infrastructure Committee (H.Rept. 109-667). The Senate did not act on H.R. 5681 before the 109th Congress adjourned in December 2006.

[22] 2000 GAO Report, pp. 19-21.

[23] 2000 GAO Report, pp. 34-35, 13.

[24] The Homeland Security Act of 2002 (P.L. 107-296) transferred the entirety of the Coast Guard from the Department of Transportation to the Department of Homeland Security. For discussion, see CRS Report RS21125, *Homeland Security: Coast Guard Operations —Background and Issues for Congress.*

[25] EPA White Paper, p. 10.

[26] "Cruising for Trouble," p. 5.

[27] The 1988 Shore Protection Act (33 U.S.C. 2601-2603) prohibits vessels from transporting municipal or commercial waste in U.S. coastal waters without a permit issued by the Department of Transportation. It was intended to minimize trash, medical debris, and potentially harmful materials from being deposited in U.S. coastal waters. However, its provisions exclude waste generated by a vessel during normal operations and thus do not apply to cruise ships.

[28] 68 *Federal Register* 53165, September 9, 2003.

[29] In 1990, Congress enacted the Non-indigenous Aquatic Nuisance Prevention and Control Act (16 U.S.C. 4701 et seq) to focus federal efforts on non-indigeous, invasive, aquatic nuisance species, specifically when such species occur in ballast water discharges. That law, as amended by the National Invasive Species Act of 1996, delegated authority to the Coast Guard to establish a phased-in regulatory program for ballast water.

[30] For information, see CRS Report RL32344, *Ballast Water Management to Combat Invasive Species.*

[31] U.S. Environmental Protection Agency, "Final Rule, Control of Emissions from New Marine Compression-Ignition Engines at or Above 30 Liters Per Cylinder," 68 *Federal Register* 9746-9789, February 28, 2003.

[32] For information, see [http://www.earthjustice.org/news/display.html?ID=53] and [http:// www.earthjustice.org/urgent/display.html?ID=158].

[33] Annex VI, which came into force internationally in May 2005, also regulates ozone-depleting emissions, sulfur oxides, and shipboard incineration, but there are no restrictions on particulate matter, hydrocarbons, or carbon monoxide.

[34] *Bluewater Network v. EPA,* D.C.Cir., No. 03-1120, June 22, 2004.

[35] For an explanation of these terms, see [http://encyclopedia.thefreedictionary.com/ Statute%20mile].

[36] In 2003, the number of cruise ship passengers in Southeast Alaska was about 800,000, with tens of thousands of crew, in addition. By comparison, the state's population is approximately 650,000. Roughly 95% of the current cruise ship traffic is concentrated in Southeast Alaska, a region with a population of approximately 73,000 people. Alaska Department of Environmental Conservation, Commercial Passenger Vessel Environmental Compliance Program, "Assessment of Cruise Ship and Ferry Wastewater Impacts in Alaska," February 9, 2004, p. 8. Hereafter, "Assessment of Impacts in Alaska."

[37] As part of its efforts to develop these vessel discharge standards, in the summer of 2004 EPA sampled wastewater from four large cruise ships operating in Alaska waters in order to evaluate the performance of various treatment systems. Results of this sampling are available at [http://www.epa.gov/owow/oceans/cruise_ships/results.html.

[38] 66 *Federal Register* 38926, July 26, 2001.

[39] "Assessment of Impacts in Alaska," pp. 33-57.

[40] Washington State Department of Ecology, Water Quality Program, "Focus on: Cruise Ship Discharges. Draft — Memorandum of Understanding (MOU)," April 10, 2004, p. 2.

[41] State of Washington. Department of Ecology. "2004 Assessment of Cruise Ship Environmental Effects in Washington." January 2005. 22 p.

[42] "Cruise Control," p. 25.

[43] EPA White Paper, p. 16.

[44] "Assessment of Impacts in Alaska," p. 12.

[45] U.S. Commission on Ocean Policy, "An Ocean Blueprint for the 21st Century" September 2004, p. 243.

[46] Friends of the Earth v. Stephen L. Johnson, D.D.C., Civ. Action No. 07-CV-0872, filed May 9, 2007.

[47] Ibid., p. 249.

[48] 2000 GAO Report, p. 34.

[49] U.S. General Accounting Office, *Coast Guard: Relationship between Resources Used and Results Achieved Needs to be Clearer*, GAO-04-432, March 2004. Also see CRS Report RS21125, *Homeland Security: Coast Guard Operations — Background and Issues for Congress.*

In: Water Pollution Issues and Developments
Editor: S. V. Thomas, pp. 25-49

ISBN: 978-1-60456-208-8
© 2008 Nova Science Publishers, Inc.

Chapter 2

MTBE IN GASOLINE: CLEAN AIR AND DRINKING WATER ISSUES[*]

James E. McCarthy and Mary Tiemann

ABSTRACT

As gasoline prices have risen in March and April 2006, renewed attention has been given to methyl tertiary butyl ether (MTBE), a gasoline additive being phased out of the nation's fuel supply. Many argue that the phaseout of MTBE and its replacement by ethanol have been a major factor in driving up prices.

MTBE has been used by refiners since the late 1970s. It came into widespread use when leaded gasoline was phased out — providing an octane boost similar to that of lead, but without fouling the catalytic converters used to reduce auto emissions since the mid-1970s. MTBE has also been used to produce cleaner-burning Reformulated Gasoline (RFG), which the Clean Air Act has required in the nation's most polluted areas since 1995. The act didn't mandate the use of MTBE (ethanol or other substances could have been used to meet the act's oxygenate requirement), but price and handling characteristics of the additive led to its widespread use.

Under the Energy Policy Act of 2005 (P.L. 109-58), the RFG program's oxygen mandate terminates on May 6, 2006, and refiners are scrambling to remove MTBE from the nation's gasoline supply by that date. The phaseout of MTBE (like its use) is not required by federal law, but gasoline refiners have focused on the May 6 date because of concerns over their potential liability for its continued presence. MTBE has contaminated drinking water in a number of states, and about half have passed legislation to ban or restrict its use. Hundreds of suits have been filed to require petroleum refiners and marketers to pay for cleanup of contaminated water supplies, the cost of which has been estimated to be in the billions of dollars. The petroleum industry has maintained that it used MTBE to meet the RFG program's oxygen mandate and therefore should not be held liable. That position could become more difficult to maintain once the oxygen mandate is removed.

To replace MTBE, refiners are switching to ethanol as swiftly as they can, leading temporarily to supply shortages and higher prices. The ethanol industry maintains that there will be sufficient ethanol to meet demand but concedes that temporary shortages

---

[*] Excerpted from CRS Report RL32787, dated April 14, 2006.

exist in some parts of the country that could affect prices until the end of June. These shortages and higher prices have led to renewed discussion by some of exempting gasoline refiners from liability for MTBE cleanup (a so-called "safe harbor" provision). Others have renewed their call for federal legislation to stimulate the construction of new refining capacity.

Besides removing the RFG program's oxygen requirement, Congress provided a major incentive to the production of ethanol in the Energy Policy Act of 2005. Under a Renewable Fuels Standard, an increasing amount of the nation's motor fuels must consist of renewable fuel, such as ethanol. The law requires 4.0 billion gallons in 2006 (a level already being achieved) and an increase of 700 million gallons each year through 2011, before reaching 7.5 billion gallons in 2012.

This report provides background regarding MTBE and summarizes the actions taken by states and Congress to address problems raised by MTBE contamination of the nation's water supplies. It will be updated if future developments warrant.

INTRODUCTION

This report provides background information concerning the gasoline additive methyl tertiary butyl ether (MTBE), discusses air and water quality issues associated with it, and reviews options available to congressional and other policy-makers concerned about its continued use. It includes a discussion of legislation in the 109th Congress.

Under the Clean Air Act Amendments of 1990, numerous areas with poor air quality were required to add chemicals called "oxygenates" to gasoline as a means of improving combustion and reducing emissions. The act had two programs that required the use of oxygenates, but the more significant of the two was the reformulated gasoline (RFG) program, which took effect January 1, 1995.[1] Under the reformulated gasoline program, areas with "severe" or "extreme" ozone pollution (124 counties with a combined population of 73.6 million) must use reformulated gasoline; areas with less severe ozone pollution may opt into the program as well, and many have. In all, portions of 17 states and the District of Columbia use reformulated gasoline (see table 1 and figure 1); about 30% of the gasoline sold in the United States is RFG.

The law required that RFG contain at least 2% oxygen by weight. Refiners could meet this requirement by adding a number of ethers or alcohols, any of which contain oxygen and other elements. Because these substances are not pure oxygen, the amount used to obtain a 2% oxygen level is greater than 2% of the gasoline blend. For example, MTBE is only 19% oxygen and, thus, RFG made with MTBE needed to contain 11% MTBE by volume to meet the 2% requirement.

By far the most commonly used oxygenate has been MTBE. In 1999, 87% of RFG contained MTBE. As restrictions on MTBE use took effect in California, New York, and Connecticut at the end of 2003, this number was reduced, but even with these state bans, 46% of RFG nationally contained MTBE in 2004.

Table 1. Areas Using Reformulated Gasoline, as of February 2005

Mandatory RFG Areas[a]
Baltimore, MD
Chicago, IL (and portions of Indiana and Wisconsin)[b]
District of Columbia (and suburbs in MD and VA)
Hartford, CT
Houston, TX
Los Angeles, CA
Milwaukee, WI[b]
New York, NY (and portions of CT and NJ)
Philadelphia, PA (and portions of DE, MD, and NJ)
Sacramento, CA
San Diego, CA
San Joaquin Valley, CA
Southeast Desert, CA
Ventura County, CA
Opt-In RFG Areas[c]
Connecticut (entire state)
Dallas / Fort Worth, TX
Delaware (entire state)
Kentucky portion of Cincinnati metropolitan area
Louisville, KY
Massachusetts (entire state)
New Hampshire portion of Greater Boston
New Jersey (entire state)
New York (counties near New York City)
Rhode Island (entire state)
St. Louis, MO
Virginia (Richmond, Norfolk - Virginia Beach - Newport News)

Source: U.S. EPA.

Notes:

a. RFG use required by the Clean Air Act. In addition to these areas, Atlanta, GA, and Baton Rouge, LA, are now also required to use RFG because they have been reclassified as severe ozone nonattainment areas; but implementation of the RFG requirement has been stayed in both areas pending the resolution of court challenges.

b. In the Chicago and Milwaukee areas, RFG has been made with ethanol rather than MTBE since 1995.

c. RFG use required by State Implementation Plan as a means of attaining the ozone air quality standard. These "opt-in" areas may opt out of the program by substituting other control measures achieving the necessary reductions in emissions.

Federal RFG & Winter OXY/RFG Programs

February, 2000

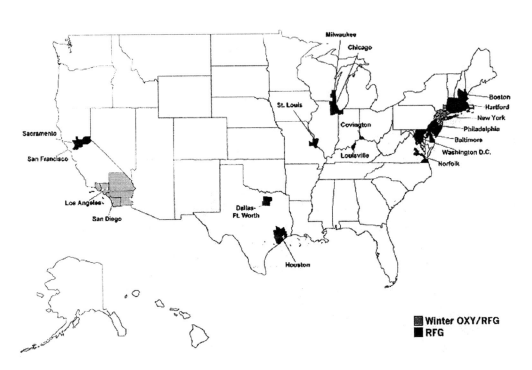

Figure 1. Federal RFG and Winter OXY/RFG Programs.

Also, MTBE has been used since the late 1970s in gasoline as an octane enhancer. MTBE use grew rapidly in the 1980s, as it replaced lead in gasoline and was used in premium fuels. As a result, gasoline with MTBE has been used virtually everywhere in the United States, whether or not an area has been subject to RFG requirements.

AIR QUALITY BENEFITS RESULTING FROM MTBE USE

State and local environmental agencies and EPA attribute marked improvements in air quality to the use of fuels containing MTBE and other oxygenates, but the exact role of oxygenates in achieving these improvements is subject to debate. In Los Angeles, which has had the worst air quality in the country, the use of reformulated gasoline was credited with reducing ground-level ozone by 18% during the 1996 smog season, compared to weather-adjusted data for the same period in 1994 and 1995. Use of RFG also reduced the cancer risk associated with exposure to vehicle emissions by 30% to 40%, according to the California EPA, largely because it uses less benzene, a known human carcinogen.[2]

Whether the oxygenates themselves should be given credit for these improvements has been the subject of debate, with the answer depending to some extent on what one assumes would replace the oxygenates if they were removed. Asked to look at the ozone-forming potential of different oxygenates used in reformulated gasoline, a National Academy of

Sciences panel concluded that "the addition of commonly available oxygenates to RFG is likely to have little air-quality impact in terms of ozone reduction."[3] An EPA advisory panel, by contrast, concluded that the use of oxygenates "appears to contribute to reduction of the use of aromatics with related toxics and other air quality benefits."[4]

Less controversy exists regarding oxygenates' role in reducing carbon monoxide emissions. Both EPA and an interagency group chaired by the White House Office of Science and Technology Policy (OSTP) have reported improvements in carbon monoxide (CO) levels due to the use of oxygenates. According to the June 1997 OSTP report, "analyses of ambient CO measurements in some cities with winter oxygenated gasoline programs find a reduction in ambient CO concentrations of about 10%."[5]

EPA also "believes that the reductions estimated in air quality studies are significant and that these reductions help to protect the public from the adverse health effects associated with high levels of CO in the air."[6] The agency based its conclusions both on its own analysis and on a report prepared for two industry groups. The latter, using hourly data for more than 300 monitoring sites gathered over a nine-year period, concluded that use of oxygenated fuels was associated with a 14% reduction in ambient CO concentrations.[7]

HEALTH-RELATED QUESTIONS

The improvements in measured air quality have not come without questions. After oxygenated fuels containing MTBE were introduced, residents in several cities complained of a variety of health effects from exposure to MTBE/RFG exhaust: headaches, dizziness, nausea, sore eyes, and respiratory irritation. Some complaints centered around the use of MTBE in cold weather; two of the principal areas noting complaints were Alaska and Milwaukee, Wisconsin. The Interagency Task Force examined these complaints and concluded:

> With regard to exposures ... experienced by the general population and motorists, the limited epidemiological studies and controlled exposure studies conducted to date do not support the contention that MTBE as used in the winter oxygenated fuels program is causing significant increases over background in acute symptoms or illnesses.[8]

Additional health effects research is being conducted by EPA, universities, and others. Under the authority of Section 211 of the Clean Air Act, EPA has requested refiners to conduct health effects studies on conventional, reformulated, and oxygenated (particularly MTBE-oxygenated) gasoline. These studies examine health effects associated with the inhalation of evaporative emissions, and several are near completion. Very little research has been done to assess the potential health risks associated with exposure to MTBE in drinking water (ingestion).

Much discussion has centered on whether MTBE has the potential to cause cancer. Although there are no studies on the carcinogenicity of MTBE in humans, EPA's Office of Research and Development (ORD) reported in 1994 that

> inhalation carcinogenicity studies in mice and rats show evidence of three types of animal tumors [testicular, liver, and kidney]. These particular studies are difficult to

interpret because of some high-dose general toxicity. Nevertheless, ORD believes the inhalation carcinogenicity evidence would support placing MTBE in Group C as a "possible human carcinogen."[9]

Also, one metabolite of MTBE (formaldehyde) is considered a probable human carcinogen, and another metabolite (tertiary butyl alcohol (TBA)) induces male rat kidney tumors.[10]

Based on animal studies, EPA has concluded that MTBE poses a potential for carcinogenicity to humans at high doses; however, because of uncertainties and limitations in the data, EPA has been unable to make a confident estimation of risk at low exposure levels.[11] The Interagency Task Force assessing oxygenated fuels concluded that the weight of the evidence supports regarding MTBE as having a carcinogenic hazard potential for humans.[12]

In 1998, the International Agency for Research on Cancer (IARC) and the U.S. National Toxicology Program determined not to list MTBE as a known human carcinogen. The IARC noted that MTBE was "not classifiable as to its carcinogenicity in humans," based on inadequate evidence in humans and limited evidence in experimental animals.[13] In 1999, California's Environmental Protection Agency determined that the MTBE carcinogenicity studies were of similar quality to studies on many other carcinogens, and established a public health goal for MTBE in drinking water based on cancer risk.[14]

Regarding noncancer effects, a California advisory committee determined that there was not clear scientific evidence to support listing MTBE as a toxic substance affecting human development or reproduction. In reviewing the research on cancer and noncancer effects, these groups generally noted that research gaps exist, and that the data were particularly limited on health effects associated with MTBE ingestion.

In response to the need for research to evaluate the potential health risks from exposure to MTBE and other oxygenates in drinking water, EPA in 1998 published a document that identified the most critical and immediate research needs. The document was intended to serve as a guide to planning future research; however, EPA has not pursued research to address the needs identified in this document.[15]

For practical purposes, the interpretation of any health risks associated with the addition of MTBE to gasoline could benefit from a comparison to the health risks associated with conventional gasoline. The Interagency Task Force, EPA, and some environmental groups have all argued that current knowledge suggests that MTBE is a less serious pollutant than the gasoline components it replaced. According to the OSTP report, the cancer risk from exposure to MTBE is "substantially less than that for benzene, a minor constituent of gasoline that is classified as a known human carcinogen; and more than 100 times less than that for 1,3-butadiene, a carcinogenic emission product of incomplete fuel combustion."[16] Such a comparison might be of limited usefulness, however, given the data gaps regarding MTBE's health effects and MTBE's ability to reach water supplies more readily than conventional gasoline.

WATER QUALITY AND DRINKING WATER ISSUES

A major issue regarding the use of MTBE concerns its detection in ground water at thousands of locations nationwide, and, usually at low levels, in various municipal drinking water supplies, private wells, and reservoirs. Although MTBE has provided air quality benefits, the inclusion of MTBE in gasoline has been a growing concern as an environmental risk since the 1980s, for several reasons. Specifically, compared to other gasoline components, MTBE (1) is much more soluble in water, (2) has a lower taste and odor threshold, (3) has a higher transport rate, and (4) often requires more time to be remediated and must be treated by more complicated and expensive treatment technologies.[17] MTBE is extremely soluble and, once released, it moves through soil and into water more rapidly than other chemical compounds present in gasoline. Once in ground water, it is slow to biodegrade and is more persistent than other gasoline-related compounds. In surface water, it dissipates more rapidly.

Studies show that most of it evaporates from the upper levels of surface water in a few weeks, while it persists longer at greater depths.[18]

The primary source of MTBE in ground water has been petroleum releases from leaking underground storage tank (UST) systems. Other significant sources include leaking above-ground storage tanks, fuel pipelines, refueling facilities, and accidental spills. The most significant source of MTBE in lakes and reservoirs appears to be exhaust from motorized watercraft, while smaller sources include gasoline spills, runoff, and ground water flow.[19]

Occurrence of MTBE in Drinking Water

Available information on the occurrence of MTBE in public drinking water supplies has increased substantially in recent years, but has been somewhat limited geographically. Although a number of serious contamination incidents have been reported, particularly in California, the available data generally do not indicate a broad presence of MTBE in drinking water supplies at levels of public health concern. However, as monitoring has increased among the states, so has the number of public water systems and private wells showing low-level detections of MTBE.

The most extensive MTBE monitoring data for drinking water are available for California, where testing for MTBE was made mandatory for most water systems in 1997. Through April 2002, some 2,957 systems had tested 9,905 sources of drinking water. MTBE was detected in 85 (0.9%) of these sources, including 54 (0.6%) of 9,234 ground water sources and 31 (4.6%) of 671 surface water sources. Overall, 53 (1.8%) of the 2,957 public water systems reported detections of MTBE in at least one of their drinking water sources, and 13 (0.4%) of the systems reported that a total of 21 (0.2%) sources of water had MTBE concentrations exceeding California's MTBE drinking water standard of 13 micrograms per liter (:g/L). As of October 2005, monitoring results had been reported for 13,620 sources. Nearly all of these results were nondetections, while 113 sources had two or more MTBE detections.[20]

In 1998, the state of Maine tested nearly 800 public water supplies and 950 randomly selected private wells and found detectable levels of MTBE in 16% of the public water supplies and 15.8% of the private wells. None of the public water supply samples exceeded the state drinking water standard of 35 :g/L, while 1% of private well samples contained MTBE concentrations above the standard. Roughly 94% of public water supply samples showed MTBE levels that were either not detectable or below 1 :g/L; the remaining 6% of samples were between 1 :g/L and 35 :g/L.[21]

Nationwide, the data on the presence of MTBE in drinking water have been more limited. In July 1999, the EPA-appointed Blue Ribbon Panel on Oxygenates in Gasoline reported that between 5% and 10% of drinking water supplies tested in high oxygenate-use areas show at least detectable amounts of MTBE, and that the vast majority of these detections have been well below levels of public health concern, with roughly 1% of detections exceeding 20

:g/L.[22]

In a study completed in 2001, the United States Geological Survey (USGS), in cooperation with EPA, assessed the occurrence of MTBE and other volatile organic compounds (VOCs) in public water supplies in 10 mid-Atlantic and northeastern states where MTBE use is common.[23] The study analyzed water from 1,194 randomly selected community water systems. The USGS reported that MTBE was detected in 8.9% of the tested water systems and was strongly associated with areas where reformulated and/or oxygenated (RFG/OXY) fuels are used. Fifteen percent of systems in RFG/OXY areas reported detecting MTBE at concentrations of 1 :g/L or more, while 3% of systems outside of RFG/OXY areas reported such detections. Most MTBE concentrations ranged from 0.5 to 5 :g/L, and less than 1% of the systems reported MTBE at levels equal to or exceeding 20 :g/L, the lower limit of EPA's drinking water advisory.[24]

A 2003 nationwide survey conducted by the American Water Works Association Research Foundation (AWWARF) reported similar results. This survey monitored sources of drinking water for 954 randomly selected community water systems (including 579 samples from groundwater-supplied systems and 375 samples from surface-water-supplied systems). MTBE was found in 8.7% of the community water system source waters, at concentrations ranging from 0.2 to 20 :g/L.[25] AWWARF also conducted a focused survey, including 451 samples collected from 134 community water systems source waters (including ground water, reservoirs, lakes, rivers, and streams) that were suspected or known to contain MTBE. The researchers found MTBE in 55.5% of the water systems.[26]

Occurrence of MTBE in Ambient Ground Water

Looking at ground water generally (not only drinking water wells), the data indicate that low levels of MTBE are found often. Nationally, the most comprehensive ground water research has been conducted by the USGS through the National Water Quality Assessment Program (NAWQA). USGS data for some 2,743 monitoring, observation, and water supply wells in 42 states (from 1993 to 1998) showed MTBE present in about 5% (145) of the wells, with MTBE levels exceeding 20 :g/L in 0.5% (12) of the wells. In all, MTBE was detected in ground water in 22 of the 42 states. The USGS further evaluated the occurrence data based on whether or not detections occurred in RFG or winter oxyfuel program areas. The researchers reported that low concentrations of MTBE were detected in 21% of ambient ground water samples in high MTBE-use areas and in 2.3% of samples in low or no-MTBE use areas.[27]

MTBE has been detected most frequently in ground water associated with leaking underground storage tank (UST) sites. The California Environmental Protection Agency has estimated that, based on monitoring information available for these sites, MTBE can be expected to be found in shallow, unused ground water at thousands of UST sites in the state, and often at high concentrations (in the parts per million range).[28] Moreover, a report by the Lawrence Livermore National Laboratory found that MTBE was not significantly degrading in the monitoring networks for these leaking UST sites.[29] The situation in other states may be similar. In a September 2000 survey of state leaking underground storage tank (LUST) programs, 31 states reported that MTBE was found in ground water at 40% or more of gasoline-contaminated sites in their states; 24 states reported MTBE at 60% to 100% of sites.[30] A 2003 update to that survey found that, averaged among the states, MTBE was found in groundwater at 60% of gasoline-contaminated sites.[31]

EPA'S RESPONSES TO MTBE OCCURRENCE IN WATER

Safe Drinking Water Act Initiatives

MTBE has not been regulated under the Safe Drinking Water Act (SDWA), but to address concerns raised by the detection of MTBE in ground water and drinking water supplies, EPA has pursued several initiatives. In December 1997, the agency issued a drinking water advisory for MTBE based on consumer acceptability (for taste and smell). EPA issues drinking water advisories to provide information on contaminants in drinking water that have not been regulated under SDWA.[32] Advisories are not enforceable, but provide guidance to water suppliers and other interested parties regarding potential health effects or consumer acceptability. While the MTBE advisory is not based on health effects, EPA notes that keeping MTBE levels in the range of 20-40 :g/L or lower for consumer acceptability reasons would also provide a large margin of safety from adverse health effects. Specifically, the advisory states that

> [c]oncentrations in the range of 20 to 40 :g/L are about 20,000 to 100,000 (or more) times lower than the range of exposure levels in which cancer or noncancer effects were observed in rodent tests. This margin of exposure is in the range of margins of exposure typically provided to protect against cancer effects by the National Primary Drinking Water Standards under the Federal Safe Drinking Water Act. This margin is greater than such standards typically provided to protect against noncancer effects. Thus, protection of the water source from unpleasant taste and odor as recommended will also protect consumers from potential health effects.[33]

In addition, EPA has taken steps that could lead to the development of an enforceable drinking water standard for MTBE. In February 1998, EPA included MTBE on a list of contaminants that are potential candidates for regulation under the Safe Drinking Water Act. Compounds on the contaminant candidate list are categorized as regulatory determination priorities, research priorities, or occurrence priorities. Because of data gaps on MTBE health effects and occurrence, EPA placed MTBE in the category of contaminants for which further occurrence data collection and health effects research are priorities. Thus, while EPA has not selected MTBE for regulation to date, the agency is pursuing research to fill the existing data gaps so that a regulatory determination may be made.

The Safe Drinking Water Act also directed EPA to publish a rule by August 1999 requiring public water systems to conduct monitoring for a list of unregulated contaminants that may require regulation. EPA included MTBE in this rule and directed large public water systems to begin monitoring for MTBE in January 2001.[34]

The occurrence data generated under the Unregulated Contaminant Monitoring Rule, combined with the results of ongoing health effects studies, are intended to provide information needed by EPA to make a regulatory determination for MTBE. Under SDWA, the next round of regulatory determinations will be made in 2006. EPA typically requires roughly three and one-half years to promulgate a drinking water regulation; thus, the earliest EPA would be expected to issue a drinking water regulation for MTBE is 2010.

Underground Storage Tank Regulation

A key EPA and state contamination prevention effort involves implementing the underground storage tank program established by the 1984 amendments to the Resource Conservation and Recovery Act (RCRA). Under this program, EPA has set operating requirements and technical standards for tank design and installation, leak detection, spill and overfill control, corrective action, and tank closure. As of 1993, all tanks were required to comply with leak detection regulations. Additionally, all tanks installed before December 1988 (when standards for new tanks took effect) were required to be upgraded, replaced, or closed by December 22, 1998.

Federal and state regulators anticipate that as tank owners and operators comply with these requirements, the number of petroleum and related MTBE leaks from UST systems should decline significantly. However, MTBE has been detected at thousands of leaking tank sites, and this additive is proving more difficult and costly to remediate than conventional gasoline. A key concern for states is that, as testing increases, it is likely that the number and scope of needed cleanups may increase as well. A 2003 state survey found that many sites have not been tested for MTBE, and most states do not plan to reopen previously closed Leaking Underground Storage Tank (LUST) sites to look for MTBE, although 32 states reported that MTBE plumes are often or sometimes longer than plumes from conventional gasoline leaks.[35] A key concern for community water suppliers and well owners is that fewer than half of the states are taking steps to ensure that MTBE and other oxygenates are not migrating beyond standard monitoring boundaries for LUST cleanup,[36] thus leaving an unknown number of MTBE plumes unremediated and ground water supplies at risk for future contamination.

In 1986, Congress created a federal response program for cleaning up releases from leaking petroleum USTs through the Superfund Amendments and Reauthorization Act, which amended RCRA Subtitle I. These provisions created the LUST Trust Fund and authorized EPA and states to use the fund to clean up underground storage tank spills and leaks in cases where tank owners or operators do not clean up sites. EPA and states use the annual trust fund appropriation primarily to oversee and enforce corrective actions performed by responsible parties. EPA and states also use fund monies to conduct corrective actions where no responsible party has been identified, where a responsible party fails to comply with a cleanup order, or in the event of an emergency, and to take cost recovery actions against parties. For FY2006, Congress provided $73 million from the LUST Trust Fund for states and EPA to administer the LUST remediation program. EPA allocates approximately 80% of the appropriated amount to the states.[37]

Since the federal underground storage tank program began, nearly 1.6 million of the roughly 2.2 million petroleum tanks subject to regulation have been closed, and, overall, the frequency of leaks from UST systems has been reduced. Through FY2005, 653,621 tanks subject to UST regulations remained in service, 452,041 releases had been confirmed, 421,924 cleanups had been initiated, and 332,799 cleanups had been completed. During FY2005, 7,421 releases were newly confirmed, compared with 8,850 in FY2004 and 12,000 in FY2003.[38]

BLUE RIBBON PANEL ON OXYGENATES IN GASOLINE

As part of its effort to gather information and focus research, in November 1998, EPA established an independent Blue Ribbon Panel on Oxygenates in Gasoline to review the broad range of issues posed by the use of MTBE and other oxygenates. The panel was established under the auspices of the Clean Air Act Advisory Committee, and its membership reflected a broad range of experts and stakeholders.[39] The panel:

- recommended that Congress act to remove the Clean Air Act requirement that 2% of RFG, by weight, consist of oxygen, in order to ensure that adequate fuel supplies can be blended in a cost-effective manner while reducing usage of MTBE;
- recommended that the winter oxygenated fuels program be continued;
- agreed broadly that use of MTBE should be reduced substantially (with some members supporting its complete phaseout), and that Congress should act to provide clear federal and state authority to regulate and/or eliminate the use of MTBE and other gasoline additives that threaten drinking water supplies;
- recommended that EPA seek mechanisms to ensure that there is no loss of current air quality benefits (i.e., no backsliding); and
- recommended a comprehensive set of improvements to the nation's water protection programs, including over 20 specific actions to enhance Underground Storage Tank, Safe Drinking Water, and private well protection programs.

The panel's numerous water protection recommendations addressed prevention, treatment, and remediation. For example, the panel recommended that EPA work with Congress to determine whether above-ground petroleum storage tanks (which generally are not regulated) should be regulated; work to enhance state and local efforts to protect lakes and reservoirs that serve as drinking water supplies by restricting use of recreational watercraft; and accelerate research for developing cost-effective drinking water treatment and remediation technologies.

The panel also suggested that EPA and others should accelerate ongoing health effects and environmental behavior research of other oxygenates and gasoline components that would likely increase in use in the absence of MTBE.

Then-EPA Administrator Carol Browner concurred with the recommendation of the Blue Ribbon Panel calling for a significant reduction in the use of MTBE. She also stated her commitment to work with Congress for "a targeted legislative solution that maintains our air quality gains and allows for the reduction of MTBE, while preserving the important role of renewable fuels like ethanol."[40]

On March 20, 2000, the former administrator announced that EPA would begin the process of issuing regulations to reduce or phase out use of MTBE. Recognizing that this process could take several years to complete, she renewed her call for congressional action to "amend the Clean Air Act to provide the authority to significantly reduce or eliminate the use of MTBE," to "ensure that air quality gains are not diminished," and to "replace the existing oxygen requirement contained in the Clean Air Act with a renewable fuel standard for all gasoline."[41]

In its few public statements on MTBE, the Bush Administration has not indicated any change in the Clinton Administration's policy, although EPA's effort to regulate MTBE using its existing authority slowed noticeably and now appear to have been terminated. Five years after EPA began the development of regulations to reduce or phase out MTBE, the agency quietly published a note in the *Federal Register* stating that its efforts to control MTBE were being "withdrawn."[42] This Administration, like the previous one, appears to have preferred a legislative solution.

STATE INITIATIVES

Among the states, California has arguably been the most active in addressing MTBE issues. Actions taken by the state legislature and the governor helped propel the issue to national prominence. Legislation signed October 8, 1997, required the state to set standards for MTBE in drinking water, and required the University of California to conduct a study of the health effects of MTBE and other oxygenates and risks associated with their use. The UC report, which was issued in November 1998, recommended a gradual phaseout of MTBE from gasoline in California.[43] Based on the report and on public hearings, Governor Davis issued a finding that "on balance, there is a significant risk to the environment from using MTBE in gasoline in California," and required the state's Energy Commission to develop a timetable for the removal of MTBE from gasoline at the earliest possible date, but not later than December 31, 2002. (This date was amended, in March 2002, to December 31, 2003.) The governor also required the California Air Resources Board (CARB) to make a formal request to U.S. EPA for a waiver from the requirement to use oxygenates in reformulated gasoline and required three state agencies to conduct additional research on the health and environmental impacts of ethanol, the most likely substitute for MTBE.

The waiver request resulted in months of negotiation between EPA and CARB, with EPA expressing skepticism that it had authority to grant a waiver under the circumstances.[44] More than two years later, on June 12, 2001, the agency finally denied California's request. Without a waiver, gasoline sold in ozone nonattainment areas in the state was required to contain another oxygenate once the MTBE ban took effect. During 2003, California's motor fuels gradually phased out MTBE in favor of ethanol.[45]

Following California's decision to phase out MTBE, at least 24 other states have acted to limit or phase out its use. The largest of these, New York, set a date of January 1, 2004, to ban MTBE. (table 2 summarizes state actions to ban MTBE.)

Table 2. State Actions Banning MTBE

State	Phaseout Date	Complete or Partial Ban?
AZ	1/1/05	Partial: no more than 0.3% (vol.) MTBE in gasoline
CA	12/31/03	Complete ban
CO	4/30/02	Complete ban
CT	1/1/04	Complete ban by 1/1/04, coordinated with NESCAUM (North East States for Coordinated Air Use Management) regional fuels task force
IL	7/24/04	Partial: may not use, sell, or manufacture MTBE as a fuel additive; may sell motor fuel containing no more than 0.5% (vol.) MTBE
IN	7/24/04	Partial: no more than 0.5% (vol.) MTBE in gasoline
IA	7/1/00	Partial: no more than trace amounts (0.5% by vol.) MTBE in motor vehicle fuel
KS	7/1/04	Partial: may not sell or deliver any motor vehicle fuel containing more than 0.5% (vol.) MTBE
KY	1/1/06	Partial: no more than trace amounts of MTBE in fuel
ME	1/1/07	Partial: no more than 0.5% (vol.) MTBE in gasoline sold
MI	6/1/03	Complete ban by 6/1/03; can be extended if determined by 6/1/02 that phaseout date is not achievable
MN	7/2/00 (partial) 7/2/05 (full)	Partial/then complete: no more than 1/3 of 1% oxygenate as of 7/2/00; complete ban as of 7/2/05. Ban also applies to ethyl tertiary butyl ether (ETBE) and tertiary amyl methyl ether (TAME)
MO	7/31/05	Partial: no more than 0.5% (vol.) MTBE in gasoline sold or stored
MT	1/1/06	Partial: no more than trace amounts in gasoline sold, stored, or dispensed
NE	7/13/00	Partial: no more than 1% (vol.) MTBE in any petroleum product
NH	1/1/07	Partial: no more than 0.5% (vol.) MTBE in gasoline sold or stored. Ban applies to other gasoline ethers and tertiary butyl alcohol (TBA)
NJ	1/1/09	Partial: no more than 0.5% (vol.) MTBE in gasoline distributed in commerce for sale in the state
NY	1/1/04	Complete ban as of 1/1/04
NC	1/1/08	Partial: no more than 0.5% (vol.) MTBE in motor fuel
OH	7/1/05	Partial: no more than 0.5% (vol.) MTBE in motor vehicle fuels
RI	6/1/07	Partial: no more than 0.5% (vol.) MTBE in gasoline sold, delivered, or imported. Ban also applies to other gasoline ethers, and TBA.
SD	7/1/01	Partial: no more than trace amounts (less than 0.5% vol.) resulting from commingling during storage or transfer
VT	1/1/07	Partial: no more than 0.5% (vol.) MTBE or other gasoline ethers in fuel products sold or stored
WA	1/1/04	Partial: may not be intentionally added to fuel or knowingly mixed in gasoline above 0.6% (vol.) Partial: no more than 0.5% (vol.) MTBE in gasoline

Source: Environmental Protection Agency, EPA 420-B-04-009, June 2004, updated July 2005 by CRS.

ALTERNATIVES TO MTBE

The major potential alternatives to MTBE are other oxygenates. Oxygenates possess several advantages, including high octane and the ability to replace toxic components of conventional gasoline.

Oxygenates that could replace MTBE include ethers, such as ethyl tertiary butyl ether (ETBE), and alcohols, such as ethanol. These other oxygenates may pose health and environmental impacts, but inadequate data make it difficult to reach definite conclusions. EPA's Blue Ribbon Panel concluded:

> The other ethers (e.g., ETBE, TAME, and DIPE) have been less widely used and less widely studied than MTBE. To the extent that they have been studied, they appear to have similar, but not identical, chemical and hydrogeologic characteristics. The Panel recommends accelerated study of the health effects and groundwater characteristics of these compounds before they are allowed to be placed in widespread use.[46]

Ethanol and other alcohols are considered relatively innocuous on their own; they generally do not persist in ground water and are readily biodegraded. However, research suggests that the presence of ethanol in a gasoline plume can extend the spread of benzene and other toxic constituents of gasoline through ground water.[47] This is largely because ethanol is likely to be degraded preferentially by microorganisms that would otherwise feed on other chemical components of gasoline, including benzene, toluene, ethylbenzene, and xylene (BTEX).

In announcing the phaseout of MTBE in his state on March 25, 1999, California's Governor Davis required three state agencies to conduct additional research on the health and environmental impacts of ethanol, the most likely substitute. In reports approved in January 2000, the agencies concluded that if ethanol were substituted for MTBE, there would be "some benefits in terms of water contamination" and "no substantial effects on public-health impacts of air pollution."[48]

A more recent article, based on the California ethanol review, focused specifically on the relative risks of ground water contamination by spills of ethanol-blended gasoline, MTBE-blended gasoline, and non-RFG gasoline. The authors concluded that

> relative to risks associated with standard formulation gasoline, *there is an increase in the risk that wells will be contaminated by RFG using either MTBE or ethanol as an oxygenate* [emphasis added]. With ethanol, the risk of contaminating wells decreases after approximately five years. However, the risk continues to grow for MTBE because of the assumption that this chemical is not degraded in the subsurface. The conservative approach used in this analysis, including the low biodegradation rates and assumption that the gasoline source areas are not remediated, results in an overstatement of the risks associated with these additives to gasoline. Nevertheless, the relative trends do favor ethanol when considering risk associated with RFG spills.[49]

The switch from MTBE to ethanol is not without technical problems, as well. Ethanol costs substantially more to produce than MTBE; and it poses challenges to the gasoline distribution system (it separates from gasoline if transported long distances by pipeline, so it must be mixed with non-oxygenated gasoline blendstock close to the market in which it is to be sold).[50] Because most ethanol is produced in the Midwest, whereas most RFG is consumed on the East and West Coasts, transportation of ethanol to markets poses logistical problems and adds cost to any gasoline-ethanol blend.

Since 1997, some refiners have discussed the possibility of making gasoline that meets the performance requirements for RFG without using oxygenates.[51] However, in the absence of congressional action, this was not permitted. Now, with the enactment of the Energy Policy Act of 2005, which ends the oxygenate requirement in May 2006 and imposes a renewable fuels requirement for gasoline, refiners generally are choosing to use ethanol to replace MTBE. Temporarily, this has led to shortages of ethanol and has contributed to higher gasoline prices in March and April 2006. Ethanol producers, represented by the Renewable Fuels Association (RFA), assert that these shortages are temporary: 500 million gallons of additional annual capacity are expected online before July, 2006, according to RFA, and another 900 million gallons by the end of the year. With additional imports from Brazilian and Caribbean suppliers, reallocation of ethanol within the marketplace, and the use of ethanol stored at terminals in anticipation of the transition, RFA says, "...virtually every refiner and gasoline analyst now acknowledges there will be sufficient ethanol supplies to meet the demand created by MTBE replacement."[52]

The increased demand for ethanol has stimulated the market for corn. Nearly 13% of the nation's corn crop was used to produce ethanol in 2004, and ethanol production has grown at least 20% since then. As much as 30% of the corn crop may be dedicated to ethanol production by 2012. Federal tax credits for ethanol blending and other state and federal legislation have played key roles in promoting this growth. (For background on ethanol and a discussion of ethanol issues, including the effect of the Energy Policy Act, see CRS Report RL33290, *Fuel Ethanol: Background and Public Policy Issues*, by Brent D. Yacobucci.)

LEGISLATION

Building on the work of earlier Congresses, the 109th Congress addressed MTBE, ethanol, and many other energy issues in H.R. 6, the comprehensive energy bill enacted in the summer of 2005. The bill passed the House April 21, 2005; a different version passed the Senate June 28, 2005.[53] Both houses — in their separate legislation — would have banned future use of MTBE in motor fuels, with some exceptions, and authorized transition assistance for MTBE producers, although the specifics of these provisions differed. The House bill would also have provided a "safe harbor" from product liability suits for MTBE producers.

Conferees on the legislation could not reach agreement on most of these provisions, so the version of H.R. 6 that emerged from conference and was signed by the President August 8, 2005 (P.L. 109-58), was stripped of many MTBE-related elements. As a result, controls on the use of MTBE and liability for cleanup of MTBE in ground water and drinking water will be left to the states and the courts respectively.

The reasons why these provisions were left out of the final version are complicated. The conferees faced time pressure as the result of a White House demand that energy legislation be delivered to the President by August 1. For that deadline to be met, the conferees needed to reach agreement on a range of issues quickly. The safe harbor and the provisions on the phaseout of MTBE, described in more detail below, were not amenable to a quick compromise. Thus, the path of least resistance was to remove them.

In the enacted version, Congress did address two issues that will affect future MTBE use. The act removes the Clean Air Act's mandate to use oxygenates (such as MTBE or ethanol) in reformulated gasoline, eliminating a major incentive for continued use of MTBE. However, the enacted bill will also require a substantial increase in the use of renewable fuels, such as the competing oxygenate, ethanol, in both conventional and reformulated gas. With ethanol use required, there will be less need for gasoline refiners to use MTBE.

Refiners began reacting to these provisions almost immediately: Valero Energy, the nation's largest petroleum refiner, announced August 2, 2005, that it will discontinue production of MTBE in May 2006, when the RFG oxygenate requirement is eliminated.[54]

The remainder of this section discusses the principal features of the House and Senate bills and how they were addressed in the enacted legislation.

Safe Harbor Provision

Perhaps the most controversial element in H.R. 6 was the House version's inclusion of a safe harbor provision protecting manufacturers and distributors of renewable fuels and fuels containing MTBE from product liability claims. The Senate bill contained a safe harbor for renewable fuels, but not for MTBE.

The effect of the House provision would have been to protect anyone in the product chain, from manufacturers to retailers, from liability for damages for contamination related to MTBE and renewable fuels, or for personal injury or property damage based on the nature of the product. The safe harbor provision would have applied retroactively to September 5, 2003, potentially barring lawsuits filed on or after that date, including those filed by the State of New Hampshire and numerous cities, towns, counties, municipal water suppliers, and schools. Prior to that date, five lawsuits had been filed. After that date, more than 150 suits have been filed on behalf of some 210 communities in 15 different states.[55]

The safe harbor provision stated that the defective products liability shield would not affect the liability of a person for environmental cleanup costs, drinking water contamination, negligence for spills, or other liabilities other than liability based upon a claim of defective product. However, MTBE manufacturers and those who blend fuels would likely have been more difficult to reach under these other bases of liability.[56]

State attorneys general, local governments, and drinking water suppliers noted that providing a products liability shield would effectively leave only gas station owners liable for cleanup, and because these businesses often have very limited resources, the effect of the safe harbor provision would have been that the burden for cleanup would fall to local communities, drinking water utilities, and the states. In light of this, the Congressional Budget Office identified the safe harbor provision as an intergovernmental and private-sector mandate in its review of the House version of H.R. 6.[57] The Attorneys General for at least 14 states, including states where RFG has been heavily used, strongly opposed the MTBE safe harbor provision. Others questioned the fairness of placing the liability burden primarily on gas station owners, who were not made aware of MTBE's exceptional contamination potential.

Oil companies and other proponents of the provision argued that a safe harbor provision was reasonable, given that the fuels were used to meet the 1990 federal oxygenated fuels and reformulated gasoline mandates, and that the key problem lay not with MTBE, but with leaking underground storage tanks, which are the primary source of MTBE contamination. Even so, MTBE producers appeared to remain concerned about potential liability exposure. MTBE production and use grew rapidly during the 1980s, and several oil companies experienced some incidents of MTBE contamination of groundwater and drinking water wells before the RFG and oxy-fuel mandates. In 1984, oil company engineers estimated that, if MTBE use in gasoline became widespread, the number of well contamination incidents would triple, and treatment costs would increase by a factor of five compared to conventional gasoline incidents.[58] In 1985, Exxon engineers "recommend[ed] that from an environmental risk point of view MTBE not be considered as an additive to Exxon gasolines on a blanket basis throughout the United States."[59]

The total costs of treating MTBE contaminated drinking water are unknown, but are expected to be in the billions. Two studies by water utilities place their best estimates of the costs, given the limited data, at $25 billion[60] and $33.2 billion.[61] A study sponsored by the American Petroleum Institute estimated that the costs of MTBE cleanup for UST sites, public wells, and residential wells that are not covered by a private party, the LUST Trust Fund, state cleanup funds, or insurance, could range from $500 million to $1.5 billion.[62]

The conference did not reach agreement on the safe harbor issue. Unable to persuade Senate conferees to accept the provision without some concessions to the local governments and water utilities that might bear the cost of cleanup (in place of MTBE producers), Representatives Barton and Bass, on behalf of the House conferees, offered to establish an $11.43 billion MTBE cleanup fund, financed by the petroleum industry, states, and federal contributions over a 12-year period.[63] Lawsuits filed by a state attorney general (i.e., New Hampshire) after September 5, 2003, would also have been exempt from the safe harbor provision. But the offer did not pick up additional support, and the safe harbor died.

Renewable Fuels Standard

Both the House and Senate versions of H.R. 6 and the enacted version of the bill amend the Clean Air Act to establish a new requirement that an increasing amount of gasoline contain renewable fuels such as ethanol. The House bill would have required that 3.1 billion gallons of renewable fuel be used in 2005, increasing to 5.0 billion gallons by 2012. (This compares to 3.4 billion gallons actually used in 2004.) The Senate bill would have required 4.0 billion gallons in 2006, increasing to 8.0 billion in 2012. The enacted bill is closer to the Senate version, requiring 4.0 billion gallons in 2006, and an increase of 700 million gallons each year through 2011, before reaching 7.5 billion gallons in 2012.

Changes to the RFG Requirements

As noted above, the enacted bill, like the earlier House and Senate versions, repeals the RFG program's 2% oxygen requirement. This step removes a major incentive for refiners to use MTBE in their fuel. The enacted bill also contains anti-backsliding provisions: gasoline refiners and importers, with some exceptions, must maintain the reduction in emissions of air toxics that they achieved in gasoline produced or distributed during 2001 and 2002.

Phase-out of MTBE and Transition Assistance

Many of the other MTBE provisions in the House- and Senate-passed bills did not make it into the enacted version. Both House and Senate would have banned the use of MTBE in motor vehicle fuel, with exceptions — the House version by December 31, 2014; the Senate, four years after the date of enactment. The conferees dropped the ban entirely.

The House and Senate bills would also have authorized funds to assist the conversion of merchant MTBE production facilities to the production of other fuel additives ($2.0 billion in the House bill, $1.0 billion in the Senate). These provisions were also dropped by the conferees.

Leaking Underground Storage Tank Issues

Both chambers addressed the issue of MTBE leaks from underground storage tanks (USTs). Adopting provisions from the House bill, Title XV, Subtitle B, of the Energy Policy Act of 2005 comprises "The Underground Storage Tank Compliance Act" (USTCA). The USTCA amends SWDA Subtitle I to add new leak prevention and enforcement provisions to the UST regulatory program and impose new requirements on states, EPA, and tank owners. The USTCA requires EPA or states that receive funding under Subtitle I to conduct UST compliance inspections every three years. It also requires states to comply with EPA guidance prohibiting fuel delivery to ineligible tanks, develop training requirements for UST operators and individuals responsible for tank maintenance and spill response, prepare compliance reports on government-owned tanks in the state, and implement groundwater protection measures for UST manufacturers and installers. The act also requires EPA to implement a strategy to address UST releases on tribal lands.

As amended in January 2006, the USTCA authorizes the appropriation of $155 million annually for FY2006 through FY2011 from the Leaking Underground Storage Tank (LUST) Trust Fund for states to use to implement the new UST leak prevention requirements and to administer state programs.[64] However, the Energy Policy Act's fuels tax extension language (§1362) prohibits the use of LUST Trust Fund appropriations for any new purposes. Thus, the Energy Policy Act significantly expands states' leak prevention responsibilities, while at the same time, it prohibits the use of the trust fund money by states to implement the new requirements, some of which have tight deadlines. States that receive funds under Subtitle I are required to implement these provisions.

The USTCA authorizes annually, from the Trust Fund, for FY2006 through FY2011, the appropriation of $200 million for the LUST clean-up program for petroleum tanks and another $200 million specifically for responding to tank leaks involving MTBE or other oxygenated fuel additives (e.g., ethanol). The Senate bill would have authorized a one-time appropriation of $200 million for the cleanup of MTBE and other ether fuels (but not ethanol) from USTs and other sources. (For a detailed comparison of the MTBE and ethanol provisions of the House and Senate bills with the enacted version, see CRS Report RL32865, *Renewable Fuels and MTBE: A Comparison of Selected Provisions in the Energy Policy Act of 2005*, by Brent D. Yacobucci, Mary Tiemann, James E. McCarthy, and Aaron M. Flynn.)

NAFTA ARBITRATION

Another MTBE issue that emerged in the wake of California's decision to phase out the use of MTBE in gasoline concerns the applicability of certain provisions in the North American Free Trade Agreement (NAFTA). Chapter 11, Article 1110, of the NAFTA requires the United States, Canada, and Mexico to treat each other's investors and investments in accordance with the principles set out in the chapter. It also allows these investors to submit to arbitration a claim that a NAFTA party has breached Chapter 11 obligations and to recover damages from any such breach.

In June 1999, the Methanex Corporation, a Canadian company that produces methanol in the United States and Canada, notified the U.S. Department of State of its intent to institute an arbitration against the United States under the investor-state dispute provisions of the NAFTA, claiming that the phaseout of MTBE ordered by the governor of California on March 25, 1999, breaches U.S. NAFTA obligations regarding fair and equitable treatment and expropriation of investments, entitling the company to recover damages which it estimated at $970 million. (Methanol is a major component of MTBE and is Methanex's only product. The California market for MTBE reportedly accounted for roughly 6% of global demand for methanol.) The 1999 Methanex claim asserted that California's phaseout was motivated by a desire to favor an MTBE competitor, ethanol, which is produced in the United States. In August 2002, an arbitration panel ordered Methanex to file a new claim more specifically relating the actions of California to the company's manufacture of methanol. Methanex did so, and a hearing was held in June 2004. In August 2005, a NAFTA arbitration panel dismissed the claim.[65]

CONCLUSION

Numerous detections in ground and surface water, and particularly in municipal and private drinking water wells, have raised significant concerns about the continued use of MTBE in gasoline. Half the states have now taken action to phase out its use, and Congress, in enacting H.R. 6, has removed the federal requirement that oxygenates (such as MTBE) be used in reformulated gasoline.

These actions may lead refiners to phase out the substance entirely. Within days of final passage of the 2005 energy bill, the nation's largest refiner, Valero, announced that it will discontinue production of MTBE. Other producers appear to be following suit.

Whether this marks the end of congressional action on MTBE remains to be seen. More than 150 suits have been filed over liability for cleanup of MTBE-contaminated water. With substantial sums of money in play, the results of this litigation will be closely watched, and may generate further pressure for congressional action.

The effects of MTBE removal on gasoline supply and price are also of concern. In mid-April 2006, gasoline prices were near record highs, and many analysts blamed the phaseout of MTBE and shortages of ethanol for a significant part of the run-up in price. Whether these factors prove transitory will bear watching

REFERENCES

[1]   The requirements for reformulated gasoline (RFG), to reduce air toxics and the emissions that contribute to smog formation, are found in Section 211(k) of the Clean Air Act. Separate requirements for oxygenated fuel, to reduce carbon monoxide formation, are contained in Section 211(m). Of the two programs, that for RFG has a much larger impact on the composition of the nation's gasoline, because RFG requirements are in effect year-round and apply to a larger percentage of the country. The Section 211(m) requirements, by contrast, are in effect during winter months only and affect a small percentage of the nation's gasoline. Ethanol has been the primary oxygenate used in winter oxygenated fuels and MTBE the primary oxygenate used in RFG, although either can be used in both fuels.

[2]   See "Reformulated Fuels Help Curb Peak Ozone Levels in California," *Daily Environment Report*, November 6, 1996, pp. A-1 and A-2.

[3]   Committee on Ozone-Forming Potential of Reformulated Gasoline, National Research Council, *Ozone-Forming Potential of Reformulated Gasoline*, May 1999, p. 5. The NAS study concluded that other characteristics of RFG, notably "lowering the Reid Vapor Pressure (RVP) of the fuel, which helps depress evaporative emissions of VOC [volatile organic compounds], and lowering the concentration of sulfur in the fuel, which prevents poisoning of a vehicle's catalytic converter," result in a reduction of about 20% in VOC emissions.

[4]   U.S. Environmental Protection Agency, Blue Ribbon Panel on Oxygenates in Gasoline, Executive Summary and Recommendations, July 27, 1999, Appendix A. Available at [http://www.epa.gov/otaq/consumer/fuels/oxypanel/blueribb.htm].

[5]   Executive Office of the President, National Science and Technology Council, *Interagency Assessment of Oxygenated Fuels*, Washington, D.C., June 1997, p. iv. Referred to hereafter as the OSTP Report. (The executive summary, recommendations, and full report are available at [http://www.ostp.gov/NSTC/html/MTBE/mtbe-top.html]). The report expressed some hesitation about its conclusions, particularly regarding the impacts of MTBE in colder weather. It also noted methodological difficulties in identifying statistically significant reductions smaller than 10%, and recommended additional research.

[6]   U.S. EPA Response to Interagency Assessment of Oxygenated Fuels, undated, p. 2.

[7]   Systems Applications International, Inc., for the Renewable Fuels Association and the Oxygenated Fuels Association, *Regression Modeling of Oxyfuel Effects on Ambient CO Concentrations*, Final Report, January 8, 1997, p. 1.

[8]   OSTP Report, p. vi. The report did suggest that "greater attention should be given to the potential for increased symptoms reporting among workers exposed to high concentrations of oxygenated fuels containing MTBE," however.

[9]   U.S. Environmental Protection Agency, *Health Risk Perspectives on Fuel Oxygenates*. Office of Research and Development, EPA 600/R-94/217, December 1994, p. 8. Detailed information is available in ORD's 1993 MTBE risk assessment, *Assessment of Potential Health Risks of Gasoline Oxygenated with Methyl Tertiary Butyl Ether (MTBE)*, EPA/600/R-93/206, at [http://www.epa.gov/ncea/pdfs/mtbe/gasmtbe.pdf].

[10] U.S. Environmental Protection Agency, Assessment of Potential Health Risks of Gasoline Oxygenated with Methyl Tertiary Butyl Ether (MTBE), EPA/600/R-93/206, p. 30.

[11] U.S. Environmental Protection Agency, *Drinking Water Advisory: Consumer Acceptability Advice and Health Effects Analysis on Methyl Tertiary-Butyl Ether (MTBE)*, EPA-822-F-97-009, December 1997, pp. 1-2, 9-10. This and other health effects information is available at [http://epa.gov/waterscience/criteria/ drinking/ mtbe.html].

[12] OSTP Report, pp. 4-26.

[13] International Agency for Research on Cancer, IARC Monographs on the Evaluation of Carcinogenic Risks to Humans and Their Supplements: Methyl tert-Butyl Ether (Group 3), World Health Organization, v. 73, 1999, pp. 339-340.

[14] California Environmental Protection Agency, *Public Health Goal for Methyl Tertiary Butyl Ether (MTBE) in Drinking Water*, Office of Environmental Health Hazard Assessment, March 1999, pp. 1-2.

[15] U.S. Environmental Protection Agency, *Oxygenates in Water: Critical Information and Research Needs*, Office of Research and Development, EPA/600/R-98/048, 1998.

[16] OSTP Report, p. vii.

[17] See, e.g., U.S. Environmental Protection Agency Memorandum from Beth Anderson, Test Rule Development Branch, re. *Division Director Briefing for Methyl tert-Butyl Ether (MTBE)*, April 1987, which notes that "[t]he tendency for MTBE to separate from the gasoline mixture into ground water could lead to widespread drinking water contamination."

[18] Arturo Keller et al., *Health and Environmental Assessment of MTBE*, Report to the Governor and Legislature of the State of California as Sponsored by SB 521, Volume I, Summary and Recommendations, University of California, November 1998, p. 35.

[19] Keller, pp. 33-34.

[20] California EPA, *MTBE in California Drinking Water,* Oct. 18, 2005. For more information, see [http://www.dhs.ca.gov/ps/ddwem/chemicals/MTBE/mtbeindex.htm].

(Micrograms per liter(:g/L) are equivalent to parts per billion (ppb) for fresh water.)

[21] Maine Department of Human Services, Department of Environmental Protection, and Department of Conservation, *The Presence of MTBE and Other Gasoline Compounds in Maine's Drinking Water*, preliminary report, October 1998, 24 pp. (Maine was not required to use RFG, but had done so voluntarily; the state opted out of the RFG program in October 1998 because of concerns over MTBE contamination of ground water and drinking water wells.)

[22] The Blue Ribbon Panel on Oxygenates in Gasoline, Executive Summary and Recommendations, July 27, 1999. Summary and full report are available at [http://www. epa.gov/otaq/consumer/fuels/oxypanel/blueribb.htm].

[23] For information on this 2001 study and other MTBE research at the USGS, see [http://sd.water.usgs.gov/nawqa/vocns/mtbe.html].

[24] Stephen J. Grady and George D. Casey, *MTBE and Other VOCs in Drinking Water in the Northeast and Mid-Atlantic Region*. Available at [http://sd.water.usgs.gov/nawqa/ vocns/dw_12state.html]. MTBE was the second most frequently detected VOC in drinking water, after trihalomethanes (disinfection byproducts), which were detected in

45% of systems tested. Chloroform, the most frequently detected trihalomethane, was found in 39% of systems.

[25] American Water Works Association Research Foundation, *Occurrence of MTBE and VOCs in Drinking Water Sources of the United States*, 2003, p. xxiii, p. 101.

[26] Ibid., p. 120.

[27] U.S. Geological Survey, data summary submitted to the EPA Blue Ribbon Panel on the Use of MTBE and Other Oxygenates in Gasoline, January 22, 1999. Available at [http://www.epa.gov/otaq/consumer/fuels/oxypanel/blueribb.htm#Presentations].

[28] California Environmental Protection Agency, *MTBE Briefing Paper*, p. 17.

[29] Anne Happel, E. H. Beckenbach, and R. U. Halden, *An Evaluation of MTBE Impacts to California Groundwater Resources*, Lawrence Livermore National Laboratory and the University of California, Berkeley, June 11, 1998, p. iv.

[30] New England Interstate Water Pollution Control Commission (NEIWPCC), *Survey of State Experiences with MTBE Contamination at LUST Sites (August 2000)*. Available at [http://www.neiwpcc.org]. The survey notes that some states began requiring testing at LUST sites in the 1980s (Maine in 1986 and Minnesota in 1987).

[31] New England Interstate Water Pollution Control Commission (NEIWPCC), *Survey of State Experiences with MTBE and Other Oxygenate Contamination at LUST Sites (August 2003)*. Available at [http://www.neiwpcc.org/Index.htm? MTBE. htm~ mainFrame].

[32] At least seven states have set health-based drinking water standards for MTBE ranging from 13 parts per billion (ppb) to 240 ppb. (Parts per billion are equivalent to :g/L.) At least five states have adopted a secondary standard (based on aesthetic qualities, i.e., taste and odor), ranging from 5 ppb to 70 ppb. At least 10 states have adopted drinking water advisory levels. At least 32 states have adopted a very wide range of ground water cleanup levels; some are guidelines, some are enforceable, and some vary depending on the use of ground water; some states apply these levels to ground-water cleanup at leaking underground storage tank sites where ground water is used for drinking water.

[33] EPA Drinking Water Advisory, p. 2.

[34] 64 *Federal Register* 50555, September 17, 1999. The law requires monitoring by all large public water systems (serving more than 10,000 people) and requires a representative sampling of smaller systems.

[35] New England Interstate Water Pollution Control Commission (NEIWPCC), Survey of State Experiences with MTBE and Other Oxygenate Contamination at LUST Sites (August 2003), Executive Summary, pp. 1-2.

[36] Ibid.

[37] For more information on the LUST program and related legislation, see CRS Report RS21201, *Leaking Underground Storage Tanks: Program Status and Issues*, by Mary Tiemann.

[38] For state-by-state information, see [http://www.epa.gov/oust/cat/camarchv.htm].

[39] A list of Blue Ribbon Panel members is provided, along with the panel report and related materials, at [http://www.epa.gov/oar/caaac/mtbe.html].

[40] Statement by former EPA Administrator Carol Browner on findings by the EPA's Blue Ribbon MTBE Panel, July 26, 1999, available on the Blue Ribbon Panel home page, previously cited.

[41] U.S. Environmental Protection Agency, "Clinton-Gore Administration Acts to Eliminate MTBE, Boost Ethanol," EPA Headquarters Press Release, March 20, 2000, pp. 7-8.

[42] U.S. EPA, Semiannual Regulatory Agenda, 70 *Federal Register* 27604, Sequence Number 3106.

[43] See Arturo Keller et al., *Health and Environmental Assessment of MTBE*, Report to the Governor and Legislature of the State of California As Sponsored by SB 521, November 1998. Available at [http://www.tsrtp.ucdavis.edu/mtberpt/homepage.html].

[44] The Clean Air Act, in Section 211(k)(2)(B), authorizes waiver of the RFG oxygenate requirement only if the Administrator determines that oxygenates would prevent or interfere with the attainment of a National Ambient Air Quality Standard. The law does not address other impacts, such as drinking water contamination.

[45] In January 2004, Governor Schwarzenegger again requested EPA to grant California a waiver from the oxygenate requirement. The governor noted that EPA's Blue Ribbon Panel concluded that a minimum oxygen content is not needed in California, and that CARB had demonstrated that the oxygen requirement is detrimental to the state's efforts to improve air quality. Governor Schwarzenegger further stated that the oxygenate requirement greatly increases fuel costs and "is no longer required to ensure substantial and sustained ethanol use in California." EPA denied Governor Schwarzenegger's request on June 2, 2005.

[46] Blue Ribbon Panel Report, p. 8.

[47] See, for example, "Ethanol-Blended RFG May Cause Small Hike in Gasoline Plume Size," *Mobile Source Report*, December 2, 1999, p. 11, or "Experts Charge Cal/EPA Rushing Approval of Ethanol in RFG," *Inside Cal/EPA*, January 14, 2000, p. 1.

[48] California Air Resources Board, Water Resources Control Board, and Office of Environmental Health Hazard Assessment, *Health and Environmental Assessment of the Use of Ethanol as a Fuel Oxygenate*, Report to the California Environmental Policy Council in Response to Executive Order D-5-99, Dec. 1999, vol. 1, Executive summary, pp. 1-22. Report is available at [http://www-erd.llnl.gov/ethanol/]).

[49] Susan Powers et al., "Will Ethanol-Blended Gasoline Affect Groundwater Quality?" *Environmental Science and Technology*, American Chemical Society, January 1, 2001, p. 28A.

[50] For additional information on ethanol, see CRS Report RL33290, *Fuel Ethanol: Background and Public Policy Issues*, Brent D. Yacobucci.

[51] In earlier versions of this report, we quoted Chevron and Tosco, two firms with large stakes in the California gasoline market, who asked permission to produce RFG without oxygenates in October and December 1997.

[52] Testimony of Bob Dinneen, President and CEO, Renewable Fuels Association, "The Impact of the Elimination of MTBE in Gasoline," Hearing, U.S. Senate, Committee on Environment and Public Works, March 29, 2006. Mr. Dinneen's testimony quotes the CEOs of Valero Energy, the nation's largest refiner, and ExxonMobil in support of his statement.

[53] Legislation that could affect MTBE use has been introduced in every Congress since the 104th. In the 108th Congress, both the House and Senate passed comprehensive energy bills (H.R. 6) that addressed MTBE. A conference report on the legislation (H.Rept. 108-375) was adopted by the House, November 18, 2003, on a vote of 246-

180. In the Senate, however, a cloture vote on the conference report, November 21, 2003, failed to achieve the 60 votes necessary to limit debate, in large part because of the MTBE safe harbor provision contained in the conference report.

[54] "Valero to Quit Making Additive," MySA.com, posted August 2, 2005, [http://www.mysanantonio.com/business/stories/MYSA080305.01E.Valero.12325438.html].

[55] Environmental Working Group. Like Oil and Water: As Congress Considers Legal Immunity for Oil Companies More Communities Go to Court Over MTBE Pollution. April 2005, at [http://www.ewg.org/reports/oilandwater/execsumm.php].

[56] For a more detailed discussion, see CRS Report RS21676, *The Safe-Harbor Provision for Methyl Tertiary Butyl Ether (MTBE)*, Aaron M. Flynn.

[57] Congressional Budget Office, "Cost Estimate for H.R. 6, the Energy Policy Act of 2005, as Introduced in the House of Representatives." Addressed to Honorable David Dreier, Chairman of the Committee on Rules, U.S. House of Representatives, April 19, 2005, 4 pp. This document is available at [http://www.cbo.gov/CESearch.htm]. The CBO determined that the MTBE and renewable fuels liability safe harbor "would impose both an intergovernmental and private-sector mandate as it would limit existing rights to seek compensation under current law.... Under current law, plaintiffs in existing and future cases may stand to receive significant amounts in damage awards, based, at least in part, on claims of defective product. Because section 1502 would apply to all such claims filed on or after September 5, 2003, it would affect more than 100 existing claims filed by local communities, states, and some private companies against oil companies. Individual judgments and settlements for similar lawsuits over the past several years have ranged from several million dollars to well over $100 million. Based on the size of damages already awarded and on information from industry experts, CBO anticipates that precluding existing and future claims based on defective product would reduce the size of judgments in favor of state and local governments over the next five years. CBO estimates that those reductions would exceed the threshold established in UMRA (Unfunded Mandates Relief Act) [$62 million] in at least one of those years."

[58] Memorandum from B. J. Mickelson to V. M. Dugan, *MTBE Contamination of Ground Water*, Exxon Oil Company, August 23, 1985, presented in *South Tahoe Public Utility District v. Atlantic Richfield Co.*, Case No. 999128 (San Fran. Super. Ct. Aug. 5, 2002).

[59] Memorandum from B. J. Mickelson to Mr. J. M. E. Mixtar, Introduction of Methyl Tertiary Butyl Ether (MTBE) in the Texas Eastern Transmission, Jacksonville, Florida; Charleston, South Carolina; and Wilmington, North Carolina Areas, Exxon Oil Company, April 19, 1985, presented in South Tahoe Public Utility District v. Atlantic Richfield Co., Case No. 999128 (San Fran. Super. Ct. Aug. 5, 2002).

[60] American Water Works Association. A Review of Cost Estimates of MTBE Contamination of Public Wells. June 21, 2005.

[61] Association of Metropolitan Water Agencies. *Cost Estimate to Remove MTBE Contamination from Public Drinking Water Systems in the United States*. June 20, 2005.

[62] American Petroleum Institute. *Analysis of MTBE Groundwater Cleanup Costs*. June 2005.

[63] See "Bass Presents MTBE Cleanup Plan," News, House Committee on Energy and Commerce [http://energycommerce.house.gov/108/News /07222005_1608.htm# Related], July 22, 2005. Additional detail can be found in numerous places, including "Barton, Bass Unveil MTBE Cleanup Plan; Petroleum Industry Refuses to Give Support," *Daily Environment Report*, July 25, 2005, p. A-9. The text of the proposal is available from CRS upon request.

[64] Technical corrections to the Energy Policy Act were enacted in P.L. 109-168 on January 10, 2006. The single substantial correction to the USTCA was the revision of the dates authorizing appropriations for Subtitle I from FY2005-FY2009 to FY2006-FY2011.

[65] U.S. Department of State. *NAFTA Tribunal Dismisses Methanex Claim.* August 10, 2005, available at [http://www.state.gov/r/pa/prs/ps/2005/50964.htm]. See also CRS Report RL31638, *Foreign Investor Protection Under NAFTA Chapter 11*, by Robert Meltz.

In: Water Pollution Issues and Developments
Editor: S. V. Thomas, pp. 51-73

ISBN: 978-1-60456-208-8
© 2008 Nova Science Publishers, Inc.

Chapter 3

ANIMAL WASTE AND WATER QUALITY: EPA REGULATION OF CONCENTRATED ANIMAL FEEDING OPERATIONS CAFO[*]

Claudia Copeland

ABSTRACT

According to the Environmental Protection Agency, the release of waste from animal feedlots to surface water, groundwater, soil, and air is associated with a range of human health and ecological impacts and contributes to degradation of the nation's surface waters. The most dramatic ecological impacts are massive fish kills. A variety of pollutants in animal waste can affect human health, including causing infections of the skin, eye, ear, nose, and throat. Contaminants from manure can also affect human health by polluting drinking water sources.

Although agricultural activities are generally not subject to requirements of environmental law, discharges of waste from large concentrated animal feeding operations CAFO into the nation's waters are regulated under the Clean Water Act. In the late 1990s, the Environmental Protection Agency (EPA) initiated a review of the Clean Water Act rules that govern these discharges, which had not been revised since the 1970s, despite structural and technological changes in some components of the animal agriculture industry that have occurred during the last two decades. A proposal to revise the existing rules was released by the Clinton Administration in December 2000. The Bush Administration promulgated final revised regulations in December 2002; the rules took effect in February 2003.

The final rules were generally viewed as less stringent than the proposal, a fact that strongly influenced how interest groups have responded to them. Agriculture groups said that the final rules were workable, and they were pleased that some of the proposed requirements were scaled back, such as changes that would have made thousands more CAFOs subject to regulation. However, some continue to question EPA's authority to issue portions of the rules. Many states had been seeking more flexible approaches than EPA had proposed and welcomed the fact that the final rules retain the status quo to a

---

[*] Excerpted from CRS Report RL31851, dated August 31, 2007.

large extent. Environmentalists contended that the rules relied too heavily on voluntary measures and fail to require improved technology.

This report provides background on the revised environmental rules, the previous Clean Water Act rules and the Clinton Administration proposal, and perspectives of key interest groups on the proposal and final regulations. It also identifies several issues that could be of congressional interest as implementation of the revised rules proceeds. Issues include adequacy of funding for implementing the rules, research needs, oversight of implementation of the rules, and possible need for legislation.

The revised CAFO rules were challenged by multiple parties, and in February 2005, a federal court issued a ruling that upheld major parts of the rules, vacated other parts, and remanded still other parts to EPA for clarification. In June 2006, EPA proposed revisions to the rules in response to the 2005 court decision; for information on the status of this proposal, see CRS Report RL33656, *Animal Waste and Water Quality: EPA's Response to the "Waterkeeper Alliance" Court Decision on Regulation of CAFOs*, which will be updated as warranted by developments.

INTRODUCTION

Agricultural operations often have been treated differently than other types of businesses under numerous federal and state laws. In the area of environmental policy, agriculture is "virtually unregulated by the expansive body of environmental law that has developed in the United States in the past 30 years."[1] Some laws specifically exempt agriculture from regulatory provisions, and some are structured in such a way that farms escape most, if not all, of the regulatory impact. The Clean Water Act (CWA), for example, expressly exempts most agricultural operations from the law's requirements, while under the Clean Air Act (CAA), most agricultural sources escape that law's regulatory programs because most of those sources do not meet the CAA's minimum emission quantity thresholds.

One exception to this general policy of exemption from environmental rules is the portion of the livestock industry that involves large, intensive animal raising and feeding operations. These facilities, which include concentrated feeding operations and feedlots, are a specialized and significant part of the livestock production process, largely separate from cropland agriculture. Certain large animal feeding operations are subject to explicit regulations under the Clean Water Act (P.L. 92-500 as amended, 33 U.S.C. 1251 *et seq.*) that are intended to restrict discharges of animal wastes which could degrade the quality of the nation's rivers, streams, lakes, and coastal waters. However, existing regulations, promulgated in the 1970s, have not been amended to reflect significant structural and technological changes in some components of the animal agriculture industry that have occurred, particularly during the last two decades. In addition, manure and waste-handling and disposal problems from intensive animal production have begun to receive attention as these facilities increase in size and the effects of these problems reach beyond the industry to affect others.[2]

In the late 1990s, the Environmental Protection Agency (EPA), the federal agency responsible for implementing the CWA, initiated a review of the existing CWA rules that govern waste discharges from large animal feeding operations. The review was part of overall Administration efforts to address problems of animal waste affecting the environment, including EPA's response to a court-ordered schedule to revise several CWA rules. A proposal to revise the existing rules for animal feeding operations was released by the Clinton

Administration in December 2000. After two years of reviewing the proposal, the Bush Administration issued final revised regulations in December 2002.

The proposed rules were controversial for a variety of reasons. Livestock and poultry groups, as well as general agriculture advocacy groups, opposed the rules, arguing that they would be too costly. Environmental groups generally supported the rules. States were divided: some favored a strengthened national approach to regulating animal waste, while many favored greater flexibility. The final revised rules adopt some elements of the proposal, modify other parts, and largely retain the structure of the previous rules. The final rules are generally viewed as less stringent than the proposal, a fact that strongly influences how interest groups have responded to them.

This report describes the revised environmental rules, the background of previous rules, the Clinton Administration proposal, and perspectives of key interest groups. It also identifies several issues that could be of congressional interest as implementation of the revised rules proceeds.

The revised CAFO rules discussed in this report were challenged by multiple parties — environmental groups and agriculture industry groups — and in February 2005, a federal court issued a ruling that upheld major parts of the rules, vacated other parts, and remanded still other parts to EPA for clarification (*Waterkeeper Alliance et al. v. EPA*, 399 F.3d 486 (2nd Cir. 2005)), leaving all parties unsatisfied to at least some extent. In June 2006, EPA proposed revisions to the CAFO rules in response to the court's decision and expects to promulgate revised regulations by June 2007. EPA's June 2006 proposal and reactions to it are not discussed here, but are discussed in CRS Report RL33656, *Animal Waste and Water Quality: EPA's Response to the "Waterkeeper Alliance" Court Decision on Regulation of CAFOs*.

LIVESTOCK PRODUCTION AND ANIMAL WASTE

There are an estimated 1.2 million farms with livestock and poultry in the United States, according to the U.S. Department of Agriculture's (USDA) 1997 Census of Agriculture. This number includes all operations that raise beef or dairy cattle, hogs, and poultry and includes both confinement and non-confinement (i.e., grazing and rangefed) production. Of these, about 238,000 are defined as animal feeding operations (AFOs, or feedlots; see box on "EPA Definitions of AFOs and CAFOs," p. 3), where livestock and poultry are confined, reared, and fed. An estimated 95% of these are small businesses: most AFOs raise small numbers of animals (i.e., fewer than 300). Concentrated animal feeding operations CAFO, which confine large numbers of animals and meet certain pollutant discharge criteria (see box, p. 3), are a small fraction of all AFOs (less than 5%), but these largest operations raise more than 40% of U.S. livestock that are reared in confined facilities. In recent years, livestock raising has become more concentrated in fewer but larger operations. From 1982 to 1997, the total number of livestock operations decreased by 24%, and total operations with confined livestock similarly fell by 27%. At the same time, the number of animals raised at large feedlots increased by 88%, and the number of large feedlots/CAFOs increased by more than 50%.[3]

EPA Definitions of AFOs and CAFOs

An Animal Feeding Operation (AFO) is a facility in which livestock or poultry are raised or housed in confinement, and where the following conditions are met: (1) animals are confined or maintained for a total of 45 days or more in any 12-month period, and (2) crops are not sustained in the normal growing season over any portion of the lot or facility (i.e., animals are not maintained in a pasture or on rangeland).

Concentrated Animal Feeding Operations CAFO are a subset of AFOs. In addition to meeting the above conditions, an AFO is a defined as a CAFO if it meets minimum size thresholds (AFOs with more than 1,000 animals are CAFOs; those with 300-999 animals may be CAFOs, depending on discharge characteristics; and those with fewer than 300 may be CAFOs in some cases) and either one of these conditions: (1) pollutants are discharged into navigable waters through a manmade ditch or similar manmade device, or (2) pollutants are discharged directly into waters of the United States that originate outside of and pass over, across, or through the facility, or otherwise come into direct contact with the confined animals. (40 C.F.R. Part 122, App. B)

By animal type, swine and poultry operations have seen the most dramatic change in the manner of production, in terms of animals being raised in confinement at very large animal feeding operations. From 1982 to 1997, there was a 12-fold increase in numbers of swine raised at large AFOs, with the greatest geographic concentration now in Oklahoma, Arkansas, North Carolina, northern Iowa, and southern Minnesota. During the same time period, poultry production at the largest operations increased 218%, with geographic concentration today in southeastern states, coastal states of Florida, Georgia, North Carolina, South Carolina; Minnesota and the surrounding areas; and western coastal states.[4]

Animal manure can be and frequently is used beneficially on farms to fertilize crops and add/restore nutrients to soil. However, the changes in animal agriculture, especially the increasing trend toward raising livestock on large feedlots, have resulted in more extensive problems associated with using and disposing of animal waste. As livestock production has become denser and more spatially concentrated, the amount of manure nutrients relative to the assimilative capacity of land available on farms for application has grown, especially in high production areas including the central northern states from New York to Nebraska, West Coast states and Arizona, and scattered areas through the Southeast.

According to USDA, in 1997, 66,000 operations had farm-level excess nitrogen (an imbalance between the quantity of manure nutrients produced on the farm and assimilative capacity of the soil on that farm) and 89,000 had farm-level excess phosphorus.[5] USDA believes that where manure nutrients exceed the assimilative capacity of a region, the potential is high for runoff and leaching of nutrients and subsequent water quality problems. Geographically, areas with excess farm-level nutrients correspond to areas with increasing numbers of confined animals, and farms with poultry accounted for about two-thirds of the farm-level excess nitrogen and over one-half of the farm-level excess phosphorus. Some of these operations can export manure to surrounding properties. Even accounting for off-site transfers, USDA believes that the number of counties with excess manure nutrients has increased by approximately 60% since 1982 and that in 1997, 165 counties had county-level excess manure nitrogen, and 374 counties had potential excess manure phosphorus. Counties with potential animal waste problems tend to be grouped together. Nearly all of the counties with excess nitrogen were in the Southeast in a region extending from Arkansas and

Louisiana to Virginia. Counties with excess phosphorus were also numerous throughout the Southeast, as well as in the Northeast (including the Delmarva Peninsula), extreme Northwest, California, and the Great Plains.[6] Poultry operations comprised 82% of the operations with farm-level excess nitrogen in those counties, and poultry, dairy, and swine operations comprised nearly 90% of those with farm-level excess manure phosphorus.[7]

ANIMAL WASTE AND THE ENVIRONMENT

Animal waste, if not properly managed, can be transported over the surface of agricultural land to nearby lakes and streams. Leaching from manure storage lagoons and percolation through the soil of fields, where animal waste is applied can contaminate groundwater resources. According to EPA, the release of waste from animal feedlots to surface water, groundwater, soil, and air is associated with a wide range of human health and ecological impacts and contributes to the degradation of the nation's surface waters.[8] Data collected for the EPA's 2000 National Water Quality Inventory identify agriculture as the leading contributor to water quality impairments in rivers and lakes and the fifth leading contributor to impairments in the nation's estuaries. Animal feeding operations are only a subset of the agriculture category, but 29 states specifically identified animal feeding operations as contributing to water quality impairment.[9]

The primary pollutants associated with animal wastes are nutrients (particularly nitrogen and phosphorus), organic matter, solids, pathogens, and odorous/volatile compounds. Animal waste also contains salts and trace elements, and to a lesser extent, antibiotics, pesticides, and hormones. Pollutants in animal waste can impact waters through several possible pathways, including surface runoff and erosion, direct discharges to surface waters, spills and other dry-weather discharges, leaching into soil and groundwater, and releases to air (including subsequent deposition back to land and surface waters). Pollutants associated with animal waste can also originate from a variety of other sources, such as cropland, municipal and industrial discharges, and urban runoff.

The most dramatic ecological impacts associated with manure pollutants in surface waters are massive fish kills. Highly publicized incidents have occurred in nearly every state — from California to Maryland. In addition, manure pollutants can seriously disrupt aquatic systems by over-enriching water (in the case of nutrients) or by increasing turbidity (in the case of solids), processes that can disrupt aquatic ecosystems. Excess nutrients cause fast-growing algae blooms that reduce the penetration of sunlight in the water column and reduce the mount of available oxygen in the water, thus reducing fish and shellfish habitat and affecting fish and invertebrates. EPA's 2000 *Water Quality Inventory* report indicates that excess algal growth alone is among the leading causes of impairment in lakes, ponds, and reservoirs.

A variety of pollutants in animal waste can also affect human health. Over 150 pathogens in livestock manure are associated with risks to humans; these include the bacteria *E. coli* and *Salmonella* species and the protozoa *Giardia* species. Contact with pathogens contained in manure during swimming or boating can result in infections of the skin, eye, ear, nose, and throat. Shellfish such as oysters, clams, and mussels can carry toxins produced by some types of algae that are associated with excess nutrients. These can affect people who eat contaminated shellfish. Further, contaminants from manure can also affect human health through drinking water sources and can result in increased drinking water treatment costs. For example, nitrogen in manure and liquid waste can be transported to drinking water as nitrates, which are associated with human health risks and which EPA has identified as the most widespread agricultural contaminant in drinking water wells. Elevated nitrate levels can cause nitrate poisoning, particularly in infants (this is known as methemoglobinemia, or "blue baby syndrome"). Nitrate contamination of private wells that has been linked to nearby livestock and poultry operations has occurred in several areas, including Delaware, the Maryland Eastern Shore, and North Carolina.

PREVIOUS CLEAN WATER ACT REGULATIONS

Since it was enacted in 1972, the Clean Water Act's predominant focus has been the control of wastewater from manufacturing and other industrial facilities and municipal sewage treatment plants, termed "point sources," which are regulated by discharge permits. As point source pollution has been brought under regulation, uncontrolled discharges in the form of runoff from "nonpoint sources" have become not only greater in absolute terms, but also proportionally a larger share of remaining water pollution problems. Nonpoint pollution occurs in conjunction with surface erosion of soil by water and surface runoff of rainfall or snowmelt from diffuse areas such as farm and ranch land, construction sites, mining and timber operations, and residential streets and yards. Most agricultural activities are considered to be nonpoint sources, since they do not discharge wastes from clearly identifiable pipes, outfalls, or similar "point" conveyances. Nonpoint sources are not subject to the permit, compliance, and enforcement regime that applies to point sources.

Under the CWA, most AFOs are considered to be nonpoint sources. However, CAFOs (large AFOs) are specifically defined in the law as point sources and are treated in a manner similar to other industrial sources of pollution, such as factories. They are subject to the act's prohibition against discharging pollutants into waters of the United States without a permit. In 1974 and 1976, EPA issued regulations defining the term CAFO for purposes of permit requirements (40 C.F.R. §122.23) and effluent limitation guidelines, specifying limits on pollutant discharges from regulated feedlots (40 C.F.R. Part 412). These regulations cover CAFOs that confine beef and dairy cattle, swine, poultry (chickens and turkeys), ducks, sheep, or horses.

Discharge permits issued pursuant to the Part 122 rules, under the act's National Pollutant Discharge Elimination System (NPDES) permit program, establish limits on the amounts and types of pollutants that can be released into waterways. Permits are issued for a fixed term, not to exceed five years, and must be renewed thereafter. NPDES permits may be issued by EPA or a state authorized by EPA to implement the NPDES program. Currently, 45 states have been authorized by EPA to administer this permit program (Oklahoma has been authorized to issue permits for most sources but not for CAFOs). States may impose additional requirements on permittees and may regulate more conduct and more types of operations than those governed by the federal NPDES rules. The two basic types of NPDES permits are individual permits, which are specifically tailored for a specific facility, and general permits, issued by a permitting authority to cover multiple facilities with similar characteristics. Because of the large number of CAFOs, EPA and states increasingly are using general permits to regulate these facilities.

EPA's regulations define a CAFO based on the length of time animals are confined, the number of animals confined, and whether or not the facility directly discharges pollutants into waters of the Untied States. In addition to criteria that define an animal feeding operation (see box on "EPA Definitions of AFOs and CAFOs," p. 3), the rules for defining a CAFO contain a three-tier structure based on the number of animal units[10] at the facility.

- The facility is a CAFO if it holds more than 1,000 animal units.
- If the facility holds from 300 to 999 animal units, the facility is a CAFO if pollutants are discharged from a manmade conveyance or are discharged directly into waters passing over, across, or through the site.
- Animal feeding operations that include fewer than 300 animal units may be designated as CAFOs if EPA or the permitting authority determines that the facility contributes significantly to water pollution.

The regulations nominally impose a zero discharge limitation on regulated operations, because they prohibit discharge of pollutants into waters of the United States, except in the event of discharges that might occur during the worst 24-hour storm in a 25-year period (termed the 25-year, 24-hour storm exception). These regulations do not specifically address discharges to surface water or leaching to groundwater that may occur from animal waste or manure which are applied to land. Nor do they address odor problems from animal agriculture operations. These topics, if regulated at all, have been subject to varied state and local authority, not federal law or regulation.

Problems with CAFO Regulation

A number of problems with the CAFO regulatory system that had existed since the 1970s were widely recognized. These problems limited its effectiveness in preventing environmental problems from livestock production.

- Less than 30% of CAFOs have CWA permits — about 4,100 out of the approximately 12,700 that meet the EPA regulatory definitions described above. One explanation is the historic emphasis by federal and state permitting authorities on regulating other large industrial and municipal dischargers rather than agricultural sources, since most of agriculture is not subject to the act. Another factor is that the 25-year, 24-hour storm exemption has allowed a large number of operations to avoid obtaining discharge permits if they discharge waste only during such a storm event.
- Some sources went unregulated because the EPA rules did not reflect changes in animal waste management technology. In particular, the 1970's rules only applied to poultry operations that have a continuous overflow watering or liquid manure handling system (i.e., "wet" systems) and thus excluded poultry CAFOs with dry manure handling systems, which predominate in this sector today. This exemption allowed more than 2,000 confined poultry operations to avoid obtaining permits.

- The federal regulations contained no requirement for plans to establish manure application rates for fields based on technical standards for nutrient management.
- CAFO inspections by federal and state regulators and compliance enforcement activities have been limited, often occurring only after citizen complaints or accidental releases following large rainfall events or equipment failures. In addition, according to the General Accounting Office (GAO), EPA's limited oversight of the states has contributed to inconsistent and inadequate implementation by states, which are the authorized permitting entities for the large majority of facilities, CAFO and other.[11]

How States Regulate AFOs and CAFOs

Since NPDES permits are the CWA vehicle for implementing the CAFO rules, and states carry out most NPDES permit activities, the nature and scope of state programs for regulating feedlots is an important consideration in evaluating overall effectiveness of current efforts. An EPA compendium of state programs for managing animal feedlots illustrates the variations and complexity of state activities.[12] According to EPA, state regulation of AFOs and CAFOs often involves both federal and state laws and regulations and several different state-level agencies, with numerous variations in approaches, requirements, and jurisdiction. Forty-five states are authorized by EPA to implement the base NPDES program to regulate CAFOs. Seven states regulate CAFOs exclusively under this authority, while 32 states administer a state NPDES CAFO program in combination with some other state permit, license, or authorization, such as a construction or operating permit. Six states, while generally authorized to implement the NPDES program, have chosen to regulate CAFOs under separate state non-NPDES programs. Further, five states are not authorized to administer the NPDES program, and EPA retains responsibility to issue CAFO permits. In three of these states, EPA permits are the sole CAFO regulation, and the other two impose some form of non-NPDES program requirement, in addition to the federally-issued permit. Substantively, state programs vary widely in defining what is a CAFO (hence, the scope of the regulatory program), permit conditions and siting requirements, details for waste management plans (if required), and enforcement procedures.

Because of the wide variability, it is difficult to say whether the glass is "half-full" or "half-empty" with regard to the adequacy of state regulatory activities. EPA concludes that state non-NPDES AFO programs are often more stringent than NPDES programs and often extend coverage to smaller classes of facilities. Further, according to EPA, the implementation of state non-NPDES programs often receives more state agency attention than implementation of NPDES programs, with several states actively choosing not to use NPDES permits. However, the GAO recently found inconsistent and inadequate implementation of CWA requirements by states that have been authorized to administer CAFO permitting. Permits do not meet all EPA requirements, and several states evaluated by GAO do not issue any type of permit to CAFOs, thereby leaving facilities and their wastes essentially unregulated.[13] In revised CAFO rules proposed in December 2000 (discussed below), EPA said that the number of non-NPDES permits issued to AFOs greatly exceeds the number of NPDES permits issued — there are nearly 20 times more non-NPDES permits.

Many would not meet the standards for approval as NPDES permits, EPA said, and because they are not NPDES permits, none meets the requirement for federal enforceability.[14]

REVISING THE CAFO REGULATIONS

In the early 1990s, environmental groups sued EPA for failure to revise existing Clean Water Act permit regulations for a number of industry categories and failure to adopt new rules for unregulated industries. Settlement of that lawsuit[15] put EPA under a court-ordered schedule to issue revised or new Clean Water Act rules for CAFOs and more than a dozen other industries. Under the consent decree, which has been modified several times, revised CAFO rules were to be proposed by December 2000 and finalized by December 15, 2002.

In response to this deadline and as part of broader efforts by EPA and the U.S. Department of Agriculture to address water quality problems associated with animal feeding operations, the Clinton Administration proposed rules to modify the existing CAFO regulations in December 2000.[16] To address shortcomings in the existing regulations, the rules proposed to clarify the conditions under which an AFO is a CAFO and is, therefore, subject to permit requirements. It proposed to increase the number of facilities required to obtain Clean Water Act permits and to restrict land application of wastes.

EPA co-proposed and asked for public comment on two alternative approaches for defining CAFOs. The first would retain the existing three-tier structure (see page 7), but with modifications and clearer criteria regarding the middle tier (1,000 Animal Units or more would be CAFOs, operations with 300 to 999 Animal Units would be CAFOs but could be exempt from permits by demonstrating no potential to discharge wastes, and fewer than 300 Animal Units would be CAFOs only if designated by the permit writer). The second option proposed a two-tier structure (500 Animal Units or more would be defined as CAFOs, fewer than 500 Animal Units would be CAFOs only if designated by the permit writer). EPA estimated that under the proposed two-tier structure, 25,590 operations would need a permit, compared with 12,700 under existing regulations. Under a revised three-tier structure, 31,930 operations would need a permit, while an additional 7,400 in the middle tier were potentially affected, but these operations were expected to be able to avoid permitting by certifying that they are not CAFOs.

In addition, permitting requirements would be extended to some livestock categories not previously regulated (i.e., dry-manure poultry operations and standalone immature swine and heifer operations). EPA also proposed to require that permitted facilities develop and implement site-specific plans which identify the amount of nutrients generated at the facility and determine rates for the application of the waste to agricultural land. Finally, it proposed a co-permitting system, in which permits would cover not just the grower or farmer, but also corporate owners (integrators) who contract out to farmers to raise the animals or poultry and exercise substantial operational control over the facility.

There was a 120-day public comment period following publication of the proposal in the *Federal Register* in January 2001, and on March 26, 2001, the EPA Administrator authorized an additional 75-day public comment period, through July 30. EPA held nine public hearings to review the proposal in the spring and early summer of 2001. Because of the change in Administrations immediately following release of the proposal, new appointees at EPA undertook a detailed and thorough review of the proposal and public comments on it before releasing final rules in December 2002.

Additional Data Considered

In November 2001, EPA published a *Federal Register* Notice of Data Availability (NODA) in which the Agency described information, data, and material received during the public comment period and subsequently concerning rule-related issues such as cost and economic impact and technology options for managing animal waste.[17] EPA said it was considering changes to certain aspects of the proposed CAFO rules. The Agency did not formally re-propose the rules, but it outlined the types of changes being considered and sought additional public comment on the specific data and issues identified in the Notice. For example, EPA said it was considering alternative definitions of what type of feedlot is a "concentrated" feedlot for certain types of livestock operations (which could result in fewer numbers of facilities being subject to regulation than under the Clinton proposal) and also was considering some alternatives that would give states the flexibility to "opt -out" of the federal regulatory program.

In July 2002, EPA published a second Notice of Data Availability that discussed three additional issues for which the Agency was considering changes to the proposal.[18] The issues were: (1) potential new regulatory thresholds for chicken operations with dry litter management practices that would lower the number of facilities defined as CAFOs; (2) potential alternative performance standards to encourage CAFOs to voluntarily install new wastewater treatment technologies and/or management practices; and (3) discussion of new financial data that EPA was considering to evaluate the economic effects of regulatory options.

Public Response

The Clinton-proposed rules were highly controversial for many reasons. Livestock and poultry groups, as well as general agriculture advocacy groups, opposed the rules, arguing that they would impose excessive economic burden on farmers and ranchers. They also criticized the proposal for taking a uniform national approach to problems that they asserted were better suited to management by state and local agencies. Environmental groups generally supported the rules (while arguing that parts should be strengthened), based on their concern that excessive nutrients and other contaminants in animal waste are polluting waterways and groundwater. While lengthy agency review of public comments on a regulatory proposal is not unusual, many in these groups feared that EPA was planning to weaken the Clinton proposal, based on discussion in the first and second NODAs. States were divided on the rules: some favored a uniform national approach to regulating animal waste pollution from the livestock industry based on strengthened EPA rules, while many favored greater state flexibility. States were concerned about diverting resources to CAFO permitting and thus undermining other water quality programs. Congress expressed some interest in the revised rules: in May 2001, a House Transportation and Infrastructure subcommittee held an oversight hearing on the proposed rules. The hearing focused on impacts and costs of the proposal on animal agriculture producers, especially small producers.

THE FINAL REVISED CAFO RULES

After nearly two years' review of the Clinton Administration proposal, EPA issued final revised CAFO regulations on December 11, 2002. The new rules were published in the *Federal Register* on February 12, 2003, with an effective date of April 14, 2003.[19] The regulations include a number of elements of the proposal and a number of modifications, with retention of much of the regulatory structure in the existing rules. Highlights include the following:

- Definition of a CAFO. The definition of what is a CAFO remains the same, unchanged from the prior rule (see box on "EPA Definitions of AFOs and CAFOs," p. 3). Also, the revised rules retain the previous three-tier structure for defining a CAFO, based on the number of animals housed at the facility. The rules retain the size thresholds for most of the regulated categories.[20] As was proposed, the final rules eliminate use of the term "animal unit" equivalents for each animal sector and replace it with the less confusing concept of numbers of animals in each sector.
- Duty to apply. The revised rules adopt an explicit duty for all CAFOs to apply for an NPDES permit, as EPA had proposed. Thus, the rules remove a permitting exemption in the previous rules that had allowed facilities which meet the definition of a CAFO, but claim to only discharge in the event of a large storm, to avoid applying for permits. However, a permit exemption can be claimed by a facility that can certify that it has no potential to discharge waste into waters of the United States.
- Poultry. As noted above, the previous rules only applied to poultry operations that have a continuous overflow watering or liquid manure handling system. The final rules include revisions, as proposed, to clarify applicability of the regulations to all types of poultry operations, regardless of the type of manure handling system. The inclusion of all poultry operations, regardless of manure handling system, brings in all large broiler and dry layer feeding operations and adds an estimated 2,198 operations to the number of regulated facilities.
- Immature animals. The final rules also regulate facilities that confine stand-alone immature animals (swine and heifers), which previously were not covered separately. As a result, 488 of these operations are now subject to regulation.
- Operations required to apply for a permit. EPA estimated that under the previous rules, 12,813 animal feeding operations were subject to regulation and should have had NPDES permits. The total includes 8,438 large facilities (more than 1,000 animals) and 4,375 medium facilities (300 to 999 facilities) which either are defined as CAFOs by size or discharge characteristics, or have been designated as CAFOs by permitting authorities. By adding all poultry operations and stand-alone, immature animal operations, the final rules are estimated to cover an additional 2,554 operations (15,437 facilities in total, consisting of 10,754 large and 4,613 medium operations).[21] The total is 34% of all large and medium animal feeding operations

and about 19% of operations of all size in the United States, based on USDA's 1997 Census of Agriculture.

- Required performance standard. Also as described above, the previous rules prohibited discharges from a CAFO except in the event of wastewater or manure overflows or runoff from a 25-year, 24-hour rainfall event. The proposed and final rules retain this design criterion without change. However, under the final rules, new sources in the swine, poultry, and veal categories must meet a more stringent design standard: storage structures must be designed and maintained to contain the runoff from a 100-year, 24-hour storm event. The final rules include a provision that was not in the proposal allowing existing CAFOs to request permit limits based on site-specific alternative technologies established by the permitting authority, to encourage innovative technologies, according to EPA. Under the new rules, alternative technology limits are required to provide pollutant control equal to or better than under the baseline rules.

- Best Management Practices. The revised rules include Best Management Practices (BMPs) for land application and animal production areas. BMPs are measures or methods that have been determined to be the most effective, practical means of preventing or reducing pollution from nonpoint sources. The requirements for land application areas are to ensure the proper application of manure, litter, and other process wastes to land that the CAFO controls. They include measures such as specified setbacks from streams, vegetated buffers, and determination of application rates, to minimize the transport of phosphorus and nitrogen from the field to surface waters, in accordance with technical standards of the permitting authority. BMPs for animal production areas also are specified, including daily and weekly inspections, maintenance of depth markers in lagoons and other impoundments to determine the design capacity, and on-site recordkeeping.

- Nutrient management plans. As part of the land application requirements, the final rules require a CAFO operator to develop a plan for managing the nutrient content of animal manure and process wastewater. The previous rules had no such requirement. The plan must be maintained on-site and available on request to EPA or the state, but it is not considered part of the facility's permit. Under this plan, manure is to be analyzed annually for nitrogen and phosphorus content, and land application areas are to be analyzed every five years for phosphorus content, to evaluate nutrient build-up in excess of amounts that crops can utilize.

- Compliance schedule. The final rules establish time frames for compliance. Operations defined as CAFOs under the previous rules are expected to already have applied for permits and, presumably, are in violation of the rules if they have not done so. Operations newly defined as CAFOs under the revised rules, such as dry litter poultry operations, must apply for permits by April 13, 2006. A new source must seek permit coverage 180 days prior to the date it commences operation. CAFOs that are existing sources are required to develop and implement nutrient management plans and other land application requirements by December 31, 2006. That date is based on EPA's belief that, by then, there will be sufficient technical experts available to develop and implement nutrient management plans. The land application and nutrient management plan requirements apply immediately to new sources. States with existing NPDES permitting programs must adopt state rule revisions to reflect the federal rules within one year. States which must amend or enact a statute to conform with the rules must make needed rule changes within two years (by April 13, 2005).

- Proposed provisions not in the final rules. Finally, the final rules omit several provisions of the proposal. In addition to not adopting reduced thresholds for defining a CAFO, EPA decided not to include requirements for co-permitting of entities that exercise "substantial operational control" over the CAFO, require zero discharge to groundwater beneath the CAFO production area where there is a direct hydrologic connection to surface water, or require that permit nutrient plans be developed by a certified expert and be re-certified every five years.

Environmental and Economic Benefits of the Rules

A number of environmental and human health benefits were expected to result from requirements of the final rules, according to EPA. These include recreational and non-use benefits from improved water quality in freshwater rivers, streams, and lakes; reduced fish kills; reduced nitrate and pathogen contamination of sources of drinking water; reduced public water treatment costs; and reduced livestock mortality from contamination of livestock drinking water.

EPA quantified the pollutant reductions associated with the final rules. It estimated that nutrient loadings (nitrogen and phosphorus) will be reduced by 23% (166 million pounds per year), sediment loadings by 6% (2.2 billion pounds), and metals discharges by 5% (one million pounds), compared with pre-regulation baseline pollutant loadings.[22] In contrast, the proposed rules estimated pollutant reductions of 179 to 187 million pounds of nutrients, 75 to 77 billion pounds of sediment, and 42 to 44 million pounds of metals (depending on which regulatory option was finalized).[23]

EPA also estimated that the environmental benefits of the final rules, such as improved surface water quality and reduced water treatment costs, will result in annual estimated economic benefits ranging from $204 to $355 million (2001 dollars).[24] Annual benefits of the proposed rules were estimated to be $146 to $163 million (1999 dollars).

Economic Costs of the Rules

The proposed and final rules also presented EPA's estimates of the costs of revised regulation. EPA estimated that the total incremental compliance costs for CAFOs is $326 million annually (pre-tax, 2001 dollars), consisting of $283 million for large CAFOs, $39 million for medium CAFOs, and $4 million for facilities that are designated as CAFOs. Federal and state permitting authorities were projected to incur $9 million per year in costs to implement the rules. Estimated annual incremental costs of the proposed rules were $831-$930 million for CAFO operators, plus $6-8 million for permitting authorities (1999 dollars).

EPA also evaluated financial effects in terms of the number of operations that will experience affordable, moderate, or stress impact because of the rules. Overall, EPA concluded that the rules are economically achievable. For the veal, dairy, turkey, and egg laying sectors, no facility closures are projected. In the beef cattle, heifer, hog, and broiler sectors, EPA's analysis showed that some existing facilities will experience financial stress. An estimated 285 facilities, or 3% of all large CAFOs, might be vulnerable to closure, according to EPA (3% of affected beef CAFOs, 9% of heifer operations, 5% of hog operations, and 1% of broiler operations).[25]

EPA estimated that about 6,200 facilities affected by the rules are small businesses, which the Small Business Administration defines in terms of average annual receipts (or gross revenue), accounting for 40% of all affected facilities. Among large CAFOs, about 2,330 operations are small businesses; most are in the broiler sector. Among medium CAFOs, about 3,870 operations are small businesses (accounting for the majority of operations in this size category), and most are in the hog, dairy, and broiler sectors. EPA's analysis further estimated that about 262 of these operations (4% of all affected small business CAFOs) are vulnerable to closure as a result of the new requirements. They are predominantly beef cattle operations.[26]

Comparing the Proposed and Final Rules

One obvious difference between the proposed and final rules was retention of the previous definition and numerical categorization of regulated facilities. Commenting on the proposal's options for either a two-tier structure or a modified three-tier structure, EPA said that it agreed with commenters, including many states, that changing to a two-tier structure would be very disruptive to ongoing programs. EPA also said that it did not adopt the proposed new set of conditions for determining when a facility in the middle of the three-tier structure (300 to 999 animals) is a CAFO because doing so would not necessarily have improved the clarity or effectiveness of the rules, as intended, but would have caused substantial permitting burdens and imposed costs on essentially all operations above 300 animals.[27]

The previous discussion concerning costs and benefits of the revised rules partially illustrates difficulties in comparing impacts of the proposed and final rules. Some differences in EPA's discussions of the two are notable, but they do not necessarily affect outcomes. For example, the 2000 proposed rules stated that of the 12,700 medium and large CAFOs that should have been subject to permits under the previous rules, but that permits had been issued for approximately 2,270 facilities. In the final rules, while continuing to acknowledge that few operations have permits, EPA stated that the number of permitted facilities is 4,100.[28]

The Notice accompanying the final rules stated, "As a result of today's action, EPA is regulating close to 60 percent of all manure generated by operations that confine animals."[29] However, the proposed rules stated that an estimated 49% of total manure would be controlled by retaining a CAFO definition threshold of 1,000 animals (as adopted in the final rules) and would increase to 64% to 72% under the regulatory options that EPA co-proposed in December 2000 which would have adopted a definition with a lower threshold.[30] The differences in estimated pollutant reductions and amounts of manure controlled under the proposed and final rules were not fully explained. Concerning amounts of manure controlled, part of the difference between the two could be explained by the final rules' inclusion of more poultry operations and stand-alone, immature animal operations than under the previous rules — except for the fact that the proposed rules also included these additional operations.

In comparing impacts of the revised requirements to a baseline, neither the proposed nor the final rules were precisely clear about what baseline was utilized. Consequently, evaluating impacts of changes is difficult. The baseline could be assuming full compliance and control of pollutant runoff from feedlots by the 12,700 operations covered by the previous rules. Alternatively, the baseline could be the partial compliance, and corresponding current water discharges, resulting from that fact that 30% or less of covered facilities are actually operating under NPDES permits. If the baseline assumed complete current compliance (which is not occurring, in fact), then the incremental pollutant reduction improvements of the revised regulatory requirements would be less than if the baseline assumed partial compliance by currently regulated facilities. There is some indication that, for estimating environmental improvements, the baseline of the proposed rules was current partial compliance with previous rules,[31] while in the final rules, the baseline was assumed to be complete compliance with the existing rules. That might explain the large estimated differences in pollutant reduction between the two; see, for example, the above discussion about estimates of reduced sediment loadings and metals discharges.[32] These differences are not satisfactorily explained or addressed in the final rules, but they are significant for evaluating the regulation. In response to inquiries about these issues, an EPA official indicated that, during review of the rules, the Agency completed more extensive modeling of previously available data to assess impacts, including disaggregation for better geographic treatment to address differences in climate, soil type, and conservation practices, and that the improved analysis contributed to the apparent differences between the proposal and final rules.[33]

REACTIONS TO THE FINAL RULES

The final rules were generally viewed as less stringent than the December 2000 proposal, a fact that strongly influenced how interest groups have responded to them. Agriculture industry groups[34] indicated that they believed the final rules were workable, and they were generally pleased that some of the proposed requirements were scaled back, including reduced definition thresholds and co-permitting of corporate owners of livestock as well as of farmers who actually raise the animals. However, some continue to question EPA's authority to issue portions of the rules. Many states, too, had been seeking more flexible approaches than EPA originally proposed, and thus welcomed the fact that the final rules retain the status quo to a large extent. Impacts on states will vary, depending on the changes in existing state programs needed to comply with the new requirements, however. Both industry and states were greatly concerned about adequacy of resources to implement the requirements. Environmentalists contended that the rules relied too heavily on voluntary measures to control runoff, instead of mandating strict compliance with national standards, and fail to require improved technology. In the weeks immediately after publication of the rules, environmental groups and several agriculture industry groups filed lawsuits challenging the rules in a number of different federal courts.

Technology Requirements

Environmental groups criticized EPA for omitting a provision in the proposal that would have required zero discharge from the CAFO's production area to ground water that has a direct hydrologic connection to surface water. A hydrologic connection refers to the interflow and exchange between surface impoundments such as lagoons and surface water through an underground corridor or ground water. The proposal would have required CAFOs to determine whether such a direct hydrologic connection exists and, if so, to monitor ground water up gradient and down gradient to ensure that zero discharge to ground water is achieved. The proposal also would have adopted a stringent zero discharge standard for regulated swine, veal, and poultry CAFOs, with no exception for chronic storm overflows. This issue was a key concern to environmentalists who point out that rural areas, where most CAFOs are located, often rely on ground water for drinking water supplies. In addition, they criticized the final rules for omitting proposed special requirements that would have restricted land application of wastes to frozen, snow-covered, or saturated soil. In the final rules, EPA explained that the proposals were rejected because pollutant discharges to surface water via ground water or as a result of application to frozen or saturated soil are highly dependent on site-specific variables, such as climate, distance to surface water, etc. Thus, a national technology-based standard is inappropriate, according to EPA.

Further, environmentalists asserted that the final rules fail to require performance standards consistent with the best available technology. The rules perpetuate that status quo, they said, because they do not require phaseout of the use of lagoons. Many environmental advocates believe that lagoons are outmoded technology that can pollute both surface and ground water as a result of weather events, human error, and system failures and, thus, are an unacceptable risk to public health and the environment. Likewise, advocates believe that sprayfields, where waste is sprayed onto crops or pastureland, pose significant risks, and many support the position that manure waste that is land applied should be injected or incorporated into the soil.

Industry groups, on the other hand, disputed environmentalists' belief that stringent national standards requiring zero discharge would encourage development of new technologies. In industry's view, the previous rules' zero discharge standard (even with the allowance for chronic storm event discharges) had virtually ensured the use of lagoons and holding ponds to store CAFO wastewater on site. Industry urged EPA to adopt final rules that would encourage alternative technologies. They argued that CAFOs — like other point sources regulated under the Clean Water Act — should be allowed to treat wastes to an established level of quality that does not impair lakes or streams and to release treated wastes to the environment. The final rules appeared to respond to industry's concern in this area: while retaining the previous rules' nominal zero discharge standard, they also allow a CAFO to request a permit based on site-specific alternative technologies established by the permitting authority that are equivalent to the baseline standard or better. EPA believed that this flexibility would encourage innovative technologies, but environmentalists believed that allowing CAFOs to "treat and release" animal waste is weaker than the previous rules and effectively allows alternative technologies to have a discharge that may harm the environment.

Air Emissions

Environmentalists also were disappointed that the final rules did not address or restrict emissions of air pollutants. AFOs can emit various pollutants, including ammonia, hydrogen sulfide, methane, volatile organic compounds, and particulate matter. Environmental impacts can vary, depending on the design and operation of the facility. Scientists generally believe that emissions present a number of issues of environmental concern but not a large public health problem, although more research on public health impacts is required. Some air emissions are important on a local scale (hydrogen sulfide, odor), and others are significant nationally or globally (ammonia, which can be redeposited to earth and contribute to water quality degradation, and the greenhouse gas methane). Industry groups pointed out that water pollution control technologies, which were the subject of the CAFO rules, do not address air emissions and that proven air abatement technologies are needed before adopting regulations.

A 2002 National Research Council report recommended developing improved approaches to estimating and measuring emissions of key air pollutants from AFOs and initiating long-term coordinated research by EPA and USDA with the goal of eliminating release of undesirable air emissions. Nitrogen emissions from production areas are substantial, the report found, and control strategies aimed at decreasing emissions should be designed and implemented now. For example, implementation of feasible management practices, such as incorporating manure into soil, that are designed to decrease emissions should not be delayed while research on mitigation technologies proceeds.[35] In the Notice accompanying the final rules, EPA estimated that the rules would not significantly alter ammonia emissions from CAFOs but will reduce hydrogen sulfide emissions and methane emissions by 12% and 11%, respectively.[36]

Resources Needed to Implement the Rules

Adequacy of resources to implement the revised regulations is an important issue for the animal agriculture industry and states, and these groups focus on the need for federal support to meet the new federal requirements. Livestock operators face costs for manure handling requirements, developing and implementing nutrient management plans, and record-keeping. A key federal financial assistance program for producers is the Environmental Quality Incentives Program (EQIP), administered by the Natural Resources Conservation Service of USDA. EQIP provides technical assistance, cost sharing, and incentive payments to assist livestock and crop producers with conservation and environmental improvements using land management and structural practices, such as site-specific nutrient management or animal waste management facilities. In the 2002 farm bill (P.L. 107-171), Congress increased funding for EQIP from $200 million to $1.3 billion per year by FY2007. Spending for this program is mandatory. Sixty percent of the available funding is to be targeted at practices relating to livestock production. EQIP funds can be used to cover 75% of the cost of measures to control manure runoff, and, under the 2002 farm bill amendments, livestock operators of all sizes including large CAFOs are eligible to receive funding. The amendments limit total payments to $450,000 per participating producer (changed from $50,000 per contract) through FY2007.

USDA, EPA, and federal agencies such as the Small Business Administration (SBA) administer a number of other assistance programs, which EPA summarized in a 2002 report.[37] The SBA, for example, administers a pollution control loan program that can be used by small and large animal feeding operations that are small businesses. Several of the EPA Clean Water Act programs described in the report, such as nonpoint source pollution management grants, can be used by AFOs, but generally not by CAFOs which are regulated as point sources under that act.

A 2003 GAO report found that neither states nor EPA are equipped to implement the program, having not made provisions for additional staffing to process permits, conduct required inspections, and take enforcement actions.[38] GAO reported that the changes will create resource and administrative challenges for states, and meeting these new demands will require additional personnel. However, most of the states reviewed by GAO cannot hire additional staff and would have to reassign personnel from other programs. EPA, too, will have to redeploy staff resources. GAO commented in the report on EPA's limited past oversight of state CAFO programs and concluded that the Agency will need to increase its oversight of state regulatory programs to ensure that the new requirements are properly adopted and carried out by states.

For state agencies that implement the NPDES permit program, the principal existing source of financial assistance is grants under Section 106 of the Clean Water Act, which states already use for various activities to develop and carry out water pollution control programs. States currently use Section 106 grants, supplemented by state resources, for standard setting, permitting, planning, enforcement, and related activities. In light of budgetary problems confronting many states, it is unclear how state agencies will find the resources needed to carry out their responsibilities under the revised rules without reducing resources for other important activities.

Other Industry Views

Fundamentally, agricultural interests emphasize that most farmers are diligent stewards of the environment, since they depend on natural resources of the land, water, and air for their livelihoods and they, too, directly experience adverse impacts on water and air quality, when they do occur. Many believe that environmental problems caused by some individual farmers do not require national solutions or standards, and most are very concerned that regulatory requirements will adversely affect the economic viability of the industry, especially compared with international competitors.

While agriculture industry groups reportedly considered the final rules workable (especially with increased resources provided by the 2002 farm bill), it was also clear that many objected to some basic elements of the regulations that were not eliminated or were changed little from the December 2000 proposal. These concerns were reflected in comments on the proposed rules. For example, livestock and general agriculture groups questioned EPA's basic authority to impose a number of the rules' requirements. These groups generally opposed eliminating the previous permitting exemption for facilities that discharge only in a large storm event, saying that such operations should not be covered by permits.

Industry also opposed imposing on CAFOs a duty to apply for permits and questioned EPA's legal authority for requiring permits from CAFOs that claim not to discharge pollutants, since in their view the Clean Water Act only requires permits for actual discharges. Some questioned EPA's finding that many CAFOs are discharging without a permit (which EPA had cited as a key reason for revising the regulations) and said that voluntary programs are working adequately to address the excess manure issue. Some objected to putting the burden on the CAFO to show that it does not discharge into waterways and argued that the CAFO should not be required to apply for a permit in the absence of evidence of an actual discharge.

Some industry commenters also argued that EPA lacks authority to include permit requirements governing land application of manure and process wastewater, because in their view runoff from land application areas is a nonpoint source discharge that is not subject to Clean Water Act permitting. EPA's view is that land application areas are integral to CAFO operations, and, because there have been significant discharges from them, non-regulatory controls alone are insufficient.

Other Views of Environmental Groups[39]

Environmental groups were critical of several other provisions in the proposal that were omitted from the final rules. Chief among these was EPA's decision not to require co-permitting of both the farmer who raises the livestock and the large companies that actually own the animals and contract with farmers. This was one of the most controversial parts of the proposed rules. Environmental advocates believe that co-permitting makes large corporations responsible for wastes produced on the farms with which they contract, while the agriculture industry said it would make corporations liable for waste management decisions over which they have no practical control.

Environmental groups also had strongly favored lowering the threshold for defining when an AFO is a CAFO, which would ensure that more operations are subject to uniform controls and enforcement.

These groups criticized changes in the final rules that they believed will limit public involvement and oversight. In particular, they said that, by not requiring that nutrient management plans be publicly developed and available, the public will not have adequate access to the plans. Many environmental advocates favored including nutrient management plans in a CAFO permit, which would make the plans an enforceable element of the permit. Agriculture industry groups argued that the plan would contain proprietary information and that making it publicly available would both discourage innovation in developing waste management technologies and could make CAFOs vulnerable to lawsuits. EPA pointed out that the final rules require CAFOs to submit annual nutrient management reports that will be public and will provide information on numbers of animals, amounts of manure generated, and how the manure is being handled. Advocates also said that, by not requiring that nutrient management plans be developed by a certified expert or be approved by the permitting authority, as had been proposed, the revised rules essentially allow farmers to write their own requirements without technical or permitting authority involvement.

ISSUES FOR CONGRESS

Implementation of the revised CAFO rules will present large challenges for those who are directly affected by the regulations — the animal agriculture industry, states, and EPA — as well as interested members of the public. Likewise, several issues of congressional interest are apparent.

- Adequacy of funding. Requests for funding assistance to help affected groups comply with the rules are expected to increase —especially by feedlot operators seeking EQIP funds. However, even at the higher EQIP contract limit provided by the 2002 farm bill ($450,000 per farmer, compared with $50,000 under prior law), the ceiling may effectively diminish some farmers' interest in the program. In addition, both states and EPA are likely to face difficulties in meeting new program and permitting responsibilities within current budgetary constraints. At issue is whether adequate resources will be provided and funding priority given as needed.

- Research needs. A large number of treatment technologies and best management practices exist for pollution prevention at animal feeding operations, as well as for handling, storage, treatment, and land application of wastes.[40] EPA believes, however, that storage lagoons and sprayfields have been and remain the most widely used technologies. Research to encourage new technologies and demonstration of technologies and practices that may pose less environmental risk could be environmentally and economically beneficial. In this regard, researchers may be interested in a program established by the 2002 farm bill that authorizes USDA to provide innovation grants to leverage federal investment in environmental protection through the use of EQIP, including demonstrating innovative nutrient management technology systems for AFOs. In addition, the National Research Council's recent report on air emissions from AFOs recommends that EPA and USDA aggressively pursue research in that area and identifies priorities for short- and long-term research programs.

- Oversight of implementation. As noted previously, GAO has been critical of EPA's past oversight of state CAFO permitting activities, and EPA has acknowledged that neither federal nor state agencies have previously given much priority to regulating feedlot wastes. At issue now will be how EPA and states demonstrate through planning and actions their commitment to implement the new requirements. USDA's commitment to supporting farmers' implementation of the rules also will be of interest.

- Is federal legislation needed? There also is the issue of whether the revised regulatory program reflects Congress' intent and expectations concerning management of animal waste and its environmental impacts. Some questions of congressional intent were raised in legal challenges brought by agriculture industry and environmental groups to the rules, such as, did Congress intend to authorize EPA to regulate land application of wastes? At the same time, some may conclude that legislation amending the Clean Water Act is needed to guide EPA, states, and industry by clarifying Congress' current view of key issues, compared with that act's enactment in 1972 — considering, for example, whether the scope of requirements should be narrowed. Alternatively, some who see gaps in parts of the final rules may favor legislation to broaden requirements — for example, concerning co-permitting or technology standards. Finally, some may believe that another legislative vehicle entirely — such as the farm bill administered by USDA — is a more appropriate tool for addressing animal waste management issues.

REFERENCES

[1]   J.B. Ruhl, "Farms, Their Environmental Harms, and Environmental Law," *Ecology Law Quarterly*, vol. 27, no. 2 (2000), pp. 263-349, 265.

[2]   For additional background, see CRS Report 98-451, *Animal Waste Management and the Environment: Background for Current Issues*, by Claudia Copeland and Jeffrey Zinn.

[3]   U.S. Department of Agriculture, Natural Resources Conservation Service, "Manure Nutrients Relative to the Capacity of Cropland and Pastureland to Assimilate Nutrients: Spatial and Temporal Trends for the United States," Publication no. nps00-579, December 2000, p. 18. Hereafter cited as USDA, "Manure Nutrients."

[4]   Ibid., pp. 44, 46.

[5]   In the agriculture context, assimilative capacity is the amount of nutrients taken up and removed at harvest for cropland and the amount that could generally be applied to pastureland without accumulating nutrients in the soil.

[6]   USDA, "Manure Nutrients," op. cit., p. 85.

[7]   Ibid., pp. 75-81.

[8]   U.S. Environmental Protection Agency, "Environmental and Economic Benefit Analysis of Final Revisions to the National Pollutant Discharge Elimination System Regulation and the Effluent Guidelines for Concentrated Animal Feeding Operations," December 2002, p. ES-6.

[9]   U.S. Environmental Protection Agency, "National Water Quality Inventory, 2000 Report," August 2002, EPA-841-R-02-001, 1 vol.

[10]   As defined by USDA, an animal unit is 1,000 pounds of live weight of any given livestock species or combination. The term varies according to animal type; one animal is not always equal to one animal unit. An EPA animal unit is equal to 1.0 beef cattle, 0.7 mature dairy cow, 2.5 pigs weighing more than 55 pounds each, 100 chickens (broilers or layers), 10 sheep or lambs, or 0.5 horses.

[11]   U.S. General Accounting Office, "Livestock Agriculture: Increased EPA Oversight Will Improve Environmental Program for Concentrated Animal Feeding Operations," January 2003, GAO-03-285, p. 7.

[12]   U.S. Environmental Protection Agency, Office of Wastewater Management, *State Compendium, Programs and Regulatory Activities Related to Animal Feeding Operations*. May 2002. Another report presents a detailed comparison of features and requirements of programs in seven states. See, Environmental Law Institute, *State Regulation of Animal Feeding Operations*, January 2003, 80 p.

[13]   GAO, op cit., pp. 7-11.

[14]   66 *Federal Register* 2969, January 12, 2001.

[15] *Natural Resources Defense Council v. Reilly*, U.S. District Court, D.C., Civ. Action No. 89-2980, April 23, 1991.

[16] 66 *Federal Register* 2959, January 12, 2001.

[17] 66 *Federal Register* 58556, November 21, 2001.

[18] 67 *Federal Register* 48099, July 23, 2002.

[19] U.S. Environmental Protection Agency, "National Pollutant Discharge Elimination System Permit Regulation and Effluent Limitation Guidelines and Standards for Concentrated Animal Feeding Operations CAFO; Final Rule," 68 *Federal Register* 7175-7274, February 12, 2003.

[20] The threshold for duck operations with dry manure-handling systems was changed from 5,000 to 30,000 animals for large operations, thus reducing the number of regulated operations from 157 under the previous rules to 25 under the final rules.

[21] U.S. Environmental Protection Agency, "Development Document for the Final Revisions to the National Pollutant Discharge Elimination System Regulation and the Effluent Guidelines for Concentrated Animal Feeding Operations," December 2002, pp. 9-3 to 9-15.

[22] 68 *Federal Register* 7239, table 7.2, February 12, 2003.

[23] 66 *Federal Register* 3116, January 12, 2001.

[24] U.S. Environmental Protection Agency, "Environmental and Economic Benefit Analysis of Final Revisions to the National Pollutant Discharge Elimination System Regulation and the Effluent Guidelines for Concentrated Animal Feeding Operations," December 2002, pp. 11-3 - 11-4.

[25] 68 *Federal Register* 7245-46, February 12, 2003.

[26] 68 *Federal Register* 7246-47, February 12, 2003.

[27] Ibid., 7189-7190.

[28] Compare 66 *Federal Register* 2969, January 12, 2001, with U.S. Environmental Protection Agency, "Development Document for the Final Revisions to the National Pollutant Discharge Elimination System Regulation and the Effluent Guidelines for Concentrated Animal Feeding Operations," December 2002, at p. 9-12.

[29] 68 *Federal Register* 7180, February 12, 2003.

[30] 66 *Federal Register* 2986, January 12, 2001.

[31] However, in the proposed rules, EPA stated that, for purposes of estimating compliance costs, it assumed that all CAFOs subject to revised regulations are currently in compliance with the existing regulatory program, even though it recognized, as a practical matter, that this is not true. EPA did not estimate the additional costs of complying with existing requirements, because it did not consider those costs part of the incremental costs of revised rules. 66 *Federal Register* 3080, January 12, 2001.

[32] US. Environmental Protection Agency, "Environmental and Economic Analysis of Proposed Revisions to the NPDES Regulation and the Effluent Guidelines for Concentrated Animal Feeding Operations," January 2001, p. 4-18.

[33] Telephone conversation, Paul Shriner, U.S. EPA, Office of Water, Office of Science and Technology, March 3, 2003.

[34] On most issues affecting agriculture, there often is a subset of interests most affected and likely to express views on legislation, regulations, etc. Their views may differ or coalesce on a given issue. The CAFO rules discussed here were of considerable interest to groups representing livestock and poultry producers, such as the National Chicken

Council, Port Producers Council, and National Cattlemen's Beef Association, as well as groups that represent agriculture as a whole, such as the Farm Bureau. In EPA's discussion of the rules (e.g., the *Federal Register* Notice accompanying the final rules), when referring to "industry," it did not distinguish among these groups, nor does this CRS report. It appears that, at least in EPA's judgment, these groups generally reflected similar interests and concerns on the CAFO rule issues. Other agriculture industry groups, representing interests of cropland producers, for example, had limited involvement in these rules.

[35] National Research Council, Board on Agriculture and Natural Resources, Board on Environmental Studies and Toxicology, "Air Emissions from Animal Feeding Operations: Current Knowledge, Future Needs, Final Report," December 2002, 241 p.

[36] 68 *Federal Register* 7242, February 12, 2003. The time period for achieving these anticipated reductions is not specified.

[37] U.S. Environmental Protection Agency, "Financial Assistance Summaries for AFOs," 2002. Available at [http://www.epa.gov/npdes/pubs/financial_assistance_ summaries. pdf].

[38] U.S. General Accounting Office, "Livestock Agriculture, Increased EPA Oversight Will Improve Environmental Program for Concentrated Animal Feeding Operations," January 2003, GAO-03-285.

[39] Within agriculture, there are some groups that reflect many interests similar to those of environmental groups, such as the Sustainable Agriculture Coalition, which promotes policies based on economically profitable, environmentally sound, family-farm based systems of agriculture and livestock production methods at small and mid-size operations that do not use animal confinement. Concerning CAFO issues, sustainable agriculture groups favored strategies based on nationally uniform standards, alternatives to large CAFO production, public accountability in issuance of CAFO permits, and legal liability for corporate owners of confined animals.

[40] Technologies are discussed extensively in: U.S. Environmental Protection Agency, "Development Document for the Final Revisions to the National Pollutant Discharge Elimination System Regulation and the Effluent Guidelines for Concentrated Animal Feeding Operations," December 2002, Chapter 8.

In: Water Pollution Issues and Developments
Editor: S. V. Thomas, pp. 75-87

ISBN: 978-1-60456-208-8
© 2008 Nova Science Publishers, Inc.

Chapter 4

ANIMAL WASTE AND WATER QUALITY: EPA'S RESPONSE TO THE *WATERKEEPER ALLIANCE* COURT DECISION ON REGULATION OF CAFOS[*]

Claudia Copeland

ABSTRACT

On June 30, 2006, the Environmental Protection Agency (EPA) proposed regulations that would revise a 2003 Clean Water Act rule governing waste discharges from large confined animal feeding operations CAFO. This proposal was necessitated by a 2005 federal court decision (*Waterkeeper Alliance et al. v. EPA*, 399 F.3d 486 (2nd Cir. 2005)), resulting from challenges brought by agriculture industry groups and environmental advocacy groups, that vacated parts of the 2003 rule and remanded other parts.

The Clean Water Act prohibits the discharge of pollutants from any "point source" to waters of the United States unless authorized under a permit that is issued by EPA or a qualified state, and the act expressly defines CAFOs as point sources. Permits limiting the type and quantity of pollutants that can be discharged are derived from effluent limitation guidelines promulgated by EPA. The 2003 rule, updating rules that had been in place since the 1970s, revised the way in which discharges of manure, wastewater, and other process wastes from CAFOs are regulated, and it modified both the permitting requirements and applicable effluent limitation guidelines. It contained important first-time requirements: all CAFOs must apply for a discharge permit, and all CAFOs that apply such waste on land must develop and implement a nutrient management plan.

EPA's proposal for revisions addresses those parts of the 2003 rule that were affected by the federal court's ruling: (1) it would eliminate the "duty to apply" requirement that all CAFOs either apply for discharge permits or demonstrate that they have no potential to discharge, which was challenged by industry plaintiffs, (2) it would add procedures regarding review of and public access to nutrient management plans, challenged by environmental groups, and (3) it would modify aspects of the effluent limitation guidelines, also challenged by environmental groups. EPA's proposal also considers modifying a provision of the rule that the court upheld, concerning the treatment of a regulatory exemption for agricultural stormwater discharges.

---

[*] Excerpted from CRS Report RL33656, dated August 31, 2007.

Public comments addressed a number of points, with particular focus on the "duty to apply" for a permit and agricultural stormwater exemption provisions of the proposal. Industry's comments were generally supportive of the proposal, approving deletion of the previous "duty to apply" provision and also EPA's efforts to provide flexibility regarding nutrient management plan modifications. Environmental groups strongly criticized the proposal, arguing that the *Waterkeeper Alliance* court left in place several means for the agency to accomplish much of its original permitting approach, but instead EPA chose not to do so. State permitting authorities also have a number of criticisms, focusing on key parts that they argue will greatly increase the administrative and resource burden on states. In July 2007 EPA extended compliance deadlines for permitted CAFOs from July 2007, subject to revised rules, to February 2009. Congress has shown some interest in CAFO issues, primarily through oversight hearings in 1999 and 2001.

INTRODUCTION

According to the Environmental Protection Agency (EPA), the release of waste from animal feedlots — the portion of the livestock industry that involves large, intensive animal raising and feeding operations — to surface water, groundwater, soil, and air is associated with a range of human health and ecological impacts and contributes to degradation of the nation's surface waters. The most dramatic ecological impacts are massive fish kills, which have occurred in a number of locations in the United States. A variety of pollutants in animal waste can affect human health in several ways, such as causing infections to the skin, eye, ear, nose, and throat. Contaminants from manure can also pollute drinking water sources. Data collected for the EPA's 2000 National Water Quality Inventory report identify agriculture as the leading contributor to water quality impairments in rivers and lakes. Animal feeding operations are only a subset of the agriculture sector, but 29 states specifically identified animal feeding operations as contributing to water quality impairment.[1] Federal efforts to control these sources of water pollution have accelerated in recent years, but they have been highly controversial.

The primary pollutants associated with animal wastes are nutrients (particularly nitrogen and phosphorus), organic matter, solids, pathogens, and odorous/volatile compounds. Animal waste also contains salts and trace elements, and to a lesser extent, antibiotics, pesticides, and hormones. Pollutants in animal waste can impact waters through several possible pathways, including surface runoff and erosion, direct discharges to surface waters, spills and other dry-weather discharges, leaching into soil and groundwater, and releases to air (including subsequent deposition back to land and surface waters). Pollutants associated with animal waste can also originate from a variety of other sources, such as cropland, municipal and industrial discharges, and urban runoff.

Although agricultural activities are generally not subject to requirements of environmental law, discharges of waste from large concentrated animal feeding operations CAFO into the nation's waters are regulated under the Clean Water Act (CWA). In the late 1990s, EPA initiated a review of the CWA rules that govern these discharges. The rules had not been revised since the 1970s, despite subsequent structural and technological changes in some components of the animal agriculture industry. A proposal to revise the existing rules was released by the Clinton Administration in December 2000. These regulatory activities and proposals have been very controversial. Agriculture industry groups have opposed permitting requirements that they consider burdensome and costly, while others, such as environmental groups, have favored more stringent national standards that require improved control technology. During this period, Congress showed some interest in CAFO issues, through oversight hearings held by House subcommittees in October 1999 and May 2001.

The Bush Administration issued final revised regulations in December 2002, which were published in the *Federal Register* in February 2003 and became effective April 14, 2003.[2] The 2003 rule was challenged by multiple parties — environmental groups and agriculture industry groups — and in February 2005, a federal court issued a ruling that upheld major parts of the rule, vacated other parts, and remanded still other parts to EPA for clarification, leaving all parties unsatisfied to at least some extent. In June 2006, EPA issued proposed revisions to the CAFO rule in response to the court's decision that have been criticized by a number of stakeholder groups. Under that proposal, EPA intended to promulgate revised rules in June 2007; the deadline for compliance with the revised rules would be July 31, 2007. However, in July, EPA extended compliance with CAFO rules to February 27, 2009.

This report describes major features of the 2003 CAFO rule. It discusses the parts of the rule that were addressed in the federal court's decision and EPA's response to the court, as presented in proposed regulatory revisions. Finally, the report provides an overview of comments on the June 2006 proposal that were submitted by several varied interest groups: the livestock and poultry industry, states, and environmentalists.

THE 2003 RULE

The CWA prohibits the discharge of pollutants from any "point source"[3] to waters of the United States unless authorized under a national pollutant discharge elimination system (NPDES) permit that is issued by EPA or a qualified state. Any discharge from a point source, even one that is unplanned or accidental, is illegal unless it is authorized by the terms of a permit. NPDES permits limit the type and quantity of pollutants that can be discharged from a facility and specify other requirements, such as monitoring and reporting. The specific discharge limitations in the permit are derived from effluent limitation guidelines and standards (ELGs) that are separately promulgated by EPA for specific categories of industrial sources. ELGs are technology-based restrictions on water pollution, because they are established in accordance with technological standards specified in the act. They vary depending upon the type of pollutant and discharge involved, and whether the point source is new or already existing.

The act expressly defines CAFOs as point sources. EPA issued NPDES permitting rules for CAFOs in 1974 (defining which animal feeding operations are subject to regulation[4]) and effluent limitation guidelines in 1976. The 2003 rule did not redefine what is a CAFO, but it revised the way in which discharges of manure, wastewater, and other process wastes from CAFOs are regulated, and it modified both the NPDES permitting requirements and applicable ELGs. Under the 2003 rule, all CAFOs are required to apply for an NPDES permit. EPA estimated that this requirement expanded the number of covered operations from about 12,800 to 15,500 — primarily the largest CAFOs, in terms of numbers of animals raised or housed on-site — or about 19% of all animal feeding operations of all size in the United

States. EPA acknowledged that prior to the revisions, permitting and enforcement had been inadequate and that only 4,000 CAFOs actually had permits.

The rule established ELGs that apply to the production areas of regulated CAFOs (including the animal confinement area, manure storage area, raw material storage area, and waste containment area) and, for the first time, to the land application area (referring to land to which manure, litter, or process wastewater is or may be applied). These ELGs are non-numerical best management practices. Discharges from a production area are subject to a performance standard requiring facilities to maintain waste containment structures that generally prohibit discharges except in the event of overflows or runoff resulting from a 25-year, 24-hour rainfall event.[5] Similarly, discharges of pollutants from land application areas must comply with ELG best management practices, such as the adoption of setback limits from surface waters or vegetative buffer strips. In addition, a permitted facility is required to submit an annual performance report to EPA and to develop and follow a plan, known as a comprehensive nutrient management plan (NMP), for handling manure and wastewater.

THE *WATERKEEPER ALLIANCE* DECISION AND EPA'S RESPONSE

The 2003 rule was challenged in court by a number of groups. The cases, brought by environmental petitioners and by farm industry petitioners, were consolidated by the Second Circuit Court of Appeals, which issued a decision on February 28, 2005 (*Waterkeeper Alliance et al. v. EPA*, 399 F.3d 486 (2nd Cir. 2005)). The ruling reflected partial victory for all of the parties, because the court upheld or did not address significant parts of the regulation (such as the definition of what is a CAFO, for regulatory purposes), but it agreed with some of the claims raised by both sets of petitioners. It vacated parts of the regulation and remanded other parts to EPA for clarification. In response to the court's ruling, EPA has proposed revisions to the 2003 rule.[6] The public comment period on this proposal concluded on August 29, 2006. EPA officials indicated in the proposal that they expected to promulgate revised regulations by June 2007. Earlier in 2006, EPA had extended compliance dates in the 2003 rule for facilities that were affected by the *Waterkeeper Alliance* decision until July 31, 2007.[7] This extension affected the date for newly defined CAFOs (facilities not defined as CAFOs as of April 14, 2003 — the effective date of the 2003 rule) to seek NPDES permit coverage and the date by which all CAFOs must develop and implement nutrient management plans.

In May 2007, EPA announced that it is still considering comments on the 2006 proposal and does not expect to complete work on a final rule until 2008. Thus, EPA extended the July 31, 2007, compliance deadline until February 27, 2009 — giving certain livestock operators another 19 months to obtain discharge permits and to develop and implement manure management plans.[8] The compliance deadline extension will not apply to new livestock operations or to existing CAFOs that were required to obtain permits prior to the 2003 rule.

The remainder of this report discusses key portions of the regulation that were affected by the court's ruling, beginning with one key issue which the court did not reject or remand. Following that is discussion of issues that EPA addressed in its proposal as a result of the litigation: (1) the "duty to apply" requirement that all CAFOs either apply for NPDES permits or demonstrate that they have no potential to discharge, which was challenged by industry plaintiffs, (2) procedures regarding review of and public access to nutrient management plans, challenged by environmental groups, and (3) aspects of the effluent limitation guidelines, also challenged by environmental groups.

Agricultural Stormwater Discharges

One issue that the court upheld concerns the rule's treatment of a regulatory exemption for agricultural stormwater discharges. This issue, which was one of the most controversial during development of the 2003 rule, arose in the context of the regulatory framework concerning the land application of manure, litter, and process wastewater. As noted above, the CWA expressly defines the term "point source" to include concentrated animal feeding operations. The same provision of the act, section 502(14), also expressly defines "point source" to *exclude* "agricultural stormwater." The court characterized this provision as "self-evidently ambiguous" and observed, "the Act makes absolutely no attempt to reconcile the two."[9] When manure and other waste is applied to land, precipitation-related runoff can transport nutrients, pathogens, and other pollutants in the waste to nearby receiving waters.

To develop the rule, EPA had to interpret the statutory inclusion of CAFOs as point sources and the agricultural stormwater exclusion consistently and to identify the conditions under which discharges from the land application area of a CAFO are point source discharges that are subject to NPDES permitting requirements, and those which are agricultural stormwater discharges and thus are not point source discharges.[10] The land application portion of the rule details requirements to ensure that animal waste is applied to land in accordance with nutrient management practices that ensure appropriate agricultural utilization of the nutrients in the waste. Under the rule as promulgated, EPA determined that when manure or process wastewater is applied in accordance with those practices, at appropriate agronomic rates, it is a beneficial agricultural production input. Where such practices have been used, any remaining discharge is agricultural stormwater which is exempt from permitting. In contrast, where such practices have not been used, EPA argued that it is reasonable to conclude that discharges of manure from a land application area have not been applied at agronomic rates, are not agricultural stormwater, and thus are subject to NPDES permitting. Under the rule, adherence to appropriate nutrient management practices eliminates any need to seek permit coverage for land application discharges or submit a land application NMP to the permitting authority.

Both groups of petitioners challenged this portion of the rule. Livestock and poultry industry plaintiffs had argued that land application runoff should be considered a point source discharge subject to permitting only if it is collected or channelized prior to discharge. In contrast, the environmental petitioners argued that the act's definition of "point source" requires regulation of all CAFO discharges, notwithstanding the statutory exemption for agricultural stormwater discharges. The court found that EPA's interpretation of the act in this regard was reasonable. The court interpreted the rule as seeking to remove liability for agriculture-related discharges primarily caused by nature, while maintaining liability for other discharges. "[W]here a CAFO has taken steps to ensure appropriate agricultural utilization of the nutrients in manure, litter, and process wastewater, it should not be held accountable for any discharge that is primarily the result of 'precipitation.'"[11] It rejected the challenges by the parties, and it upheld this portion of the rule.

Although the court did not direct EPA to revise this provision, the agency stated in the Preamble to the June proposal that it is considering adding a requirement that would apply to runoff from CAFO fields that are otherwise unpermitted because they do not discharge or propose to discharge (and thus are considered to be agricultural stormwater). Under this addition, such CAFOs that do not have permits would still be required comply with any more prescriptive nutrient management technical standards for land application (field-specific standards, for example) that have been established by the permitting authority (the state or EPA), in addition to the practices specified in the EPA rule.

Duty to Apply for a Permit

The 2003 rule explicitly required all CAFOs to apply for an NPDES permit, or to demonstrate to the permitting authority that they have no potential to discharge. EPA's policy rationale for this "duty to apply" provision was based on its "presumption that most CAFOs have a potential to discharge pollutants into waters of the United States."[12] However, farm industry plaintiffs argued that, unless there is a discharge of a pollutant, CAFOs and other point sources are neither statutorily obligated to comply with EPA regulations, nor are they obligated to seek or obtain an NPDES permit. The *Waterkeeper Alliance* court ruled in support of these plaintiffs and held that EPA exceeded its authority under the CWA in ordering all CAFOs to apply for a permit, finding that the law requires permits only where there is an actual discharge, not just a potential to discharge.

In its proposal to revise the regulation, EPA would replace the "duty to apply" requirement of the 2003 rule with a requirement that all CAFOs that "discharge or propose to discharge" must seek coverage under an NPDES permit. A similar requirement for all point sources already exists under other parts of EPA regulations that were not affected by the *Waterkeeper Alliance* decision (40 C.F.R. §122.21(a)(1)). The proposal deletes the 2003 rule's provision allowing CAFOs to demonstrate that they have no potential to discharge, saying that such a designation would be irrelevant because the proposal requires only those CAFOs that discharge or propose to discharge to seek coverage under a permit. EPA estimated that the change in the "duty to apply" provision means that 25% fewer CAFOs would ultimately receive permits and that CAFO operators will experience a $15.5 million per year reduction (or 26%) in administrative burden, compared with the 2003 rule.

Nutrient Management Plans

The 2003 rule mandated that NPDES permits for all CAFOs that land apply animal waste include a new requirement that the permittee develop and implement a nutrient management plan that includes minimum elements specified in the rule, such as ensuring adequate storage of manure, litter, and process wastewater, and preventing direct contact of confined animals with waters of the United States. CAFOs were to develop and implement an NMP by the same date that the rule required them to comply with the rule's land application provisions (generally December 31, 2006, under the original rule; after the *Waterkeeper Alliance* decision, EPA extended the deadline to July 31, 2007). The rule provided that NMPs would be retained on-site at the CAFO. It must be available to EPA or the permitting authority, but it is not considered part of the facility's permit.

The environmental plaintiffs argued that the NMP part of the rule was unlawful under the Clean Water Act and the Administrative Procedure Act[13] because it failed to require that the terms of the NMP be included in the NPDES permit (inclusion in the permit would make the NMP enforceable by the government and private citizens) and because it allowed permitting authorities to issue permits in the absence of any meaningful government or public review of this aspect of the permit. They also argued that the permitting aspects of the rule violate the Clean Water Act's public participation requirements by effectively shielding the plans from public scrutiny and comment. The court agreed with the environmental plaintiffs on these points and vacated these portions of the rule.

In response, EPA proposes to require that CAFOs seeking permit coverage submit an NMP as part of its permit application and that the permitting authority make the plan available for review prior to developing the facility's permit. The permitting authority would be responsible for reviewing the NMP for completeness and sufficiency. The terms of the NMP (such as the minimum elements described above) would become terms and conditions of the permit, as required by the court. In its proposal, EPA distinguishes between NMP *terms,* which must be incorporated as enforceable conditions of the permit following the public review process, and the plan as a whole, which must be submitted to the permitting authority for review. The NMP as a whole, EPA says, will include underlying data, calculations, and other information such as technical standards that provide a basis for the facility-specific requirements.

EPA allows permitting authorities to issue two types of permits: either individual facility-specific permits, or general permits to cover multiple facilities without the need to receive individual permit applications from facilities in advance of developing the permit. In the 2003 rule, EPA indicated that it expected that most permitting authorities would utilize general permits, as a way of minimizing regulatory burden. The *Waterkeeper Alliance* ruling required EPA to expressly address public participation in review of NMPs, since they must be included in a permit. In the case of individual permits, existing NPDES rules already establish procedures for public participation. Thus, because the NMP would be part of the individual permit application, it would be subject to existing rules requiring public participation, and no rule changes were needed.

EPA's response to the *Waterkeeper Alliance* ruling does contain new provisions for public participation in review of NMPs for those facilities intending to be covered by a general permit, because there is no provision in existing rules that explicitly addresses incorporation of site-specific NMP requirements into a general permit. The proposal includes mechanisms so that general permits for CAFOs can be modified, once issued, to include the terms of an NMP applicable to a specific CAFO and to provide an opportunity for public review of a CAFO's Notice of Intent (including the entire NMP) to be covered by a general permit, before the CAFO actually receives coverage under the general permit. The proposal gives the permitting authority (state or EPA) discretion as to how best to provide public notification and comment in the context of general permits.

Aspects of the Effluent Limitation Guidelines for CAFOs

Specific effluent limitations contained in individual NPDES permits are dictated by the terms of more general effluent limitations guidelines promulgated by EPA that typically specify the maximum allowable levels of pollutants that may be discharged by facilities within an industrial category or subcategory using specific technologies. While the limits are based on the performance of specific technologies, they do not generally require the industry to use these technologies, but rather allow the industry to use any effective alternatives to meet the pollutant limits. As noted above, in the 2003 rule, EPA established non-numerical effluent limitation guidelines for the production areas of CAFOs, and did so for four subcategories of the CAFO industry. The environmental petitioners challenged several aspects of the ELGs, and the *Waterkeeper Alliance* court upheld parts of their claims. In this portion of the decision, the court remanded the rule to EPA with instruction to present additional analysis and justification, so as to clarify its decisionmaking rationale.

Standards for New Sources of Swine, Poultry, and Veal Operations

The CWA requires EPA to promulgate New Source Performance Standards (NSPS) for new, as opposed to already existing, sources of pollution, based on what is determined to be the best available demonstrated control technology. The 2003 rule dictated that new sources in this subcategory meet a waste management standard of no discharge, except in the event of manure runoff and precipitation from a 100-year, 24-hour rainfall event.[14] The rule also allowed a less restrictive alternative performance standard (a 25-year, 24-hour storm standard) for those facilities that will voluntarily use new technologies and management practices that perform as well as or better than the baseline ELGs at reducing pollutant discharges to surface waters from the production area. The court held that EPA had not provided adequate statutory and evidentiary basis for these portions of the rule and had not justified its decision to allow compliance through an alternative standard. In its proposal to revise the rule, EPA deleted the provision allowing CAFOs to meet the no discharge standard through the use of a 100-year, 24-hour rain event containment structure, thus effectively prohibiting all discharge of manure, litter, and process wastewater from the production area for new sources in this subcategory. EPA also proposes to delete the voluntary superior performance standards provision, since the baseline for all new facilities in this subcategory will now be no discharge.

Technology for Pathogen Control

An effluent limitation guideline establishes the degree of pollutant reduction that is attainable by industrial sources through the application of various levels of technology. The CWA requires that ELGs be based on standards that are progressively more stringent: (1) best practicable control technology currently available (BPT), the minimum technological requirement, (2) best control technology for conventional pollutants (BCT), and (3) best available technology economically achievable (BAT), representing the best control measures that have been developed or are capable of being developed within the industrial category. The act required existing sources to meet BPT by July 1, 1977, and BAT by July 1, 1983. BCT is not an additional limitation, but it replaces BAT for control of a group of pollutants that are naturally occurring in the aquatic environment, are biodegradable, and are the traditional and primary focus of wastewater control. Five pollutants are presently considered conventional pollutants; one of these, the pathogen fecal coliform, is associated with manure discharges from CAFOs. Point sources that discharge conventional pollutants are required to meet the BCT standard, but the act requires that, in establishing BCT, EPA must conduct a "cost reasonableness" test of attaining more stringent pollutant control than BPT.

In the 2003 rule, EPA said that the ELG requirements of the rule were not specifically designed to reduce pathogens in animal waste but may, in EPA's view, achieve some incidental reductions of pathogens. The environmental plaintiffs argued that EPA had not presented adequate evidence to justify establishing a BCT standard for pathogens that is no more stringent than the rule's BPT standard. The court upheld this complaint and ruled that EPA must make an affirmative finding that the BCT-based ELGs adopted in the rule do in fact represent the best control technology for reducing pathogens. In its June proposal to revise the 2003 rule, EPA retains the BCT standard promulgated previously and provides a lengthy narrative discussion and cost analysis justifying its rationale.

Water Quality-Based Effluent Limitations

While technology-based NPDES permits derived from EPA's ELGs may result in meeting state water quality standards for individual waterbodies, the effluent guidelines program is not specifically designed to ensure that the discharge from each facility meets the water quality standards for that particular waterbody. For this reason, the CWA requires permitting authorities to establish water quality-based effluent permit limitations (WQBELs), where necessary to attain and maintain water quality standards, that specify discharge limitations that are more stringent than the national ELGs. Where WQBELs are necessary, they are established without consideration of treatment technologies or cost. In the 2003 rule, EPA included no requirements concerning WQBELs, saying that it did not expect that WQBELs will be established for CAFO discharges from land application areas since, as described above, any precipitation-related discharges from those areas will be considered agricultural stormwater, which is exempt from NPDES permitting.

The environmental plaintiffs challenged EPA's failure to justify the lack of WQBELs for other than agricultural stormwater discharges. They also charged that the 2003 rule bars states from promulgating WQBELs. The *Waterkeeper Alliance* court partly upheld these complaints and directed EPA on remand to explain whether or not, and why, WQBELs are needed to assure that CAFO discharges will not interfere with the attainment and maintenance of water quality standards. The court also found that the Preamble to the 2003 rule is ambiguous about whether states may promulgate WQBELs for discharges other than agricultural stormwater, and it ordered EPA to clarify this issue. In the June proposal, EPA restated its view that precipitation-related discharges from land application areas are statutorily exempt from any effluent limitations, including WQBELs, because they are agricultural stormwater, but it clarified that WQBELs can be applied in appropriate cases to further limit discharges from CAFO production areas and with respect to non-precipitation-related land application discharges. This reasoning applies to state-issued as well as EPA-issued permits. Further, EPA said that it is possible that a state, acting under its own regulatory authorities, could impose additional requirements that are broader than the federal NPDES program, if they so choose. Whether many states will do so, however, is unclear.

PUBLIC RESPONSE TO EPA'S PROPOSAL

Several hundred public comments on EPA's June 2006 regulatory proposal were submitted by individual citizens, environmental advocacy groups, state agencies (environmental, public health, and agricultural departments), individual livestock and poultry producers, and groups that represent livestock and poultry producers.[15]

Public comments addressed a number of general and specific technical points, with particular focus on the "duty to apply" and agricultural stormwater exemption provisions of the proposal. Industry's comments were generally supportive of the proposal, approving deletion of the previous "duty to apply" provision and also EPA's efforts to provide flexibility regarding nutrient management plan modifications — especially to limit review and public participation requirements to only those changes that are substantial. Environmental groups, on the other hand, strongly criticized the proposal, arguing that the *Waterkeeper Alliance* court left in place several means for the agency to accomplish much of its original permitting approach, but instead EPA chose not to do so. State environmental and resource agencies, the primary implementers of CWA permitting, also have a number of criticisms. They focus on key parts that they argue will greatly increase the administrative and resource burden on states.

Duty to Apply

Both state permitting authorities and environmental groups are unhappy with EPA's deletion of the requirement that all CAFOs must apply for an NPDES permit. They concur that in doing so, EPA would change the entire permitting program from one that is pro-active to one that is reactive, because it "would allow CAFO operators to decide whether their situation poses enough risk of getting caught having a discharge to warrant the investment of time and resources in obtaining a permit."[16] Although EPA estimates that 25% fewer CAFOs will seek permit coverage, states argue that this overestimates the number that will voluntarily get permits, because under EPA's proposed revisions, there is virtually no incentive to seek a permit. Further, states contend that any cost savings that CAFOs will experience will be shifted to permitting authorities which will be placed in a more adversarial position of first proving that a facility has a discharge and then taking an enforcement action. As one state observed, the number of CAFOs, permitted or not, is the same, and EPA expects states to inspect those that don't apply for permit coverage, as well as process permits for those that do.[17] Overall, states believe that the administrative burden on states of EPA's proposal to delete the "duty to apply" requirement will be greater than under the 2003 rule, not less, as EPA concluded.

Agriculture industry commenters have very different concerns about this aspect of EPA's proposal. They challenged the "duty to apply" provision of the 2003 rule, and the court upheld their argument that the CWA only requires facilities that actually discharge to seek permit coverage. Industry groups fundamentally disagree with any presumption that CAFOs do discharge pollutants, contrary to EPA's position in support of the 2003 rule or environmentalists' contentions.[18] Thus, they object to EPA's attempts to get CAFOs to voluntarily seek permits and the specific addition of a permit requirement for those that "propose to discharge" (see page 6). According to this view, EPA may not lawfully establish permitting requirements based on speculation as to possible future CAFO discharges. Any "duty to apply" triggered by accidental discharges could arise (if at all) only after an actual discharge has occurred and should be limited to facilities that accidentally discharge and fail after a reasonable time to identify the cause and take appropriate corrective measures.[19] One of EPA's rationales for promulgating the 2003 rule was recognition that large numbers of unpermitted CAFOs were discharging wastes that contribute to water quality impairments.[20] Critics of industry's position on this issue contend that allowing CAFOs to self-regulate, self-report accidental releases, and then possibly seek permit coverage will likely perpetuate those same conditions.

Agricultural Stormwater Exemption

Industry groups endorse EPA's proposal regarding agricultural stormwater, which assumes that where land application is conducted in accordance with the rule's nutrient management standards, stormwater runoff is exempt from NPDES permitting. However, these groups strongly object to EPA's suggestion in the Preamble to the rule that it is also considering requiring CAFOs to comply with additional technical standards established by a permitting authority (see page 6), because they maintain that such a change would unlawfully narrow the exemption.

Environmentalists, on the other hand, argue that this portion of the proposal would unlawfully allow CAFOs to self-regulate, as it fails to require them to get permits in order to claim the exemption. States express a similar view, contending that neither a state nor EPA can take enforcement action against an unpermitted CAFO to comply with technical or other standards. One state observed that EPA's proposal represents "a circular arrangement that would be quite difficult to enforce and administer," and that courts will be skeptical of enforcement cases against facilities that are exempt from regulation.[21]

CONCLUSION

While there is no overall agreement in the views of these varied interest groups, they do concur on at least one point: EPA should provide much more clarity and guidance on such key concepts as criteria or circumstances defining the need for a CAFO to seek permit coverage, what terms in a nutrient management plan should be included in a permit, and what constitutes a substantial change to a NMP (since non-substantial changes could be incorporated in a permit without time-consuming review). The Preamble to the proposal offers some examples on these points, but the public comments indicated that considerable uncertainty still exists about issues that are fundamental to implementation of the rule.

Further, agriculture industry groups and states generally agreed on one other issue. As previously noted, EPA had expected to promulgate a final revised rule by June 2007. The 2006 proposal did not include an extension of the July 31, 2007, deadline for compliance with the rule, apparently assuming that states have already adopted provisions of the 2003 rule and would simply need to rescind provisions of the vacated rule and replace them with language of a revised rule. States considered that date "unrealistic and unattainable," because most states likely stopped their rulemaking adoption of the 2003 rule during the *Waterkeeper Alliance* challenge.[22] Industry groups argued that one month is not enough time for CAFOs to decide whether to apply for a permit, prepare the permit application, and prepare or update their NMPs to meet the new regulatory requirements. Thus, many of their comments urged EPA to extend the compliance deadline.

Because of delays in completing work on a final rule, in May 2007, EPA proposed to extend the compliance deadline from July 2007 to February 27, 2009, as discussed above. In reaction, environmental advocates objected to the proposed extension, asserting that it would further delay the time when states will issue needed permits to CAFOs. Some states said that EPA's delay will complicate the work of state regulators who are anxious to have the rule finalized, since until that time, they lack clear direction from EPA about what is required of their permit programs.

Finally, because of the differing perspectives on EPA's proposal, one can anticipate that whatever revised regulation emerges from the current process will be challenged. Some of the discussions in the public comments echoed criticisms that were made of the 2003 rule and seem to preview legal critiques that are likely to be raised in future challenges. Thus, it is nearly as difficult to estimate when the issues discussed here will ultimately be resolved, as it is to estimate how they will be resolved.

REFERENCES

[1]   U.S. Environmental Protection Agency, "National Water Quality Inventory, 2000 Report," August 2002, EPA-841-R-02-001, 1 vol.

[2]   U.S. Environmental Protection Agency, "National Pollutant Discharge Elimination system Permit Regulation and Effluent Limitation Guidelines and Standards for Concentrated Animal Feeding Operations CAFO; Final Rule," 68 *Federal Register* 7175-7274, February 12, 2003. For additional information on the rule, see CRS Report RL31851, "Animal Waste and Water Quality: EPA Regulation of Concentrated Animal Feeding Operations CAFO," by Claudia Copeland.

[3]   Under the act, point sources are defined as any discernible, confined, and discrete conveyance, such as any pipe, ditch, channel, or conduit from which pollutants are or may be discharged. In contrast, nonpoint source pollution, which is not regulated by NPDES permits, is any source of water pollution that is not associated with a discrete conveyance, including precipitation runoff from fields, forest lands, or mining and construction activities.

[4]   An animal feeding operation (AFO) is a facility in which livestock or poultry are raised or housed in confinement for a total of 45 days or more in any 12-month period and animals are not maintained in a pasture or on rangeland. CAFOs are a subset of AFOs. In addition to meeting the confinement criteria, an AFO is a CAFO if it meets minimum size thresholds (those with more than 1,000 animals are CAFOs; those with fewer animals may be defined as CAFOs in some cases).

[5]   This is a rainfall event with the probability of recurrence once in 25 years (or a 4% chance of being exceeded in a 24-hour period in any single year). The amount of precipitation that constitutes a 25-year, 24-hour rainfall event varies by location.

[6]   U.S. Environmental Protection Agency, "Revised National Pollutant Discharge Elimination System Permit Regulation and Effluent Limitation Guidelines for Concentrated Animal Feeding Operations in Response to Waterkeeper Decision; Proposed Rule," 71 *Federal Register* 37744-37787, June 30, 2006.

[7]   U.S. Environmental Protection Agency, "Revised Compliance Dates for National Pollutant Discharge Elimination System Permit Regulation and Effluent Limitation Guidelines for Concentrated Animal Feeding Operations," 71 *Federal Register* 6978-6984, February 10, 2006.

[8]   U.S. Environmental Protection Agency, "Revised Compliance Dates Under the National Pollutant Discharge Elimination System Permit Regulations and Effluent Limitations Guidelines and Standards for Concentrated Animal Feeding Operations," 72 *Federal Register* 40245-40250, July 24, 2007.

[9]   Waterkeeper Alliance et al. v. EPA, 399 F.3d at 507.

[10] Production areas such as feedlots and lagoons are not eligible for the agricultural stormwater exemption, because they involve the type of industrial activity that originally led Congress to single out CAFOs as point sources. See 68 *Federal Register* 7198.

[11] Waterkeeper, p. 509.

[12] 71 Federal Register at 37748.

[13] The Administrative Procedure Act, 5 U.S.C. §§701-706, contains provisions that govern federal agency rulemaking proceedings.

[14] This is a statistical event defined as the amount of rainfall that has a 1% chance of being exceeded in a 24-hour period in any given year (or once in 100 years).

[15] Materials in the EPA docket for this rulemaking, No. EPA-HQ-OW-2005-0037, including EPA documents and public comments on the proposal, can be found at [http://www. regulations.gov/fdmspublic/component/main].

[16] Natural Resources Defense Council, Sierra Club, Waterkeeper Alliance, Comments on the revised CAFO regulation, August 29, 2006, p. 9.

[17] Ohio Department of Agriculture, Ohio Environmental Protection Agency, Ohio Department of Natural Resources, Comments on the revised CAFO regulation, undated, p. 6.

[18] National Pork Producers Council, United Egg Producers, American Farm Bureau Federation, National Council of Farmer Cooperatives, National Corn Growers Association, "Comments on Proposed Post-*Waterkeeper* CAFO NPDES Regulations," August 29, 2006, p. 38.

[19] Id., p. 14.

[20] See 68 *Federal Register* 7179-7181,

[21] Illinois Environmental Protection Agency, Comments on the revised CAFO regulation, August 29, 2006, p. 4.

[22] Association of State and Interstate Water Pollution Control Administrators, Comments on revised CAFO regulation, August 29, 2006, p. 4.

In: Water Pollution Issues and Developments
Editor: S. V. Thomas, pp. 89-102

ISBN: 978-1-60456-208-8
© 2008 Nova Science Publishers, Inc.

Chapter 5

ALLOCATION OF WASTEWATER TREATMENT ASSISTANCE: FORMULA AND OTHER CHANGES*

Claudia Copeland

ABSTRACT

Congress established the statutory formula governing distribution of financial aid for municipal wastewater treatment in the Clean Water Act (CWA) in 1972. Since then, Congress has modified the formula and incorporated other eligibility changes five times. Federal funds are provided to states through annual appropriations according to the statutory formula to assist local governments in constructing wastewater treatment projects in compliance with federal standards. The most recent formula change, enacted in 1987, continues to apply to distribution of federal grants to capitalize state revolving loan funds (SRFs) for similar activities

The current state-by-state allotment is a complex formulation consisting basically of two elements, state population and "need." The latter refers to states' estimates of capital costs for wastewater projects necessary for compliance with the act. Funding needs surveys have been done since the 1960s and became an element of distributing CWA funds in 1972. The Environmental Protection Agency (EPA) in consultation with states has prepared 13 clean water needs surveys since then to provide information to policymakers on the nation's total funding needs, as well as needs for certain types of projects.

This report describes the formula and eligibility changes adopted by Congress since 1972, revealing the interplay and decisionmaking by Congress on factors to include in the formula. Two types of trends and institutional preferences can be discerned in these actions. First, there are differences over the use of "need" and population factors in the allocation formula itself. Over time, the weighting and preference given to certain factors in the allocation formula have become increasingly complex and difficult to discern. Second, there is a gradual increase in restrictions on types of wastewater treatment projects eligible for federal assistance.

Crafting an allotment formula has been one of the most controversial issues debated during past reauthorizations of the Clean Water Act. The dollars involved are significant, and considerations of "winner" and "loser" states bear heavily on discussions of policy

---

* Excerpted from CRS Report RL31073, dated March 9, 2007.

choices reflected in alternative formulations. This is likely to be the case again, when Congress moves to reauthorize the act. In the 109th Congress, legislation to extend CWA infrastructure financing was approved by the Senate Environment and Public Works Committee (S. 1400, S.Rept. 109-186). It included changes to the allotment process. However, the bill did not receive further action, partly because of controversies over the proposed allocation formulas. Because the current allocation formula is now 20 years old, and because needs and population have changed, the issue of how to allocate state-by-state distribution of federal funds is likely to be an important topic in debate over water infrastructure legislation. This report will be updated as developments warrant.

INTRODUCTION

Congress established the statutory formula governing distribution of financial aid for municipal wastewater treatment in the Clean Water Act (CWA) in 1972. Since then, Congress has modified the formula and incorporated other eligibility changes five times, actions which have been controversial on each occasion. Federal funds are provided to states through annual appropriations according to the statutory formula to assist local governments in constructing wastewater treatment projects in compliance with federal standards. Congress has appropriated more than $75 billion since 1972. The formula originally applied to the act's program of grants for constructing such projects. That grants program was replaced in the law in 1987 by a new program of federal grants to capitalize state revolving loan funds (SRFs) for similar activities. The most recent formula change, also enacted in 1987, continues to apply to federal capitalization grants for clean water SRFs.

The current state-by-state allotment is a complex formulation consisting basically of two elements, state population and "need." The latter refers to states' estimates of capital costs for wastewater projects necessary for compliance with the act. Funding needs surveys have been done since the 1960s and became an element for distributing CWA funds in 1972. The Environmental Protection Agency (EPA), in consultation with states, has prepared 13 clean water needs surveys since then to provide information to policymakers on the nation's total funding needs, as well as needs for certain types of projects. Legislation to fund water infrastructure projects is on the agenda of the 108th Congress. In part because the allocation formula is 16 years old and needs and population have changed, the issue of state-by-state distribution of funds is likely to be an important topic when legislation is considered.

This report describes the formula and eligibility changes adopted by Congress since 1972, revealing the interplay and decisionmaking by Congress on factors to include in the formula. Two types of trends and institutional preferences can be discerned in these actions. First, there are differences over the use of need and population factors in the allocation formula itself. During the 1970s, the Senate strongly favored reliance on use of population factors in the allocation formula, while the House strongly advocated a needs-based approach. During the 1980s, the period when categorical eligibilities were restricted in order to emphasize water quality benefits, the Senate favored needs as the basis for grants distribution, while the House position generally was to retain formulas used in prior years, which incorporate both needs and population elements. When population has been used as a factor, differences have occurred over whether a current or future year population estimate is appropriate, but there is no clear trend on this point.

Second, there have been gradual increases in restrictions on types of wastewater treatment projects eligible for federal assistance. Beginning with a limitation that denied use of federal funds for stormwater sewer projects in 1977, debate over categorical eligibility has had two elements. One has been fiscal: a desire to not fund types of projects with the highest costs and often the most unreliable cost estimates. The other focus has been environmental: a desire to use federal resources to assist projects which benefit water quality protection most directly. While some of these eligibility restrictions presented Congress with rather straightforward choices, others have been more complex. Some continue to be debated, such as whether certain types of projects should be fully eligible for federal aid or should be the responsibility of state and local governments.

FORMULA AND OTHER CHANGES

The following table provides a generalized summary of the components of the allocation formula since 1972. Details discussed below should be consulted, because a summary table such as this cannot fully reflect factors such as "hold harmless" or "minimum share" provisions frequently included in the state-by-state distribution scheme to protect states with small allocations or to minimize potential disruptions when formula changes were adopted. The term "total needs" refers to funding needs identified by states for all categories of projects and water quality activities eligible for assistance. The term "partial needs" refers to a subset of eligible project categories, primarily construction or upgrades to comply with the act's minimum requirement that municipalities achieve secondary treatment of wastewater.

Table 1. Needs and Population Components of CWA Allocation Formula

Fiscal Year	Total Needs	Partial Needs	Population
Pre-1973	—	—	100%
1973-1974	100%	—	—
1975-1976	50% [b]	50% [a]	—
Talmadge-Nunn Act, P.L. 94-369	—	50% [a]	50% [c]
1977	25% [b]	50% [a]	25% [d]
1978-1982	25% [b]	50% [a]	25% [e]
1983-1986	12.5% [b]	50% [f]	37.5% [e]
1987-present [g]	?	?	?

[a] Project categories I, II, and IVB(see pages 4-5) [b] Excluding separate stormwater sewers [c] 1990 population [d] 1975 population [e] Population year cannot be determined [f] Backlog needs only, categories I, II, IIIA, and IVB (see page 10) [g] Precise factors included in the formula are unclear; see text for discussion.

Grants Allocation before P.L. 92-500 (Pre-1973)

Prior to enactment of the Federal Water Pollution Control Act Amendments in 1972 (FWPCA, P.L. 92-500), the federal government administered a comparatively small program of aid for constructing municipal wastewater treatment plants.[1]

Under the prior program, assistance was allocated to states on the basis of population. There was no statutory formula. Nor was there a systematic process for the federal government or states to estimate and report on funding needs for sewage treatment. Needs surveys had been developed by the Conference of State Sanitary Engineers, which reported generally (but not rigorously) on estimated construction costs of municipal waste treatment facilities planned by communities to meet water quality standards or other standards or enforcement requirements. They lacked both consistent definitions of objectives and consistent reporting requirements. Moreover, these surveys tended to be based on needs of larger municipalities, so needs in small or rural communities were underrepresented.

The first funding needs survey undertaken by the federal government was published in 1968, in response to a general requirement in the 1966 Clean Water Restoration Act for an annual report on "the economics of clean water," but it was a considerably more modest effort than followed enactment of P.L. 92-500. These early documents reported state-by-state and national total needs over a given period of time but did not estimate or report needs for particular categories of waste treatment projects, such as secondary treatment. Annual surveys were published each year through 1974; Congress then changed the reporting requirement to biennial.

P.L. 92-500 and the Formula for 1973-1974

In P.L. 92-500 Congress provided the first statutory formula, governing state-by-state allocations in fiscal years 1973 and 1974. It was entirely needs-based and contained no categorical limitations. Despite weaknesses of the prior surveys, they were the only tool available to guide Congress when the decision was made in the 1972 legislation to move away from a population-based distribution of grants. The 1972 survey estimated total needs, from 1972 through 1976, to be $18.1 billion. Estimated construction costs for the first three years of that period were reported to be $14.6 billion. The rationale for changing to a needs basis for grants allocation despite limitations of available needs information was explained in the House Public Works Committee's report on the 1972 legislation.[2]

> This needs formula is a sound basis for allotting funds since our experience to date clearly demonstrates that there is no necessary correlation between the financial assistance needed for waste treatment works in a given State and its population.
> The Committee is fully aware that at the present time there is no satisfactory estimate of the total funds required by the States for construction of publicly owned treatment works... However [the 1972 Needs Survey] report does provide some measure of the relative needs of the various States and in the absence of any better measure has been incorporated in the bill for the determination of the State allotments for the fiscal years 1973 and 1974.

The Senate favored retaining population as the basis for grants allocation, and the available public records — Committee reports and Senate debates — give no indication whether an alternative approach, such as one based on needs, was considered.

The 1972 FWPCA incorporated a statutory formula for distributing grants that was derived from the 1972 survey for the period 1972 through 1974. It covered reported needs in the 50 states and territories, with little categorical restriction. Some limitation was included on use of federal funds for new collector sewers (which collect and carry wastewater from an individual house or business to a major, or interceptor, sewer that conveys the wastewater to a treatment facility). In addition, eligibility for funds was limited to communities in existence when P.L. 92-500 was enacted and could only be provided if the treatment plant had sufficient existing or planned capacity to treat sewage collected by such sewers.

Section 205(a) of the FWPCA cross-referenced a table in a House Public Works and Transportation Committee Print that identified each state's percentage share under the legislation.[3] The percentages would apply to total grant amounts made available through annual congressional appropriations. The statute provided that this distribution formula would apply for two years; in section 516 of the act, EPA was directed to prepare a new needs survey that would govern distribution in FY1975.

The 1973 Needs Survey

In response to the 1972 statutory directive, EPA undertook a new method of preparing the needs survey, and the 1973 Needs Survey was the first effort to report and evaluate needs for categories of waste treatment projects, as well as state and national totals. This survey reported costs for the following categories:

I — Secondary treatment required by the 1972 act
II — Treatment more stringent than secondary required by water quality standards
III — Rehabilitation of sewers to correct infiltration and inflow
IV — New collector and interceptor sewers
V — Correction of overflows from combined stormwater and sanitary sewers (CSOs)

This original categorization was subsequently refined. Category III was subdivided to include category IIIA — correction of infiltration and inflow in existing sewers; and category IIIB — replacement or rehabilitation of structurally deteriorating sewers. Category IV was subdivided to include category IVA — new collector sewers; and category IVB — new interceptor sewers. Needs surveys have continued to be based on this same categorical arrangement since the mid-1970s.

However, from an initial estimate of $63 billion in the 1973 survey, the survey figure for wastewater treatment and collection system projects went to a high of $342 billion in 1974, dropped to $96 billion in 1976, rose to $106 billion in 1978, $120 billion in 1980, declined to $80 billion in 1990, and was assessed at $67 billion in 2000, the thirteenth and most recent survey. Since the 1992 survey, states also have assessed needs for projects to address nonpoint pollution from sources such as agriculture, silviculture, and urban runoff. In the 2000 survey, needs for these types of projects were an additional $14 billion. Over time, inconsistencies and variations in the surveys have been ascribed to several factors, including the lack of precision with which needs for some project categories could be assessed and the desire of state estimators to use the needs survey as a way of keeping their share of the federal allotment as high as possible.[4]

Formula Applicable in 1975 and 1976

In December 1973, Congress enacted P.L. 93-243, Waste Treatment Fund Allocations, providing the section 205(a) allocation formula for FY1975. As enacted, the formula was based on EPA's November 1973 Needs Survey, with a formula that split the difference between total needs and partial needs. The formula was: one-half of amounts reported in the 1973 Needs Survey for all categories (secondary treatment, more stringent than secondary, sewer rehabilitation to correct infiltration and inflow, new collector sewers, new interceptor sewers, and CSO correction, but not separate stormwater sewers), and one-half of amounts just for categories including secondary treatment, more stringent than secondary and new interceptor sewers. The formula also included a hold harmless provision, under which no state would receive less in construction grant funds than it was allotted under the previous formula.

Use of the partial needs categories was based on EPA's recommendation to the Congress that the allocation formula should only include the costs of providing treatment works to achieve secondary treatment (the basic national treatment requirement mandated in the 1972 act), treatment more stringent than secondary as required by water quality standards, and eligible new interceptor sewers, force mains, and pumping stations (categories I, II and IVB, respectively). These were the core categories representing projects to comply with the basic water quality objectives of the Clean Water Act. EPA's basis for this recommendation was the Agency's assessment that the data for the other categories, as reported by the states, were limited and considerably less reliable than for these three categories.

In the 1973 survey, EPA reported that total needs nationwide were $60.1 billion (1973 dollars), but that reported costs probably underestimated actual expenditures — by half — due to underreporting of CSO needs and failure of states to report all needs in categories I and II. EPA reported that estimates from only 15 states included cost surveys of all communities in the state; data from the remaining 35 states represented all urban areas plus a sample of communities of less than 10,000 persons located outside urban areas.

The Senate Committee on Public Works found that EPA's recommendations would lead to inequities affecting a number of states. In its version of legislation to establish an allocation formula for 1975 (S. 2812), it recommended distribution based 75% on partial needs and 25% on 1972 population (i.e., the ratio of a state's 1972 population compared to the population of all states).

The formula recommended by the House, in its version of the legislation (H.R. 11928), was the same as the version finally agreed to: one-half partial needs, and one-half total needs, based on the 1973 EPA Needs Survey. The House Committee's actions were explained by the chairman of the Public Works and Transportation Committee.[5]

> The Environmental Protection Agency proposed two tables for allocation of the grant funds to the States. One was based on all of the needs of the States...The other table was based on only part of the needs...The committee heard testimony from several States, some of which would receive more funds under one table and some of which would receive more under the other table. In addition, some States found that under the needs concept they would receive less than they had previously when funds were allocated on the basis of population. The primary reason for this appears to be that these States have not yet accurately identified their true needs for wastewater treatment facilities.
>
> The committee is very much committed to the allocation of funds on the basis of need. After much consideration, we determined that the most equitable solution would be to allocate the funds for the next 2 fiscal years on the basis of 50 percent of each of the two tables, with no State receiving less than its allocation of 1972. While some States may receive a little less under the committee's solution, all States will benefit greatly in the long run.

Although the House-passed bill called for a two-year allocation formula, the enacted legislation applied only to FY1975. Nevertheless, the formula continued to apply through FY1976, because Congress did not enact legislation to modify it until 1977.[6]

1977 Allocation and Appropriations

Appropriations in FY1977 were provided under two appropriations acts, the Public Works Employment Appropriations Act of 1976 and the Fiscal Year 1977 supplemental appropriations act, each using a different allocation formula.

Allotment under the Public Works Employment Act (Talmadge-Nunn Act)

The 1973 allocation legislation, P.L. 93-243, required EPA to prepare a new, comprehensive needs survey no later than September 3, 1974, and directed that it include all of the categories included in the 1973 survey, plus costs to treat separate storm water flows. In response, the next wastewater needs survey (the 1974 survey) was transmitted to Congress in February 1975.[7] Based on that survey, EPA recommended that future formulas focus on needs reported for categories I, II, and IVB. This recommendation came from the Agency's conclusion that data and cost estimates for other categories submitted in prior surveys had been of poor and inconsistent quality and had resulted in an inequitable allocation formula, as expressed by EPA Administrator Russell Train.[8]

> There is serious doubt, however, that we will be able to provide accurate estimates of the total national needs, or of needs for each State, which would form an equitable basis for allocation of construction grant funds. Even categories I, II, and IV(b) will be very difficult to refine for purposes of allocation because of the large variations in approach used by the States in estimating needs in these categories.
>
> I believe that the fundamental differences in reported cost estimates for the construction of publicly owned wastewater treatment facilities highlighted by the last two surveys confirms our concerns about basing the allocation of Federal funds on "needs," at least as they are currently reported.

Congress adopted EPA's recommendation to limit the use of "total needs" in connection with the allotment formula that governed distribution of $700 million in authorized monies under the Public Works Employment Act of 1976, P.L. 94-369, but in so doing, it reintroduced a population factor. This act, commonly referred to as the Talmadge-Nunn Act, authorized funds for a number of public works programs, including wastewater treatment construction, in order to counter unemployment conditions in certain regions of the country. Under the statutory language, the wastewater treatment monies authorized in P.L. 94-369 were to be allocated just to the 33 states and four territories that had received inequitable allocations as a result of the prior two needs surveys.[9] The action in this legislation is significant, because it restored population as a factor in the construction grants allocation formula. The formula in P.L. 94-369 was used to govern the distribution of $480 million in FY1977 construction grants to the 33 states and 4 territories identified in that act.[10] The formula provided under P.L. 94-369 was 50% partial needs, as reported in the 1974 needs survey, and 50% 1990 projected population.

1977 Supplemental Appropriations Allocation

The second portion of funds provided in Fiscal Year 1977, totaling $1 billion, was governed by the formula that Congress enacted in the FY1977 supplemental appropriations act, P.L. 95-26. That legislation directed that construction grants allocation be according to the 25-50-25 formula contained in the table on page 16 of S.Rept. 95-38,[11] which was 25% total needs from the EPA 1974 needs survey, 50% partial needs from the 1974 survey, and 25% 1975 population. The needs factors used in this formula were the same as had been in use since FY1975 (derived from the 1974 needs survey), but the population basis was different — population in 1975, rather than projected 1990 population, as under the formula that applied in 1976 under Talmadge-Nunn.

1978-1982 Allocation

The next Clean Water Act amendments that addressed the allocation formula were in the 1977 amendments (P.L. 95-217); these amendments provided the distribution formula for FY1978 through FY1981. The final version of the formula was based 25% on total needs (excluding costs of treating separate stormwater flows), 50% on partial needs (categories I, II, and IVB), and 25% on population. The resulting distribution, on a percentage basis, was summarized in tables included in a House Public Works and Transportation Committee print; the allocation provided in table 3 from that report is referenced in section 205(a) of the Clean Water Act, as amended by P.L. 95-217. (As discussed below, this same formula was subsequently extended to 1982.)

Documents in the legislative history do not indicate clearly either which year's needs survey or which population year were reflected in the final formula. The formula provided in the House version of the 1977 legislation (H.R. 3199) contained a ratio similar to the final version and was based on data from EPA's 1974 needs survey and 1990 estimated population (the factors also used under the Public Works Employment Act of 1976).[12]

The Senate version of the legislation, S. 1952, contained a formula based on 1975 population and needs reported in the 1976 survey for categories I, II, III, IVB, and V. The committee formula utilized the higher of the two percentages each state would receive under the two formulas and then reduced the total (which added up to 117.34%) to 100%. In addition, no state would receive less than one-half of one percent of total funds.

Although the 25-50-25 ratio in the final formula was the same as under the House bill, the state-by-state percentages were not identical, so it appears that, although conferees endorsed the basic House approach, they made some changes, as well. Neither the conference report nor House and Senate debates on the final legislation provides sufficient explanation to determine which population year (1990 or 1975) or needs survey (1974 or 1976) was used in the final allocation formula.

Categorical Restrictions

Beyond the question of which categories should be included for purposes of the allocation formula, the 1977 amendments presented the first explicit restrictions on categories eligible for federal grant assistance. Based on provisions in the Senate bill (the House version had no similar provisions), the 1977 amendments made one categorical restriction. The legislation prohibited use of federal funds for projects to control pollutant discharges from separate storm sewer systems, category VI in the EPA needs survey. The concerns here were fiscal (the 1974 survey estimated category VI costs at $235 billion, or double all other costs in total) and environmental. The committee sought to assure that federal funds would be used for facilities most critical to reducing pollutant discharges, according to the report on the Senate bill, S. 1952.[13]

> The cost of controlling stormwater is substantial even after consideration of other options such as land use controls which may be more cost-effective in some situations. The Federal share for stormwater projects is beyond the reach of the limitations of the Federal budget. It is, furthermore, a cost for which water quality benefits have not been sufficiently evaluated, particularly since stormwater discharges occur on an episodic basis during which water use is minimal.

Senate-proposed restrictions on new collector sewer systems and rehabilitation of existing collectors were not included in the final 1977 amendments. Like its proposal concerning stormwater sewers, the committee had contended that the costs of all such projects were excessive, while the water quality benefits were less significant than other core projects, such as constructing secondary treatment plants.

1983-1986 Allocation

P.L. 97-117, passed in 1981, contained the formula governing distribution of construction grant funds from 1982 through 1985. It was subsequently extended through 1986. These amendments included a number of eligibility restrictions, as well.

The House bill, H.R. 4503, proposed to extend the existing formula through FY1982 only. The position of the House Public Works Committee was that it would address multi-year funding issues in a comprehensive review of the Clean Water Act in 1982.

In S. 1716, the Senate adopted a new formula based on 1980 population; backlog needs for categories I, II and IVB, as reported in the 1980 needs survey; plus a minimum state share and "hold harmless" provisions to protect states in order to alleviate disruption of state programs, by minimizing potential loss of funds under a new formula. Backlog needs, used for the first time in connection with this legislation, were defined as facility requirements to meet the needs of the 1980 population — rather than 20 years' future growth, as had been customary in previous needs surveys and allotment formulas. The Senate formula would apply through FY1984. EPA was directed to conduct a new needs survey placing greater emphasis on public health and water quality needs; that survey would be the basis for allocation beginning in FY1985.

As enacted, P.L. 97-117 incorporated the House formula for 1982. For 1983 through 1985, the legislation used the average of the House formula and the Senate formula for 1984 — which was 1980 population, backlog partial needs (for categories I, II, IVB *and* IIIA), and a hold harmless provision that no state would receive less than 80% of what it would have received under the 1977 amendments formula. These four categories were those which were to be fully eligible for federal grants, under categorical restrictions included in the legislation (see below).

Because of delays in enacting a reauthorization bill in the mid-1980s, Congress extended
this formula through 1986, as well.

Categorical Restrictions

The 1981 legislation put in place several eligibility restrictions intended to restructure the
grants program. The Senate Committee explained the rationale in its report on the
legislation.[14]

> The members of this Committee, the Administration, and the majority of the
> witnesses who came before the Committee agree that the time has come to provide
> priority funding to those parts of the program which provide the greatest water quality
> benefit. The Committee bill reflects this principle. In the future, only treatment facilities
> and the necessary interceptor sewers associated with those plants will be eligible for
> Federal assistance.

Two broad points were made by those who advocated restrictions: (1) current budgetary
problems made it necessary to focus limited federal resources on the highest priority
environmental problems; and (2) the Administration believed that the federal government's
funding responsibilities had largely been met, and remaining water quality needs were local,
not national in scope. Based on these issues, the Reagan Administration proposed a number
of program changes that Congress endorsed with some modifications:

- The Administration recommended eliminating eligibility for new collector sewers,
  sewer rehabilitation, infiltration and inflow correction, and combined sewer overflow
  projects. The 1981 amendments retained full eligibility for infiltration and inflow
  projects, on the basis that they can reduce the need for additional sewage treatment
  plant capacity. The amendments made the other categories generally ineligible for
  federal grants, *but* allowed governors to use up to 20% of their annual allotment for
  such projects. The general prohibition on use of federal funds for separate storm
  sewer projects, established in the 1977 legislation, was continued.
- The Administration recommended eliminating eligibility for reserve capacity to meet
  future population growth and recommended that the allotment formula be based only
  on backlog needs. The legislation provided that, after October 1, 1984, no grant
  would be made for reserve capacity in excess of that needed when an actual
  construction grant is awarded and in no event in excess of needs existing on October
  1, 1990.
- The Administration recommended eliminating "hold harmless" and minimum
  allocation provisions of the formula which were not related to water quality benefits.
  Congress did not adopt these recommendations.

Finally, although not part of the Administration's recommendations, the enacted
legislation reduced the federal share for eligible projects from 75% to 55%, to extend the
availability of limited federal funds to more projects.

Adoption of the Current Allotment Formula

In the 1987 amendments, P.L. 100-4, Congress adopted the allocation formula that is currently in effect. Unlike the 1981 legislation, Congress did not make fundamental changes in eligibility — there were no further limitations on types of projects eligible for federal assistance. The prohibition on federal funding for separate storm sewers was continued. The bigger policy issues debated in this legislation concerned establishing state revolving funds as the future funding mechanism, thus replacing the previous construction grants program. Congress directed that the act's statutory allotment formula would govern the new SRF program (Title VI of the act) and also would continue to govern construction grants allotment during the transition from the old funding program to the new one in 1991.

Nowhere in the legislative history of Congress' final action on the 1987 amendments is there a clear statement about the weighting or factors that went into the final allocation formula — it is even difficult to guess. The conference report on the final legislation merely states: "The conference substitute adopts a new formula for distributing construction grant funds and the state revolving loan fund capitalization grants among states for fiscal years 1987 through 1990. The allotment formula for FY1986 is the same as under current law."[15]

It is clearer, however, where the two houses began. During consideration of the legislation, the House favored retaining the formula adopted in 1981. The Senate proposed an entirely new formula.

The Senate formula was based on partial needs (year 2000 needs — not backlog needs, as in the 1981 formula) reported in the 1984 needs survey for the 4 categories which are fully eligible for federal funds: I, II, IIIA (made eligible in 1981), and IVB. As reported by the committee, the formula was essentially based on needs for these categories. There was no *explicit* population factor — but an implicit population factor was incorporated in reverse, because 21 small states were allotted a slightly larger share in order to be able to maintain viable programs, according to the committee report.[16] In addition, the formula included an 80% hold harmless provision for 11 large states that were expected to experience greater changes in eligibility because of the revised formula, compared with the average.[17]

The formula adopted by the Senate, after debate on S. 1128, was different still: it provided that the full extent of formula changes would apply to the last three of the five years covered by the reauthorization and that a modified version would govern during the first two years. The two-year modified version gave the large states an 85% hold harmless by holding down the amount of increased share that the smaller states would receive — so that large states would lose less, and smaller states would gain less, at least in the first two years.

Accordingly, the Senate formula was essentially needs-based, with an unquantifiable population factor apparently included, as well. It was merged — in ways that are not clear from available public documents — with the House formula, which had total needs, partial needs, and 1980 population factors.

CONCLUDING THOUGHTS

Since adoption of the allocation formula that has governed distribution of Clean Water Act assistance since 1987,[18] EPA and the states have produced four updated needs surveys (in 1988, 1992, 1996, and 2000). In addition, updated population information became available through two subsequent decennial Censuses (in 1990 and 2000). Although population changes have occurred during that time, and needs for water quality projects also have changed, none of this more recent information is reflected in the currently applicable distribution formula.

Also since 1987, Congress has on several occasions considered CWA reauthorization legislation that would have modified that formula. In the 104th Congress, the Senate Environment and Public Works Committee approved a bill (S. 2093), and in the 105th Congress, the House passed a comprehensive reauthorization bill (H.R. 961). In the 107th Congress, House and Senate committees approved legislation to reauthorize water infrastructure financing programs in the act (H.R. 3930, S. 1961). These bills included revised allocation formulas, as well as other program changes. Neither received further action.

In the 108th Congress, legislation to extend CWA infrastructure financing was approved by a House subcommittee (H.R. 1560) in July 2003. It included changes to the allotment process: it would have extended the current statutory formula for two years, and thereafter, a portion of funds would be allocated by that formula and a portion would be distributed based on each state's proportional needs of total needs. In October 2004, the Senate Environment and Public Works Committee reported a similar bill (S. 2550, S.Rept. 108-386). Like H.R. 1560, S. 2550 would have modified the allotment formula, moving towards a target allotment based entirely on needs, but with several complex factors intended to moderate the potential for substantial loss or gain of funds under the target, compared with the current statutory allotment formula. (For information, see CRS Report RL32503, *Water Infrastructure Financing Legislation: Comparison of S. 2550 and H.R. 1560*, by Claudia Copeland and Mary Tiemann.) Neither measure received further action, in part because of controversies over the allotment formulas in the bills.

In the 109th Congress, the Senate Environment and Public Works Committee approved S. 1400, the Water Infrastructure Financing Act (S.Rept. 109-186). Among other provisions, S. 1400 would have revised the CWA formula for allocation of SRF monies. This bill was similar to S. 2550 from the 108th Congress, although the proposed allocation formula was different from that in the earlier bill. As in the previous congress, no further action occurred on this bill.

In 1996, Congress amended the Safe Drinking Water Act and established a drinking water state revolving loan fund program modeled after the clean water SRF. However, in that act (P.L. 104-182), Congress directed that drinking water SRF capitalization grants be allotted among the states by EPA based on the proportional share of each state's needs identified in the most recent national drinking water needs survey, not according to a statutory allotment formula.

It is also worth noting that, while the clean water and drinking water SRFs represent significant amounts of federal financial assistance, Congress has provided other assistance, as well, in the form of grants earmarked in EPA appropriations acts for specific communities, both small and large. In recent years, congressional appropriators have dedicated an increasing portion of annual water infrastructure assistance as earmarked special project grants which are not subject to any statutory or other allotment formula. For example, for FY2005 (P.L. 108-447), Congress appropriated $1.09 billion for clean water SRF capitalization grants, $843 million for drinking water SRF capitalization grants, and $402 million in earmarked grants for projects in 669 listed communities. Since the first of these earmarks in EPA appropriations in FY1989, Congress has provided $6.5 billion for special project grants.

Crafting an allotment formula has been one of the most controversial issues debated during past reauthorizations of the Clean Water Act. The dollars involved are significant, and considerations of "winner" and "loser" states bear heavily on discussions of policy choices reflected in alternative formulations. This is likely to be the case again, when Congress reauthorizes the wastewater infrastructure funding portions of the act. In part because the current allocation formula is now 20 years old, the issue of how to allocate state-by-state distribution of federal funds is likely to be an important topic during debate on water infrastructure financing legislation, and elements of funding need, population, and possibly other factors are likely to again be debated.

REFERENCES

[1]   The Water Pollution Control Act of 1948 (P.L. 80-845) started the trickle of federal aid to municipal wastewater treatment authorities that grew in subsequent years. It authorized loans for treatment plant construction. With each successive statute in the 1950s and 1960s, federal assistance to municipal treatment agencies increased. A grant program replaced the loan program; the amount of authorized funding went up; the percentage of total costs covered by federal funds was raised; and the types of project costs deemed grant-eligible were expanded.

[2]   U.S. Congress. Senate. Committee on Public Works. *A Legislative History of the Water Pollution Control Act Amendments of 1972.* January 1973. Serial No. 93-1. 93rd Congress, 1st session. p. 780.

[3]   U.S. Congress. House. Committee on Public Works and Transportation. *Estimated Construction Cost of Sewage Treatment Facilities Planned for the Period Fiscal Years 1972-74.* Committee Print 92-50. 92nd Congress, 2d sess. Table 3, p. 3.

[4]   Water Pollution Control Federation (now, the Water Environment Federation). *The Clean Water Act with Amendments.* 1982. p. 14.

[5]   Congressional Record, vol. 119, part 32, p. 42259.

[6]   In FY1976, construction grant funds provided to states resulted from the release of $9 billion originally authorized in P.L. 92-500 which the Administration had impounded. The withheld sums amounted to $3 billion for each of the Fiscal Years 1973, 1974, and 1975. In 1975, the Supreme Court affirmed a lower court decision requiring the allotment of these funds (*Train v. City of New York*, 420 U.S. 35 (1975)), and EPA did so in February 1975. A portion of the released funds (representing the $3 billion from FY1973 and the $3 billion from FY1974) was allocated according to the allotment formula specified in P.L. 92-500 that had applied to other funds previously distributed for those years. The remaining $3 billion (representing withheld FY1975 funds) were allocated on the basis of the requirements of P.L. 93-243, that is, one-half partial needs, one-half total needs. See U.S. Environmental Protection Agency. Amendment to Final Construction Grant Regulations. 40 *Federal Register* 40, Feb. 27, 1975, p. 8349.

[7]   The 1974 needs survey also reported rough estimates of construction costs for a new category: costs for treatment and/or control of stormwaters that are not part of

combined stormwater and sanitary sewer systems. This is now category VI. In the 1974 survey, this category alone was estimated to be $235 billion, but EPA said that this estimate was unreliable.

[8]   U.S. Senate. Committee on Public Works. Subcommittee on Environmental Pollution. *The Environmental Protection Agency's 1974 Needs Survey.* Hearing, 93rd Congress, 2d session. September 11, 1974. Serial No. 93-H53. p. 15.

[9]   The States were: Alabama, Alaska, Arizona, Arkansas, Colorado, Florida, Georgia, Hawaii, Idaho, Iowa, Kansas, Kentucky, Louisiana, Maryland, Mississippi, Missouri, Montana, Nebraska, Nevada, New Mexico, North Carolina, North Dakota, Oklahoma, Oregon, South Carolina, South Dakota, Tennessee, Texas, Utah, Washington, West Virginia, Wisconsin, Wyoming, plus Guam, Puerto Rico, American Samoa, and Trust Territory of the Pacific.

[10] P.L. 94-447, Public Works Employment Appropriations Act of 1976.

[11] U.S. Congress. Senate. Committee on Environment and Public Works, *Public Works Employment Act of 1977, report to accompany S. 427.* 95th Congress, 1st session. S.Rept. 95-38. 36 pp.

[12] During debate on H.R. 3199, the House rejected an amendment offered by Congressman Ottinger which would have struck the population factor from the formula and retained the prior fundamental reliance on needs.

[13] U.S. Congress. Senate. Committee on Environment and Public Works. *A Legislative History of the Clean Water Act.* 95th Congress, 2d session. Serial No. 95-14. p. 672.

[14] U.S. Congress. Senate. Committee on Environment and Public Works. Legislative History of the Water Quality Act of 1987, including Public Law 97-440; Public Law 97-117, Public Law 96-483; and Public Law 96-148. 100th Congress, 2d session. p. 2471.

[15] Ibid., p. 791.

[16] These 21 states were Alaska, Arkansas, Colorado, Delaware, District of Columbia, Hawaii, Idaho, Kansas, Maine, Montana, Nebraska, Nevada, New Mexico, North Dakota, Oklahoma, Oregon, Rhode Island, South Dakota, Utah, Vermont, and Wyoming. Ibid., pp. 1479-82.

[17] The 11 states were California, Illinois, Indiana, Maryland, Michigan, Minnesota, New York, Ohio, Pennsylvania, Virginia, and West Virginia. Ibid.

[18] In 1995, three districts of the U.S.-administered United Nations Trust Territory of the Pacific Islands, which previously had been eligible for Clean Water Act funds, completed the process of becoming Freely Associated States with status as sovereign states by adopting a Compact of Free Association. As of FY1999, the Trust Territory, which had been receiving 0.1295% of available funds, is no longer eligible for grants under the act. EPA made an administrative adjustment to allotment totals for all other recipients for FY2000 and onwards to reflect this change.

In: Water Pollution Issues and Developments   ISBN: 978-1-60456-208-8
Editor: S. V. Thomas, pp. 103-109   © 2008 Nova Science Publishers, Inc.

Chapter 6

ARSENIC IN DRINKING WATER: REGULATORY DEVELOPMENTS AND ISSUES[*]

Mary Tiemann

ABSTRACT

In 2001, the Environmental Protection Agency (EPA) promulgated a new regulation for arsenic in drinking water, as required by 1996 Safe Drinking Water Act Amendments. The rule set the legal limit for arsenic in tap water at 10 parts per billion (ppb), replacing a 50 ppb standard set in 1975, before arsenic was classified as a carcinogen. When issuing the rule, the EPA projected that compliance could be costly for some small systems, but many water utilities and communities expressed concern that the EPA had underestimated the rule's costs significantly. The arsenic rule was to enter into effect on March 23, 2001, and public water systems were given until January 23, 2006, to comply. Subsequently, the EPA postponed the rule's effective date to February 22, 2002, to review the science and cost and benefit analyses supporting the rule. After completing the review in October 2001, the EPA affirmed the 10 ppb standard. The new standard became enforceable for water systems in January 2006.

Since the rule was completed, Congress and the EPA have focused on how to help communities comply with the new standard. In the past several Congresses, numerous bills have been offered to provide more financial and technical assistance and/or compliance flexibility to small systems; however, none of the bills has been enacted.

BACKGROUND

Sources of arsenic in water include natural sources, particularly rocks and soils, and also releases from its use as a wood preservative, in semi-conductors and paints, and from mining and agricultural operations. Elevated levels of arsenic are found more frequently in ground water than in surface water. Because small communities typically rely on wells for drinking

---

[*] Excerpted from CRS Report RS20672, dated July 20, 2007.

water, while larger cities typically use surface-water sources, arsenic tends to occur in higher concentrations more frequently in water used by small communities.

In the United States, the average arsenic level measured in ground-water samples is less than or equal to 1 part per billion (ppb, or micrograms per liter [µg/L]); however, higher levels are not uncommon. Compared with the rest of the United States, Western states have more water systems with levels exceeding 10 ppb; levels in some locations in the West exceed 50 ppb. Parts of the Midwest and New England also have some water systems with arsenic levels exceeding 10 ppb, but most systems meet the new standard. When issuing the rule, the EPA estimated that roughly 4,000 (5.5%) of the regulated water systems, serving a total of 13 million people, were likely to exceed the 10 ppb standard.

The previous drinking water standard for arsenic, 50 ppb, was set by the U.S. Public Health Service in 1942. EPA adopted that level and issued an interim drinking water regulation for arsenic in 1975. This standard was based on estimated total dietary intake[1] and non-cancer health effects. In 1986, Congress amended the Safe Drinking Water Act (SDWA), converted all interim standards to National Primary Drinking Water Regulations, and included arsenic on a list of 83 contaminants for which the EPA was required to issue new standards by 1989. EPA's extensive review of arsenic risk assessment issues had caused the agency to miss the 1989 deadline. As a result of a citizen suit, the EPA entered into a consent decree with a new deadline for the rule of November 1995. The EPA continued work on risk assessment, water treatment, analytical methods, implementation, and occurrence issues but, in 1995, decided to delay the rule in order to better characterize health effects and assess cost-effective removal technologies for small utilities.

ARSENIC AND THE 1996 SDWA AMENDMENTS

In the 1996 SDWA Amendments (P.L. 104-182), Congress directed the EPA to propose a new drinking water standard for arsenic by January 1, 2000, and to promulgate a final standard by January 1, 2001. Congress also directed the EPA to develop a comprehensive research plan for arsenic to support the rulemaking effort and to reduce the uncertainty in assessing health risks associated with low-level exposures to arsenic. The EPA was required to conduct the study in consultation with the National Academy of Sciences. In 1996, the EPA requested the National Research Council (NRC) to review the available arsenic toxicity data base and to evaluate the scientific validity of EPA's risk assessments for arsenic.

The NRC issued its report in 1999 and recommended that the standard be reduced, but it did not recommend a particular level. The NRC affirmed that the available data provided ample evidence for EPA's classification of inorganic arsenic as a human carcinogen, but that EPA's dose-response assessment, which was based on a Taiwan study, deserved greater scrutiny. The NRC explained that the data in the study lacked the level of detail needed for use in dose-response assessment. The Council also reported that research suggested that arsenic intake in food is higher in Taiwan than in the United States, further complicating efforts to use the data for arsenic risk assessment. Based on findings from three countries where individuals were exposed to very high levels of arsenic (several hundreds of parts per billion or more), the NRC concluded that the data were sufficient to add lung and bladder cancers to the types of cancers caused by ingestion of inorganic arsenic; however, the NRC noted that few data address the risk of ingested arsenic at lower concentrations, which would be more representative of levels found in the United States.[2] The NRC concluded that key studies for improving the scientific validity of risk assessment were needed, and recommended specific studies to the EPA.

EPA'S FINAL ARSENIC RULE

In June 2000, EPA published its proposal to revise the arsenic standard from 50 ppb to 5 ppb and requested comment on options of 3 ppb, 10 ppb, and 20 ppb. The EPA stated that the proposal relied primarily on the NRC analysis and some recently published research, and that it would further assess arsenic's cancer risks before issuing the final rule. As proposed, the standard would have applied only to community water systems. Non-transient, non-community water systems (e.g., schools with their own wells) would have been required only to monitor and then report if arsenic levels exceeded the standard. In the final rule, published on January 22, 2001 (66 *FR* 6976), the EPA set the standard at 10 ppb and applied it to non-transient, non-community water systems, as well as community water systems. The agency gave the water utilities five years to comply (the maximum amount of time allowed under SDWA). The EPA estimated that 3,000 (5.5%) of the 54,000 community water systems, and 1,100 (5.5%) of the 20,000 non-transient, non-community water systems, would need to take measures to meet the standard.[3]

Standard-Setting Process

In developing standards under the Safe Drinking Water Act, the EPA is required to set a maximum contaminant level goal (MCLG) at a level at which no known or anticipated adverse health effects occur and that allows an adequate margin of safety. (The EPA sets the MCLG at zero for carcinogens [as it did for arsenic], unless a level exists below which no adverse health effects occur.) The EPA must then set an enforceable standard, the MCL, as close to the MCLG as is "feasible" using the best technology, treatment, or other means available (taking costs into consideration).[4] EPA's determination of whether a standard is feasible typically has been based on costs to large water systems (serving more than 50,000 people). Less than 2% of community water systems (roughly 750 of 54,000 systems) are this large, but they serve roughly 56% of all people served by community systems.[5]

Variances and Exemptions

Congress has long recognized that the technical and cost considerations associated with technologies selected for large cities often are not applicable to small systems. In the 1996 amendments, Congress expanded SDWA variance and exemption provisions to address small system compliance concerns.

The *small system variance* provisions require that for each rule establishing an MCL, the EPA must list technologies that comply with the MCL and are affordable for three size categories of small systems. If the EPA does not list affordable compliance technologies for small systems, then it must list variance technologies. A variance technology need not meet the MCL, but must be protective of public health. If the EPA lists a variance technology, a state then may grant a variance to a small system, allowing the system to use a variance technology to comply with a regulation. The EPA has not identified variance technologies for arsenic or any other standards because, based on its current affordability criteria, the EPA has determined that affordable compliance technologies are available for all standards. Thus, small system variances are not available.

Congress took issue with EPA's assessment that small system variance technologies were not merited for the arsenic standard, and in 2002, directed the EPA to review the criteria it uses to determine whether a compliance treatment technology is affordable for small systems. In March 2006, EPA proposed three options for revising its affordability criteria (71 *FR* 10671). Under the current affordability criteria, EPA considers a treatment technology affordable unless the average compliance cost exceeds 2.5% of the area's median household income. Based on this measure, EPA determined that affordable technologies are available for all SDWA standards. The proposed options under consideration are well below the current level: 0.25%, 0.50%, and 0.75% of an area's median household income. The revised criteria, as proposed, are also expected to address how to ensure that a variance technology would be protective of public health. According to EPA, the final criteria would apply only to the new Stage 2 Disinfectants/Disinfection Byproducts Rule and future rules, and not to the arsenic rule.

Exemptions potentially offer a source of compliance flexibility for small systems. States may grant temporary exemptions from a standard if, for certain reasons (including cost), a system cannot comply on time. The arsenic rule gives systems five years to comply with the new standard; an exemption allows another three years for qualified systems. Systems serving 3,300 or fewer persons may receive up to three additional two-year extensions, for a total exemption duration of nine years (a total of 14 years to achieve compliance). In the final rule, EPA noted that exemptions will be an important tool to help states address the number of systems needing financial assistance to comply with this rule and other SDWA rules (66 *FR* 6988). However, to grant an exemption, the law requires a state to hold a public hearing and make a finding that the extension will not result in an "unreasonable risk to health." Because the exemption process is complex, states have seldom granted them. State officials have noted that "unreasonable risk to health" has never been defined, and that states must make a separate finding for each system. Many states have indicated that they plan to grant few or no exemptions.

Balancing Costs and Benefits

When proposing a rule, the EPA must publish a determination as to whether or not the benefits of the standard justify the costs. If the EPA determines that costs are not justified, then it may set the standard at the level that maximizes health risk reduction benefits at a cost that is justified by the benefits. The EPA determined that the "feasible" arsenic level (for large systems) was 3 ppb, but that the benefits of that level did not justify the costs. Thus, the EPA proposed a standard of 5 ppb. Also, the EPA proposed to require non-transient, non-community water systems (e.g., schools) only to monitor and report (as opposed to treating), largely because of cost-benefit considerations. In setting the standard at 10 ppb, EPA cited SDWA, stating that this level "maximizes health risk reduction benefits at a cost that is justified by the benefits." The EPA applied the final rule to schools and similar non-community systems.

Anticipated Benefits and Costs

In the final rule, the EPA estimated that reducing the standard to 10 ppb could prevent roughly 19 to 31 bladder cancer cases and 5 to 8 bladder cancer deaths each year. The agency further estimated that the new standard could prevent 19 to 25 lung cancer cases and 16 to 22 lung cancer deaths each year, and provide other cancer and non-cancer health benefits that were not quantifiable.

Regarding the cost of meeting the 10 ppb standard, the EPA estimated that for systems that serve fewer than 10,000 people, the average cost per household could range from $38 to $327 per year. Roughly 97% of the systems that were expected to exceed the standard are in this category, and most of these systems serve fewer than 500 people. For larger systems, projected water cost increases range from $0.86 to $32 per household. The EPA estimated the total national, annualized cost for the rule to be about $181 million.

EPA's Science Advisory Board (SAB) raised concerns about the rule's economic and engineering assessment, and concluded that several cost assumptions were likely to be unrealistic and other costs seemed to be excluded. The SAB also suggested that the EPA give further thought to the concept of affordability as applied to this standard.[6] Many municipalities and water system representatives also disagreed with the agency's cost estimates. The American Water Works Association (AWWA), while supporting a stricter standard, estimated that the new rule will cost $600 million annually and require $5 billion in capital outlays. The AWWA attributed differences in cost estimates partly to the costs of handling arsenic-contaminated treatment residuals and the estimated number of wells affected. The AWWA projected that the rule could cost individual households in the Southwest, Midwest, and New England as much as $2,000 per year.[7]

Arsenic Rule Review

The EPA issued the final rule on January 22, 2001. In March 2001, the Administrator delayed the rule for 60 days, citing concerns about the science supporting the rule and its estimated cost to communities. On May 22, 2001, the EPA delayed the rule's effective date until February 22, 2002, but did not change the 2006 compliance date for water systems (66 *FR* 28342). At EPA's request, the NRC undertook an expedited review of EPA's arsenic risk analysis and recent health effects research, the National Drinking Water Advisory Council (NDWAC) reassessed the rule's cost, and the SAB reviewed its benefits. The EPA also requested public comment on whether the data and analyses for the rule support setting the standard at 3, 5, 10, or 20 ppb (66 *FR* 37617). The NRC concluded that "recent studies and analyses enhance the confidence in risk estimates that suggest chronic arsenic exposure is associated with an increased incidence of bladder and lung cancer at arsenic levels in drinking water below the current MCL of 50 µg/L."[8] The NDWAC reported that the EPA produced a credible cost estimate, given constraints and uncertainties, and suggested ways to improve estimates. The SAB offered ways to improve the benefits analysis. In October 2001, the EPA concluded that 10 ppb was the appropriate standard and announced plans to provide $20 million for research on affordable arsenic removal technologies to help small systems comply.

LEGISLATIVE ACTION

Since the arsenic standard was revised, Congress repeatedly has expressed concern over the potential cost of the regulation, especially to small and rural communities. The 107th Congress directed EPA to review its affordability criteria and how the small system variance and exemption programs should be implemented for arsenic (P.L. 107-73, H.Rept. 107-272, p. 175). Congress directed EPA to report on its affordability criteria, administrative actions, potential funding mechanisms for small system compliance, and possible legislative actions. In 2002, EPA submitted its report to Congress, *Small Systems Arsenic Implementation Issues*, on actions EPA was undertaking to address these directives. Major activities included reviewing the small system affordability criteria and variance process, and developing and implementing a small community assistance plan to improve access to financial and technical assistance, improve compliance capacity, and simplify the use of exemptions. The EPA also has offered technical assistance to small systems, sponsored research on low-cost arsenic treatment technologies, and issued *Implementation Guidance for the Arsenic Rule*, which includes guidance to help states grant exemptions.

The 108th Congress again expressed concern over the financial impact that the new standard could have in many communities. In the conference report for the omnibus appropriations act for FY2005 (P.L. 108-447, H.Rept. 108-792), Congress provided $8.2 million for arsenic removal research. Conferees also expressed concern that many rural communities would be unable to meet the new requirements which could pose a "huge financial hardship." The conferees directed the EPA to report on the extent to which communities were being affected by the rule and to propose compliance alternatives and make recommendations to minimize costs. This report is pending.

In the 109th Congress, legislative efforts focused on helping economically struggling communities comply with the arsenic rule and other drinking water regulations. Various bills were introduced to provide more regulatory flexibility, financial assistance, and technical assistance, especially to small systems. The EPA's uncompleted FY2007 funding bill, as reported by the Senate Committee on Appropriations (H.R. 5386, S.Rept. 109-275), would have required the EPA to make available at least $11 million for small system compliance assistance. Companion bills, H.R. 2417 and S. 689, would have required the EPA to establish a grant program to help qualified communities comply with standards, delay enforcement of the arsenic rule until states implement the grant program, and prevent EPA from enforcing a standard during the grant application process. H.R. 4495 proposed to give small systems two more years to comply with the arsenic rule. S. 1400 (S.Rept. 109-186), the Water Infrastructure Financing Act, would have increased funding authority for the drinking water state revolving fund (DWSRF) program and established a grant program for priority projects, including projects to help small systems comply with standards. S. 2161 would have prevented enforcement of regulations for small systems unless the EPA had identified a variance technology and sufficient DWSRF funds were made available. S. 2161 also proposed to require EPA to develop new affordability criteria related to small system variances. S. 41 and H.R. 1315 would have required states to grant small community water systems exemptions from regulations for naturally occurring contaminants in certain cases. None of the bills was enacted. In the 110th Congress, H.R. 2141 (a bill identical to H.R. 1315 in the past Congress) has been introduced to require states to grant to eligible small systems exemptions from the requirements of regulations for arsenic and other naturally occurring contaminants.

REFERENCES

[1]   Food is a significant source of arsenic. The National Research Council estimates that, in the United States, arsenic intake from food is comparable to drinking water containing 5 ppb arsenic.

[2]   National Research Council, *Arsenic in Drinking Water*, National Academy of Sciences, National Academy Press, Washington, D.C., 1999, pp. 7, 22.

[3]   See EPA's *Technical Fact Sheet: Final Rule for Arsenic in Drinking Water*, available online at [http://www.epa.gov/safewater/arsenic/regulations.html].

[4]   For a more detailed discussion of these and other SDWA provisions, see CRS Report RL31243, *Safe Drinking Water Act: A Summary of the Act and Its Major Requirements*.

[5]   SDWA does not discuss how the EPA should consider cost in determining feasibility; thus, the EPA has relied on legislative history for guidance. The Senate report for the 1996 amendments states that "[f]easible means the level that can be reached by large regional drinking water systems applying best available treatment technology. ... This approach to standard setting is used because 80% of the population receives its drinking water from large systems and safe water can be provided to this portion of the population at very affordable costs."(U.S. Senate, *Safe Drinking Water Amendments Act of 1995*, S.Rept. 104-169, November 7, 1995, p. 14.) Systems serving 10,000 or more people serve about 80% of the population served by community water systems.

[6]   Science Advisory Board, Arsenic Proposed Drinking Water Regulation: A Science Advisory Board Review of Certain Elements of the Proposal, EPA-SAB-DWC-01-001, December 2000, p. 4.

[7]   AWWA, January 17, 2001. See [http://www.awwa.org/Advocacy/pressroom/pr/010111.cfm].

[8]   National Research Council, *Arsenic in Drinking Water: 2001 Update*, NAS, p. 14.

In: Water Pollution Issues and Developments
Editor: S. V. Thomas, pp. 111-118

ISBN: 978-1-60456-208-8
© 2008 Nova Science Publishers, Inc.

Chapter 7

PERCHLORATE CONTAMINATION OF DRINKING WATER: REGULATORY ISSUES AND LEGISLATIVE ACTIONS[*]

Mary Tiemann

ABSTRACT

Perchlorate is the explosive component of solid rocket fuel, fireworks, road flares, and other products. Used mainly by the Department of Defense (DOD) and related industries, perchlorate occurs naturally and is present in some organic fertilizer. This soluble, persistent compound has been detected in sources of drinking water for more than 11 million people. It also has been found in milk, fruits, and vegetables. Concern over the potential health risks of perchlorate exposure has increased, and some states and Members of Congress have urged the Environmental Protection Agency (EPA) to set a drinking water standard for perchlorate. The EPA has not determined whether to regulate perchlorate and has cited the need for more research on health effects, water treatment techniques, and occurrence. Related issues have involved environmental cleanup and water treatment costs, which will be driven by federal and state standards. Interagency disagreements over the risks of perchlorate exposure led several federal agencies to ask the National Research Council (NRC) to evaluate perchlorate's health effects and EPA's risk analyses. In 2005, the NRC issued its report, and the EPA adopted the NRC's recommended reference dose (i.e., the expected safe dose) for perchlorate exposure. However, new studies raise more concerns about potential health risks of low-level exposures, particularly for infants. Perchlorate bills in the 110th Congress include S. 150 and H.R. 1747, which direct the EPA to set a standard. This report reviews perchlorate water contamination issues and developments.

---

[*] Excerpted from CRS Report RS21961, dated April 4, 2007.

BACKGROUND

Ammonium perchlorate is the key ingredient in solid fuel for rockets and missiles; other perchlorate salts are used to manufacture products such as fireworks, air bags, and road flares. Uncertainty about the health effects of perchlorate has slowed efforts to establish drinking water and environmental cleanup standards for it. However, because of perchlorate's persistence in water and ability to affect thyroid function, concern has escalated with the detection of perchlorate in water in at least 33 states. In the absence of a federal standard, states have begun to adopt their own measures. Massachusetts set a drinking water standard of 2 parts per billion (ppb, or micrograms per liter [:g/L]) in 2006, and California has proposed a 6 ppb standard. Several states have issued health goals or advisory levels ranging from 1 ppb in Maryland (advisory level) and New Mexico (drinking water screening level) to 51 ppb in Texas (industrial cleanup level).

Occurrence

Perchlorate has been used heavily by the Department of Defense (DOD) and its contractors, and perchlorate contamination of water has been found most often near weapons and rocket fuel manufacturing facilities and disposal sites, research facilities, and military bases. Fireworks, road flares, and other manufacturing facilities and construction sites also have been sources of contamination. Perchlorate also occurs naturally (in West Texas, for example) and is present in organic fertilizer imported from Chile.[1] In 1997, a new test method lowered the detection limit for perchlorate in drinking water from 400 parts per billion to 4 ppb, prompting several states to begin testing. Within two years, perchlorate was detected in drinking water sources for more than 11 million people in the Southwest and in surface and ground water in scattered locations across the country. Contamination has been found most often in ground water; however, it has been detected at low levels in the Colorado River, a major source of drinking water and irrigation water for Arizona, California, and Nevada.[2] Perchlorate also has been detected in dairy milk in various states, especially California and Texas.

In 1999, the EPA required public water systems to monitor for perchlorate under the Unregulated Contaminant Monitoring Rule (UCMR) to determine the frequency and levels at which it is present in public water supplies nationwide. The UCMR required monitoring by all water systems serving more than 10,000 persons and by a representative sample of smaller systems. Of some 3,700 water systems tested, perchlorate was detected in 153 systems in 26 states and 2 commonwealths.[3] In 14 systems, perchlorate levels exceeded EPA's preliminary remediation goal of 24.5 ppb.[4] The EPA also has reported perchlorate contamination at 65 DOD facilities, 7 other federal facilities, and 37 private sites. In California, where most perchlorate releases have been identified, perchlorate has been detected in 456 sources of drinking water (including closed wells).

Monitoring also has been undertaken to assess the presence of perchlorate in foods. In 2004, the Food and Drug Administration (FDA) tested 500 samples of foods, including vegetables, milk, and bottled water for perchlorate. Samples were taken in areas where water was thought to be contaminated. The FDA found perchlorate in roughly 90% of lettuce samples (average levels ranged from 11.9 ppb to 7.7 ppb for lettuces in four states), and in 101 of 104 bottled milk samples (with an average level of 5.7 ppb across 14 states).[5] This research is relevant to the EPA's standard-setting efforts, as EPA would take into account other exposures to perchlorate when setting a drinking water standard.

Health Effects

Perchlorate is known to disrupt the uptake of iodine in the thyroid, and health effects associated with perchlorate exposure are expected to parallel those caused by iodine deficiency. Iodine deficiency decreases the production of thyroid hormones, which help regulate the body's metabolism and growth. A key concern is that impairment of thyroid function in pregnant women can affect fetuses and infants and can result in delayed development and decreased learning capability. Several human studies have indicated that thyroid changes occur in humans at significantly higher concentrations of perchlorate than the amounts typically observed in water supplies.[6] However, a new study by the Centers for Disease Control and Prevention (CDC) of a representative sample of the U.S. population found that environmental exposures to perchlorate have an effect on thyroid hormone levels in women with iodine deficiency. No effect was found in men. Fully 36% of the 1,111 women in this study were found to be iodine deficient, and the median level of urinary perchlorate measured in the women was 2.9 ppb.[7]

EPA REGULATION OF PERCHLORATE

The EPA has taken steps toward establishing a standard for perchlorate in drinking water but has not made a determination to regulate it. Under the Safe Drinking Water Act (SDWA, §1412(b)), the EPA must establish a standard for a contaminant if the Administrator determines that it occurs at a frequency and level of public health concern and that its regulation presents a meaningful opportunity for reducing health risks. In 1997, when a better detection method became available for perchlorate and detections increased, scientific information was limited. In 1998, the EPA placed perchlorate on the list of contaminants that were candidates for regulation, but concluded that information was insufficient to determine whether perchlorate should be regulated under the SDWA. The EPA listed perchlorate as a priority for further research on health effects and treatment technologies and for collecting occurrence data. In 1999, the EPA required water systems to monitor for perchlorate under the Unregulated Contaminant Monitoring Rule to determine the frequency and levels at which it is present in public water supplies nationwide. The EPA recently determined that it had collected sufficient data and that further monitoring was not needed (72 *Federal Register* 367, January 4, 2007).

Perchlorate Risk Assessment

In 1992, and again in 1995, EPA issued draft reference doses (RfDs) for perchlorate exposure. An RfD is an estimate (with uncertainty spanning perhaps an order of magnitude) of a daily oral exposure that is not expected to cause any adverse, non-cancer health effects during a lifetime. In developing an RfD, the EPA incorporates factors to account for sensitive subpopulations, study duration, inter-and intraspecies variability, and data gaps. The draft RfDs range of 0.0001 to 0.0005 milligrams per kilogram (mg/kg) body weight per day translated to a drinking water equivalent level of 4 ppb-18 ppb. The EPA takes the RfD into account when setting a drinking water standard; it also considers costs, the capabilities of monitoring and treatment technologies, and other sources of perchlorate exposure, such as food.

The EPA continued to assess perchlorate risks, and its 1999 draft risk characterization resulted in a human risk benchmark of 0.0009 mg/kg per day (with a 100-fold uncertainty factor), which converted to a drinking water equivalent level of 32 ppb. However, the EPA determined that the available health effects and toxicity database was inadequate for risk assessment. In 1999, the EPA issued an *Interim Assessment Guidance for Perchlorate*, which recommended that EPA risk managers use the earlier reference dose range and drinking water equivalent level (DWEL) of 4-18 ppb for perchlorate-related assessment activities at hazardous waste sites.

In 2002, the EPA completed a draft risk assessment that concluded that the potential human health risks of perchlorate exposures include effects on the developing nervous system and thyroid tumors, based on rat studies that observed benign tumors and adverse effects in fetal brain development. The document included a draft RfD of 0.00003 mg/kg per day, which translated to a drinking water equivalent level of 1 ppb. This document was controversial, both for its implications for cleanup costs and for science policy reasons. (For example, some peer reviewers expressed concern over the EPA's risk assessment methodology and reliance on rat studies.) The DOD, water suppliers, and other commentors expressed concern that the draft RfD could lead to unnecessarily stringent and costly cleanups of perchlorate releases at federal facilities and in water supplies. In 2002, a federal interagency perchlorate working group convened to discuss perchlorate risk assessment, research and regulatory issues, and related agency concerns. Working group members included the DOD, the EPA, the Department of Energy, the National Aeronautics and Space Administration, the Office of Science and Technology Policy, the Council on Environmental Quality, and the Office of Management and Budget.

NRC Perchlorate Study

To resolve some of the uncertainty and debate over perchlorate's health effects and the 2002 draft risk assessment, the interagency working group asked the National Research Council (NRC) to review the available science for perchlorate and EPA's draft assessment. The NRC was asked to comment and make recommendations. The NRC Committee to Assess the Health Implications of Perchlorate Ingestion issued its review in January 2005 and suggested several changes to EPA's draft risk assessment. The committee concluded that because of key differences between rats and humans, studies in rats are of limited use for quantitatively assessing human health risk associated with perchlorate exposure. Although the committee agreed that thyroid tumors found in a few rats were likely perchlorate treatment-related, it concluded that perchlorate exposure is unlikely to lead to thyroid tumors in humans. The committee noted that, unlike rats, humans have multiple mechanisms to compensate for iodide deficiency and thyroid disorders. Also, the NRC found flaws in the design and methods used in the rat studies. The committee concluded that the animal data selected by the EPA should not be used as the basis of the risk assessment.

The committee also reviewed the EPA's risk assessment model. It agreed that the EPA's model for perchlorate toxicity represented a possible early sequence of events after exposure, but it did not think that the model accurately represented possible outcomes after changes in thyroid hormone production. Further, the committee disagreed with the EPA's definition of a change in thyroid hormone level as an adverse effect. Rather, the NRC defined transient changes in serum thyroid hormone as biochemical events that might precede adverse effects, and identified hypothyroidism as the first adverse effect.

Because of research gaps regarding perchlorate's potential effects following changes in thyroid hormone production, the committee made the unusual recommendation that EPA use a *nonadverse effect* (i.e., the inhibition of iodide uptake by the thyroid in humans) rather than an adverse effect as the basis for the risk assessment. The committee explained that "[i]nhibition of iodide uptake is a more reliable and valid measure, it has been unequivocally demonstrated in humans exposed to perchlorate, and it is the key event that precedes all thyroid-mediated effects of perchlorate exposure."[8] Based on the use of this point of departure, the reliance on human studies, and the use of an uncertainty factor of 10 (for intraspecies differences), the NRC's recommendations led to an RfD of 0.0007 mg/kg per day. The committee concluded that this RfD should protect the most sensitive population (i.e., the fetuses of pregnant women who might have hypothyroidism or iodide deficiency) and noted that the RfD was supported by clinical studies, occupational and environmental epidemiologic studies, and studies of long-term perchlorate administration to patients with hyperthyroidism.[9] In addition, the NRC identified data gaps and research needs. The committee has received some criticism for the extent to which it relied on a small, short-term human study, and debate over perchlorate's health risks continues.

EPA's Response

In 2005, the EPA adopted the NRC recommended reference dose of 0.0007 mg/kg per day, which translates to a drinking water equivalent level of 24.5 ppb. The DWEL is the concentration of a contaminant in water that is expected to have no adverse effects; it is intended to include a margin of safety to protect the fetuses of pregnant women who might have a preexisting thyroid condition or insufficient iodide intake. The DWEL is based on the assumption that all exposure would come from drinking water. If the EPA were to develop a drinking water standard for perchlorate, it would adjust the DWEL to account for other sources of exposure, such as food.

In January 2006, the EPA's Superfund office issued guidance adopting the NRC reference dose and the DWEL of 24.5 ppb as the recommended value to be considered as the preliminary remediation goal (PRG) to guide perchlorate assessment and cleanup at Superfund sites. In March, the EPA's Children's Health Protection Advisory Committee

(CHPAC) wrote to the EPA Administrator that the PRG does not protect infants, who are highly susceptible to neurodevelopmental toxicity and may be more exposed to perchlorate than fetuses. The CHPAC noted that perchlorate is concentrated in breast milk and that nursing infants could receive daily doses greater than the RfD if the mother is exposed to 24.5 ppb perchlorate in tap water. The committee recommended that the Superfund office lower the PRG and that the Office of Water develop a drinking water standard for perchlorate and, in the interim, issue a drinking water health advisory that takes into account early life exposures. CHPAC's assessment, combined with the new CDC study and ongoing data gaps, could further complicate EPA regulatory efforts.

DEPARTMENT OF DEFENSE

The DOD has the largest number of identified sites with perchlorate contamination and has allotted significant resources to address this problem. Through FY2006, the DOD had spent roughly $88 million on perchlorate-related research activities, including $64 million on treatment technologies, $9.5 million on health and toxicity studies, and $11.6 million on pollution prevention. Additional funds have been spent on testing and cleanup.

Although remediation has proceeded at some sites, cleanups typically are driven by drinking water standards or other established cleanup standards. With no federal standard and just one promulgated state standard, cleanup goals and responsibilities have been ambiguous. In January 2006, following EPA's establishment of a reference dose and DWEL for perchlorate, the DOD adopted a policy that establishes 24 ppb as the level of concern to be used in managing perchlorate releases, unless a more stringent federal or state standard has been promulgated. The policy applies broadly to DOD installations and former military lands. The policy directs the services to test for perchlorate when it is reasonably expected that a release has occurred. If perchlorate levels exceed 24 ppb, a site-specific risk assessment must be conducted; if the assessment indicates that the perchlorate could result in adverse health effects, then the site must be prioritized for risk management.[10] The DOD uses a relative risk site evaluation framework across DOD to evaluate the risks posed by one site relative to other sites. The DOD uses this framework to help prioritize environmental restoration work and to allocate resources among sites.

CONGRESSIONAL ACTIONS

The 109th Congress targeted some funding for perchlorate cleanup in conference reports for various appropriations acts, including DOD and EPA appropriations acts for FY2006 (P.L. 109-148 and P.L. 109-54, respectively). In the conference report for the Department of Health and Human Services FY2006 appropriations act (P.L. 109-149), conferees encouraged support for studies on the long-term effects of perchlorate exposure. The conference report for the FDA's FY2006 funding act (P.L. 109-97) directed the FDA to continue conducting surveys of perchlorate in food and to report to Congress. An array of bills addressed cleanup of perchlorate-contaminated water in California. The House passed two such bills, H.R. 186

and H.R. 18, but no action occurred in the Senate. Other bills called for EPA to issue a standard for perchlorate in drinking water.

In the 110th Congress, perchlorate is back on the agenda. Responding to the EPA decision to not require further monitoring for perchlorate as an unregulated contaminant, S. 24 (Boxer) would require community water systems to test for perchlorate and disclose its presence in annual consumer confidence reports. S. 150 (Boxer) and H.R. 1747 (Solis) would direct the EPA to establish a drinking water standard for perchlorate.

REFERENCES

[1] Purnendu K. Dasgupta, et al., "Perchlorate in the United States: Analysis of Relative Source Contributions to the Food Chain," *Environmental Science and Technology*, v. 40, n. 21, Nov. 21, 2006, p. 6608-6614. This study suggests that although Chilean fertilizer is a much smaller source of perchlorate than oxidizers, the fertilizer may have a proportionally greater impact as a source of perchlorate in the food chain because it is applied directly to crop land. This article reports that processing methods have reduced the perchlorate content of Chilean fertilizer in recent years.

[2] A key, but diminished, source of perchlorate in the Colorado River has been a facility in Nevada, where perchlorate production began in 1951. Since 1997, Nevada and the EPA have worked with Kerr McGee to control the source of releases. From January 2004 through June 2005, only three of the monthly samples had detectable levels of perchlorate. U.S. EPA, Region 9, *Perchlorate Monitoring Results: Henderson, Nevada to the Lower Colorado River*, June 2005.

[3] EPA, Federal Facilities Restoration and Reuse, *Known Perchlorate Releases in the U.S.*, Mar. 25, 2005, at [http://www.epa.gov/fedfac/documents/ perchlorate_l inks.htm#occurrences].

[4] U.S. Government Accountability Office, Perchlorate: A System to Track Sampling and Cleanup Results is Needed, GAO-05-462, May 2005, p. 3.

[5] The FDA test results are available online at [http://www.cfsan.fda.gov/ ~dms/clo4data.html].

[6] Michael A. Kelsh et al., "Primary Congenital Hypothyroidism, Newborn Thyroid Function, and Environmental Perchlorate Exposure Among Residents of a Southern California Community," *Journal of Occupational Environmental Medicine*, 2003, p. 1117.

[7] Benjamin C. Blount, James L. Pirkle, et al., "Urinary Perchlorate and Thyroid Hormone Levels in Adolescent and Adult Men and Women Living in the United States," Centers for Disease Control and Prevention, in *Environmental Health Perspectives*, October 2006.

[8] National Research Council, *Health Implications of Perchlorate Ingestion*, Committee to Assess the Health Implications of Perchlorate Ingestion, National Academy of Sciences, 2005, p. 9.

[9] Ibid., p. 10.

[10] For more information, see the DOD perchlorate website regarding policy and guidance, [http://www.denix.osd.mil/denix/Public/Library/MERIT/Perchlorate/efforts/policy/inde x.html].

In: Water Pollution Issues and Developments
Editor: S. V. Thomas, pp. 119-125
ISBN: 978-1-60456-208-8
© 2008 Nova Science Publishers, Inc.

Chapter 8

LEAKING UNDERGROUND STORAGE TANKS: PREVENTION AND CLEANUP[*]

Mary Tiemann

ABSTRACT

To address a nationwide water pollution problem caused by leaking underground storage tanks (USTs), Congress created a leak prevention, detection, and cleanup program in 1984. In 1986, Congress established the Leaking Underground Storage Tank (LUST) Trust Fund to help the Environmental Protection Agency (EPA) and states oversee LUST cleanup activities and pay the costs of remediating leaking petroleum USTs where owners fail to do so. Despite progress in the program, challenges remain. A key issue has been that state resources have not met the demands of administering the UST leak prevention program. States have long sought larger appropriations from the trust fund to support the LUST cleanup program, and some have sought flexibility to use the fund to administer and enforce the UST leak prevention program. Another issue has involved the detection of methyl tertiary butyl ether (MTBE) in groundwater at many LUST sites. This gasoline additive was used widely to reduce air pollution from auto emissions. However, MTBE is very water soluble and, once released, tends to travel farther than conventional gas leaks, making it more likely to reach water supplies and more costly to remediate. For more than a decade, Congress considered various bills to broaden the use of the trust fund and strengthen the leak prevention program.

The Energy Policy Act of 2005 (P.L. 109-58) added new leak prevention provisions to the UST program, imposed new requirements on states, and authorized EPA and states to use LUST Trust Fund appropriations for both cleanup and prevention activities. A key now issue concerns the level of resources available to states to meet the new mandates. Some stakeholders have called for increased trust fund appropriations, while EPA has asked Congress to modify UST inspection requirements. This report reviews the LUST program, legislative changes made by P.L. 109-58, and related developments.

---

[*] Excerpted from CRS Report RS21201, dated July 12, 2007.

BACKGROUND

In the 1980s, EPA determined that many of the roughly 2.2 million underground storage tanks (USTs) in the United States, most of them storing petroleum, were leaking. Many other tanks were nearing the end of their useful life expectancy and were expected to leak in the near future. Approximately 50% of the U.S. population relies on ground water for their drinking water, and states were reporting that leaking underground tanks were the leading source of groundwater contamination.

In 1984, Congress responded to this environmental and safety threat and established a leak prevention, detection, and cleanup program for USTs containing chemicals or petroleum through amendments to the Solid Waste Disposal Act (42 U.S.C. 6901 et seq., also known as the Resource Conservation and Recovery Act (RCRA)). Subtitle I directed EPA to establish operating requirements and technical standards for tank design and installation, leak detection, spill and overfill control, corrective action, and tank closure. The universe of regulated tanks was extremely large and diverse, and included many small businesses. Consequently, EPA phased in the tank regulations over a 10-year period (1988 through 1998). Strict standards for new tanks took effect in 1988, and all tanks were required to comply with leak detection regulations by late 1993. All tanks installed before 1988 had to be upgraded (with spill, overfill, and corrosion protection), replaced, or closed by December 22, 1998.

In 1986, Congress established a response program for leaking petroleum USTs through the Superfund Amendments and Reauthorization Act (P.L. 99-499), which amended Subtitle I of RCRA. The amendments authorized EPA and states to respond to petroleum spills and leaks, and created the Leaking Underground Storage Tank (LUST) Trust Fund to help EPA and states cover the costs of responding to leaking USTs in cases where the UST owner or operator does not clean up a site. EPA and the states primarily use the annual LUST Trust Fund appropriation to oversee and enforce corrective actions performed by responsible parties. They also use the funds to conduct corrective actions where no responsible party has been identified, where a responsible party fails to comply with a cleanup order, in the event of an emergency, and to take cost recovery actions against parties. EPA and states have been successful in getting responsible parties to perform most cleanups. In these cases, the cleanup costs typically have been paid for by a state fund (discussed below), the responsible party, and/or private insurance.

State Funds

The 1986 law also directed EPA to establish financial responsibility requirements to ensure that UST owners and operators are able to cover the costs of taking corrective action and compensating third parties for injuries and property damage caused by leaking tanks. As mandated, EPA issued regulations requiring most tank owners and operators selling petroleum products to demonstrate a minimum financial responsibility of $1 million. Alternatively, owners and operators could rely on state assurance funds to demonstrate financial responsibility, saving them the cost of purchasing private insurance.

Most states established financial assurance funds. Unlike the federal LUST Trust Fund, state funds often are used to reimburse financially solvent tank owners and operators for some or all of the costs of remediating leaking tank sites. Revenues for state funds typically have been generated through gas taxes and tank fees and, collectively, these funds have provided more cleanup funds than the LUST Trust Fund. A June 2006 survey of states showed that, cumulatively, states had collected and spent roughly $14.18 billion through their funds. During 2006, state funds collected $1.48 billion in annual revenues and spent a total of $1.03 billion, while outstanding claims against state funds reached $1.32 billion. Ten states have made a transition to private insurance, but 20 states have extended their fund's original sunset date to address the backlog of leaking tanks. (See *2006 State Financial Assurance Funds Survey* at [http://www.astswmo.org].)

LUST TRUST FUND: FUNDING AND USES

The LUST Trust Fund is funded primarily through a 0.1 cent-per-gallon motor fuels tax that began in 1987. The Energy Policy Act of 2005 (P.L. 109-58, H.R. 6) extended the tax through March 2011. During FY2006, the tax generated $197 million, and the trust fund earned roughly $99 million in interest on the balance in the fund. By the end of June 2007, the fund's total net assets had reached $2.84 billion. From October 2007 through June 2007, the fund had earned $98.5 million in interest.

Congress has annually appropriated funds from the trust fund to support the LUST response program. EPA roughly estimates that the average cost of cleaning up a leaking tank site is $125,000, and through March 2007, 110,985 releases still needed remediation. Although EPA expects that private parties will pay for most cleanups, states estimate that it will cost $12 billion to remediate at least 54,000 tank sites that lack viable owners.[1]

To support the LUST cleanup program, Congress provided from the trust fund $69.4 million for FY2005 and $72 million for FY2006.[2] For FY2007, the House approved and the Senate Appropriations Committee recommended $72.8 million, as requested. The continuing resolution providing FY2007 appropriations (P.L. 110-5, H.J.Res. 20) generally funded EPA programs at the FY2006 level, which included $72 million from the LUST Trust Fund. The request for FY2008 is $72.5 million.

In recent years, EPA has allocated approximately 81% (roughly $58 million) of the annual trust fund appropriation to the states in the form of cooperative agreements and 4% to support LUST-eligible activities on Indian lands. EPA has used the remaining 15% for its program responsibilities. The Energy Policy Act of 2005 (P.L. 109-58, §1522) requires EPA to allot least 80% of the LUST Trust Fund appropriation to the states. Under cooperative agreements with EPA, the states receive grants to help cover the cost of administering the LUST program. States use most of their LUST program grants to hire staff for technical oversight of corrective actions performed by responsible parties. They typically use about one-third of the LUST money they receive for cleaning up abandoned tank sites and undertaking emergency responses.

EPA uses its portion of the appropriation to oversee cooperative agreements with states, implement the LUST corrective action program on Indian lands, and support state and regional offices. EPA priorities in the LUST program include reducing the backlog of confirmed releases; promoting better and less expensive cleanups; providing assistance to

Indian tribes; assisting with the cleanup of more complicated sites, especially sites contaminated with MTBE; and implementing the Energy Policy Act provisions.

PROGRAM ACCOMPLISHMENTS AND ISSUES

EPA reports that since the federal underground storage tank program began, more than 1.6 million of the roughly 2.2 million petroleum tanks subject to regulation have been closed and, overall, the frequency and severity of leaks from UST systems have been reduced significantly. As of mid-year FY2007, 637,612 tanks remained in service and subject to UST regulations, 468,331 releases had been confirmed, 439,450 cleanups had been initiated, and 357,346 cleanups had been completed. The backlog of sites requiring remedial action dropped to 110,985 sites. During FY2006, 8,361 releases were newly confirmed, compared with 7,421 in FY2005, 7,850 in FY2004 and 12,000 in FY2003.[3] Nearly 14,500 corrective actions were completed in FY2006.

Methyl Tertiary Butyl Ether (MTBE)

In the 1990s, as states and EPA were making solid progress in addressing tank leaks, another problem emerged. The gasoline additive MTBE was being detected at thousands of LUST sites and in numerous drinking water supplies, usually at low levels. Gasoline refiners had relied heavily on MTBE to produce gasoline that contained oxygenates, as required by the 1990 Clean Air Act Amendments as a way to improve combustion and reduce emissions. Once released into the environment, however, MTBE moves through soil and into water more rapidly than other gasoline components, and it is more difficult and costly to remediate than conventional gasoline. Because of its mobility, MTBE is more likely to reach drinking water supplies than conventional gas leaks. Although MTBE is thought to be less toxic than some gasoline components (such as benzene), even small amounts can render water undrinkable because of its strong taste and odor. Also, in 1993, EPA's Office of Research and Development concluded that the data support classifying MTBE as a possible human carcinogen.[4] Although EPA has not done so, at least seven states have set drinking water standards for MTBE, and many states have established cleanup standards or guidelines. At least 25 states have enacted limits or bans on the use of MTBE in gasoline.

At least 42 states require testing for MTBE in ground water at LUST sites. In a 2000 survey, 31 states reported that MTBE was found in ground water at 40% or more of LUST sites in their states; 24 states reported MTBE at 60% to 100% of sites. An update of this survey found that many sites had not been tested for MTBE and that most states did not plan to reopen closed sites to look for MTBE.[5] The total cost of treating MTBE-contaminated drinking water is unknown. Two studies by water utilities place their estimates of the costs, given limited data, at $25 billion and $33.2 billion.[6]

Implementation and Compliance Issues

EPA estimated that by FY2001, 89% of USTs had upgraded tank equipment to meet federal requirements. However, the Government Accountability Office (GAO) reported that because of poor training of tank owners, operators, and other personnel, about 200,000 (29%) USTs were not being operated or maintained properly, thus increasing the risk of leaks and ground water contamination. GAO also reported that only 19 states physically inspected all their tanks every three years (the minimum EPA considers necessary for effective tank monitoring) and that, consequently, EPA and states lacked the information needed to evaluate the effectiveness of the tank program and take appropriate enforcement actions.[7] Among its initiatives to improve compliance, EPA revised the definition of compliance ("significant operational compliance") to place greater emphasis on the proper operation and maintenance of tank equipment and systems. Through mid-year FY2007, EPA reports that 75% of recently inspected UST facilities were in compliance with the release *prevention* requirements, 69% were in compliance with the leak *detection* requirements, and 62% of facilities had complied with the combined requirements.

LEGISLATION

The 109th Congress addressed LUST and MTBE issues in the Energy Policy Act of 2005 (P.L. 109-58, H.R. 6). The act revised the UST leak prevention and cleanup programs (Title XV, Subtitle B) and extended the 0.1 cent-per-gallon motor fuels tax that finances the LUST Trust Fund through March 2011 (§ 1362).

MTBE in the Energy Policy Act

The House version of H.R. 6 had included a retroactive safe harbor provision to protect manufacturers and distributors of fuels containing MTBE or renewable fuels from product liability claims. This provision was opposed by water utilities, local government associations, and many states. Opponents argued that providing a liability shield would effectively leave gas station owners liable for cleanup, and as these businesses often have few resources, the effect of the provision would have been that the burden for cleanup would fall to local communities, water utilities, and the states. Proponents argued that a safe harbor was merited because MTBE was used heavily to meet federal clean air mandates. They further argued that the focus should be placed on preventing leaks from USTs, which have been the main source of MTBE contamination.[8] Ultimately, the conferees dropped the safe harbor provision and a provision to ban MTBE. P.L. 109-58 also repealed the Clean Air Act oxygenated fuel requirement that had prompted greater use of MTBE.

The Underground Storage Tank Compliance Act

Title XV, Subtitle B, of P.L. 109-58 comprises "The Underground Storage Tank Compliance Act" (USTCA). The USTCA amended SWDA Subtitle I to add new leak prevention and enforcement provisions to the UST regulatory program and imposed new requirements on states, EPA, and tank owners. The USTCA requires EPA, and states that receive funding under Subtitle I, to conduct compliance inspections of tanks at least once every three years. It also requires states to comply with EPA guidance prohibiting fuel delivery to ineligible tanks, to develop training requirements for UST operators and individuals responsible for tank maintenance and spill response, and to prepare compliance reports on government-owned tanks in the state. Additionally, states must require either that new tanks located near drinking water wells are equipped with secondary containment, or that UST manufacturers and installers maintain evidence of financial responsibility to provide for the costs of corrective actions. (USTCA implementation information and documents are available online at [http://www.epa.gov/oust/fedlaws/epact_05.htm].)

The USTCA authorizes the appropriation of $155 million annually for FY2006 through FY2011 from the LUST Trust Fund for states to use to implement new and existing UST leak prevention requirements and to administer state programs. However, the energy act's tax extension language (§1362) prohibited the use of trust fund appropriations for any new purposes. Thus, while the Energy Policy Act greatly expanded state responsibilities, it prohibited the use of LUST Trust Fund money to support implementation of these mandates. To address this issue, the 109th Congress amended the Internal Revenue Code to allow the trust fund to be used for prevention and inspection activities (P.L. 109-433, H.R. 6131). The USTCA also includes new authorizations of appropriations to hasten the cleanup of leaking tanks and related MTBE contamination. It authorizes LUST Trust Fund appropriations of $200 million annually for FY2006 through FY2011 for EPA and states to administer the LUST cleanup program, and another $200 million annually for FY2006 through FY2011, specifically for addressing MTBE and other oxygenated fuels leaks (e.g., ethanol).

The 110th Congress

Underground storage tank issues continue to receive congressional attention. A key concern involves the level of resources available to states to meet the new USTCA mandates, and some stakeholders have called for increased LUST Trust Fund appropriations. Despite the USTCA's funding authority, however, no increase in Trust Fund appropriations was sought or approved for FY2007; Congress provided the requested $72.3 million for LUST cleanup activities.

For FY2008, the President requested $72.46 million from the LUST Trust Fund for cleanup activities, and another $22.3 million through the State and Tribal Assistance Grants (STAG) account for leak prevention activities. The Senate Committee on Appropriations similarly recommended $72.49 million from the trust fund and $22.5 million from the STAG account (S.Rept. 110-91). In contrast, the House report for EPA's FY2008 funding bill, H.R. 2643 (H.Rept. 110-187) notes that the Energy Policy Act authorized the prevention grants to be funded from the LUST Trust Fund. The House-passed bill would provide $117.9 million from the trust fund for cleanup and leak prevention (including tank inspection) activities. This amount includes $10 million more than requested for LUST cooperative agreements, and $15.7 million more for UST grants authorized by the Energy Policy Act (which, when combined with the funds moved from the STAG account, would provide a total of $35.5 million for prevention activities). Noting this increase in UST funding, the House rejected the President's request that Congress revise the state inspection requirements under the USTCA.

Another UST issue concerns storage tank infrastructure. The Energy Policy Act's renewable fuel mandates, and congressional interest in increasing the use of such fuels, have presented new technical issues for USTs and for fuel storage and delivery infrastructure generally. Ethanol, for example, is more corrosive than gasoline and may increase the risk of leaks in tank systems. In February, the House passed H.R. 547 (H.Rept. 110-7) to require EPA to establish a program to research and develop materials that could be added to biofuels to make them more compatible with existing infrastructure used to store and deliver petroleum-based fuels. (For more information on this issue, see CRS Report RL33928, *Ethanol and Biofuels: Agriculture, Infrastructure, and Market Constraints Related to Expanded Production* by Brent Yacobucci and Randy Schnepf.)

REFERENCES

[1]   Government Accountability Office, Leaking Underground Storage Tanks: EPA Should Take Steps to Better Ensure the Effective Use of Public Funding for Cleanups, GAO-07-152, 2007.

[2]   For FY2006, Congress provided another $15 million in supplemental appropriations for cleaning up releases from tanks damaged by hurricanes Katrina and Rita.

[3]   For state-by-state information, see [http://www.epa.gov/oust/cat/camarchv.htm].

[4]   U.S. Environmental Protection Agency, Assessment of Potential Health Risks of Gasoline Oxygenated with Methyl Tertiary Butyl Ether (MTBE), EPA/600/R-93/206, 1993.

[5]   The New England Interstate Water Pollution Control Commission's 2000 Survey of State Experiences with MTBE Contamination at LUST Sites, and the 2003 Survey of Oxygenates at LUST Sites.

[6]   Respectively, the American Water Works Association, A Review of Cost Estimates of MTBE Contamination of Public Wells, June 21, 2005, and the Association of Metropolitan Water Agencies, Cost Estimate to Remove MTBE Contamination from Public Drinking Water Systems in the United States, June 20, 2005.

[7]   U.S. GAO, Environmental Protection: Improved Inspections and Enforcement Would Better Ensure the Safety of Underground Storage Tanks, GAO-01-464, May 2001, pp. 2-6. Also see Environmental Protection: More Complete Data and Continued Emphasis on Leak Prevention Could Improve EPA's Underground Storage Tank Program, GAO-06-45, November 2005.

[8]   For more information on LUST and MTBE provisions in P.L. 109-58, see CRS Report RL32865, Renewable Fuels and MTBE: A Comparison of Selected Provisions in the Energy Policy Act of 2005 (P.L. 109-58 and H.R. 6), by Brent D. Yacobucci et al.

In: Water Pollution Issues and Developments
Editor: S. V. Thomas, pp. 127-265

ISBN: 978-1-60456-208-8
© 2008 Nova Science Publishers, Inc.

SPECIAL BIBLIOGRAPHY OF BOOKS

Marlene Randall

1988 Oregon statewide assessment of nonpoint sources of water pollution / Planning & Monitoring Section, Water Quality Division, Oregon Department of Environmental Quality. Published/Created: Portland, Or. (811 SW 6th Ave., Portland 97204): The Section, [1988] Related Names: Oregon. Dept. of Environmental Quality. Water Quality Division. Planning & Monitoring Section. Description: viii, 182, [105] p.: maps (some col.); 28 x 44 cm." Subjects: Water quality--Oregon. Nonpoint source pollution--Oregon. LC Classification: TD224.O7 A15 1988 Dewey Class No.: 363.73/942/09795 20 Geographic Area Code: n-us-or

1991 doser study in Maryland coastal plain: use of lime doser to mitigate stream acidification: final report / Lenwood W. Hall ... [et al.]. Published/Created: Annapolis, MD: State of Maryland, Dept. of Natural Resources, Tidewater Administration, Chesapeake Bay Research and Monitoring Division, [1992] Related Names: Hall, Lenwood W. Maryland. Chesapeake Bay Research and Monitoring Division. Description: 1 v.: ill., maps; 28 cm. Subjects: Fish habitat improvement--Maryland. Aquatic liming--Maryland. Fishes--Effect of water quality on--

Maryland. Acid pollution of rivers, lakes, etc.--Maryland. Water acidification--Maryland. LC Classification: SH157.8 .A18 1992 Geographic Area Code: n-us-md

1995-1997 Maryland Biological Stream Survey results for selected small watersheds / prepared for Maryland Department of Natural Resources; prepared by Nancy E. Roth ... [et al.]. Portion of Title: Maryland Biological Stream Survey results for selected small watersheds Published/Created: Annapolis, Md.: Maryland Dept. of Natural Resources, [2000] Related Names: Roth, Nancy E. Maryland. Dept. of Natural Resources. Description: 1 v. (various pagings): ill.; 28 cm. Subjects: Stream ecology--Maryland. Water--Pollution--Maryland. Acid deposition--Environmental aspects--Maryland. Environmental monitoring--Maryland. LC Classification: QH105.M3 A14 2000

1996 update, industrial water, and wastewater market outlook. Portion of Title: Industrial water and wastewater market outlook Published/Created: Dodgeville, Wis., U.S.A. (3541 Norwegian Hollow Rd., Dodgeville 53533): W.T. Lorenz & Co., [c1996] Related Names: William T. Lorenz

& Co. Description: 280 p.; 29 cm.
Subjects: Water pollution control industry-
-United States. Water pollution control
equipment industry--United States. Water
purification equipment industry--United
States. LC Classification: HD9718.5.W363
U517 1996 Geographic Area Code: n-us---

1997 Red River Flood Groundwater
Rehabilitation Program / by Rick Lemoine
... [et al.]. Variant Title: Nineteen ninety-
seven Red River Flood Groundwater
Rehabilitation Program Published/Created:
[Winnipeg]: Manitoba Environment, 1998.
Related Names: Lemoine, Rick. Manitoba.
Manitoba Environment. Description: iv,
130, [12] leaves; 28 cm. LC Classification:
TD224.M25 A16 1998 Canadian Class
No.: COP.MA.2.1998-50 Dewey Class
No.: 363.739/4/0971274 22 National
Bibliography No.: C98-803317-8

1997 update: implementing the Clean Air Act:
April 10, 1997: an ABA satellite seminar /
co-sponsored by the American Bar
Association, Section of Natural Resources,
Energy, and Environmental Law, Center
for Continuing Legal Education; Air and
Waste Management Association; U.S.
Environmental Protection Agency.
Published/Created: [Chicago]: American
Bar Association, c1997. Related Names:
American Bar Association. Description:
474 p.; 28 cm. Subjects: Water--Pollution--
Law and legislation--United States. LC
Classification: KF3790.Z9 A135 1997
Geographic Area Code: n-us---

1998 update: the Clean Water Act: May 12,
1998, an ABA satellite seminar / co-
sponsored by the American Bar
Association, Section of Natural Resources,
Energy, and Environmental Law, Center
for Continuing Legal Education, Water
Environment Federation; in cooperation
with U.S. Environmental Protection
Agency. Portion of Title: Clean Water Act
Published/Created: [Chicago]: The
Association, c1998. Related Names:

American Bar Association. Description: 1
v. (various pagings): ill., forms; 28 cm.
Subjects: Water--Pollution--Law and
legislation--United States. LC
Classification: KF3790.Z9 A135 1998
Dewey Class No.: 344.73/046343 21
Geographic Area Code: n-us---

2002 list of Connecticut water bodies not
meeting water quality standards / State of
Connecticut, Department of Environmental
Protection. Portion of Title: List of
Connecticut water bodies not meeting
water quality standards Published/Created:
Hartford, CT: State of Connecticut, Dept.
of Environmental Protection, [2002?]
Related Names: Connecticut. Dept. of
Environmental Protection. Related Titles:
Connecticut waterbodies not meeting water
quality standards, 1998. Description: i, 28,
38, 12 p.: map; 28 cm. LC Classification:
TD224.C8 A6152 2002 Dewey Class No.:
363.739/4/09746 22

2003 watershed protection plan update for the
Wachusett Reservoir watershed.
Published/Created: [Boston, Mass.]:
Commonwealth of Massachusetts,
Executive Office of Environmental Affairs,
Dept. of Conservation and Recreation,
Division of Water Supply Protection,
Bureau of Watershed Management, 2003.
Related Names: Massachusetts. Dept. of
Conservation and Recreation. Bureau of
Watershed Management. Description: vi,
223 p.: ill. (some col.), col. maps; 28 cm.
LC Classification: TD224.M4 A5763 2003

208 phase 1 results: ranking of 128 sub-basins
for water pollution potential based on
intensity of land-disturbing activities:
selection of five priority sub-basins for
initial detailed studies / North Carolina
Department of Natural Resources and
Community Development, Division of
Environmental Management.
Published/Created: [Raleigh, N.C.] (P.O.
Box 27687, Raleigh 27611): The Division,
[1978] Related Names: North Carolina.

Division of Environment Management. Related Titles: Two hundred eight phase one results. Description: 74 p. in various pagings, [1] folded leaf of plates: ill. (some col.); 28 cm. LC Classification: TC424.N8 A43 1978 Dewey Class No.: 363.7/3942/09756 19 Geographic Area Code: n-us-nc

208 planning: what's it all about? / Edited by Frank L. Cross, Jr. Published/Created: Westport, Conn.: Technomic Pub. Co., c1976. Related Names: Cross, Frank L. Description: 74 p.: ill.; 28 cm. ISBN: 0877622108 Notes: Includes bibliographical references. Subjects: United States. Laws, statutes, etc. Federal water pollution control act amendments of 1972. Water quality management--United States. Series: Environmental monograph series; v. 8 LC Classification: TD223 .T85 Dewey Class No.: 363.6 Geographic Area Code: n-us---

208: the transition from planning to management. Published/Created: Washington: Urban Land Institute, 1977. Related Names: Urban Land Institute. Related Titles: Environmental comment. Description: 23 p.: ill.; 28 cm. Notes: Special issue of Environmental comment. Subjects: Water--Pollution--Law and legislation--United States. Water quality management--United States. LC Classification: KF3790.A2 T87 Dewey Class No.: 614.7/72/0973 Geographic Area Code: n-us---

305(b) technical report for Oklahoma / prepared by the Oklahoma State Department of Health; coordinated by the Pollution Control Coordinating Board. Portion of Title: Three hundred five (b) technical report for Oklahoma Serial Key Title: 305(b) technical report for Oklahoma Published/Created: Oklahoma City: The Department Related Names: Oklahoma. State Dept. of Health. Oklahoma. Pollution Control Coordinating Board. Description:

v.: ill., maps; 28 cm. Current Frequency: Biennial ISSN: 0740-9923 Notes: " ... produced as a result of the Clean water act, Public Law 92-500, Section 305(b) ..." Description based on: Water years 1980-1981. SERBIB/SERLOC merged record Subjects: Water-supply--Oklahoma--Periodicals. Water--Pollution--Oklahoma--Periodicals. LC Classification: TD224.O5 A57 Dewey Class No.: 363.7/3942/09766 19

A Competitive assessment of the U.S. water resources equipment industry / prepared by Capital Goods and International Construction Sector Group, International Trade Administration, Department of Commerce. Published/Created: Washington, D.C.: The Administration: For sale by the Supt. of Docs., U.S. G.P.O., 1985. Related Names: United States. International Trade Administration. Capital Goods and International Construction Sector Group. Related Titles: Competitive assessment of the U.S. water resources industry. Description: xii, 119 p.: ill.; 28 cm. Notes: Cover title: A competitive assessment of the U.S. water resources industry. Shipping list: 85-1067-P. "October 1985." S/N 003-009-00468-3 Item 231-B-1 Bibliography: p. 117-119. Subjects: Water purification equipment industry--United States. Water pollution control equipment industry--United States. Water purification equipment industry. Water pollution control equipment industry. Market surveys. Competition, International. LC Classification: HD9718.5.W383 U632 1985 Dewey Class No.: 338.4/76281/0973 19 Government Document No.: C 61.2:W 29/2 Geographic Area Code: n-us---

A Compilation of state water quality standards for marine waters / United States Environmental Protection Agency, Office of Water Planning and Standards. Published/Created: Washington, D.C.: U.S.

Govt. Print. Off., 1978. Related Names: United States. Environmental Protection Agency. Office of Water Planning and Standards. Description: 626 p. in various pagings; 28 cm. Notes: Cover title. Subjects: Water quality--Standards--United States. Marine pollution--Law and legislation--United States. LC Classification: TD223 .C675 Dewey Class No.: 628.1/686/162 Geographic Area Code: n-us---

A Compilation of state water quality standards for marine waters. Published/Created: Washington, DC: U.S. Environmental Protection Agency, Office of Water Regulations and Standards, 1982. Related Names: United States. Environmental Protection Agency. Office of Water Regulations and Standards. Description: 1 v. (various pagings): ill., 1 map; 28 cm. Notes: Cover title. "June 1982." Item 431-I-1 (microfiche) Subjects: Water quality--Standards--United States. Marine pollution--Law and legislation--United States. LC Classification: TD223 .C675 1982 Dewey Class No.: 363.7/39462/0973 19 Government Document No.: EP 1.2:M 33 Geographic Area Code: n-us---

A comprehensive program for control of water pollution: Missouri drainage basin / adopted by U.S. Dept. of Health, Education & Welfare, Public Health Service. Published/Created: [Washington DC: Division of Water Pollution Control, Missouri Drainage Basin Office] 1953. Related Names: United States. Division of Water Supply and Pollution Control. Description: 123p.: ill. Notes: Cover-title: Missouri drainage basin: a cooperative state-federal report on water pollution. June 1953. Subjects: Water--Pollution--Missouri River Valley. Water resources development--Missouri River Valley. Missouri River Watershed. Series: Public health service publication 317 Water

pollution series #56 LC Classification: TD420 .U58 no. 56

A Comprehensive study of Texas watersheds and their impacts on water quality and water quantity. Published/Created: Temple, Tex. (P.O. Box 658, Temple 76503): Texas State Soil and Water Conservation Board, [1991] Related Names: Texas State Soil and Water Conservation Board. Description: viii, 208 p.: ill., maps; 28 cm. Notes: Cover title. "January, 1991." Includes bibliographical references. Subjects: Water quality--Texas. Watersheds--Texas. Agricultural pollution--Texas. LC Classification: TD224.T4 C64 1991 Dewey Class No.: 363.73/94/09764 20 Geographic Area Code: n-us-tx

A Contingent valuation study of lost passive use values resulting from the Exxon Valdez oil spill: a report to the Attorney General of the State of Alaska / Richard T. Carson ... [et al.]. Published/Created: [Juneau, Alaska?: Distributed by the State of Alaska Attorney General's Office, 1992] Related Names: Carson, Richard T. Description: 2 v.: col. ill.; 28 cm. Contents: v. 1. Report -- v. 2. Appendices A-D. Notes: "November 10, 1992." Includes bibliographical references. Subjects: Exxon Valdez (Ship) Water--Pollution--Economic aspects--Alaska--Prince William Sound. Oil spills--Economic aspects--Alaska--Prince William Sound. Tankers--Accidents. Natural resources surveys--Alaska--Prince William Sound. Contingent valuation. LC Classification: HC107.A46 P83 1992 Geographic Area Code: n-us-ak

A guide for best management practice (BMP) selection in urban developed areas / sponsored by Environmental and Water Resources Institute of ASCE; produced by Urban Water Infrastructure Management Committee's Task Committee for Evaluating Best Management Practices. Published/Created: Reston, Va.: American Society of Civil Engineers, c2001.

Description: vii, 51 p.; 22 cm. ISBN:
0784405573 (pbk.)

A Guide to marine pollution. Compiled by
Edward D. Goldberg. Published/Created:
New York, Gordon and Breach Science
Publishers [c1972] Description: x, 168 p.
illus. 24 cm. ISBN: 0677125003

A guide to protecting coastal waters through
local planning. Published/Created: Raleigh,
N.C.: Division of Coastal Management,
North Carolina Dept. of Natural Resources
and Community Development, [1986]
Description: 63 p.: ill.; 28 cm.

Abstracts of refining literature. Other Title: API
abstracts of refining literature Serial Key
Title: Abstracts of refining literature
Abbreviated Title: Abstr. refin. lit.
Published/Created: Washington, American
Petroleum Institute. Related Names:
American Petroleum Institute. Description:
v. 28 cm. v. L8- Jan. 1961- Continues: API
technical abstracts 0096-5073 (DLC)sn
89003958 Continues in part: Petroleum
refining and petrochemicals literature
abstracts Abstracts of air and water
conservation literature Abstracts of
transportation and storage literature
Abstracts of petroleum substitutes
literature ISSN: 0003-0422 Cancelled
ISSN: 0091-214X Notes:
SERBIB/SERLOC merged record Vols.
for 1971-<June 1972> called also
Combined edition and include separately
paged sections: Petroleum refining and
petrochemicals literature abstracts,
Abstracts of air and water conservation
literature, Abstracts of transportation and
storage literature, and abstracts of
petroleum substututes literature, which are
also published separately. Subjects:
Petroleum--Refining--Abstracts--
Periodicals. Petroleum--Abstracts--
Periodicals. Air--Pollution--Abstracts--
Periodicals. Water--Pollution--Abstracts--
Periodicals. LC Classification: TP690.A1
A82 Dewey Class No.: 665/.53/08

Acidic deposition and aquatic ecosystems:
regional case studies / Donald F. Charles,
editor; Susan Christie, technical editor.
Published/Created: New York: Springer-
Verlag, c1991. Related Names: Charles,
Donald F. (Donald Franklin), 1949-
Christie, Susan. Description: xii, 747 p.:
ill., maps; 27 cm. ISBN: 0387973168
(New York: acid-free paper) 3540973168
(Berlin: acid-free paper) Notes: Includes
bibliographical references and index.
Subjects: Acid deposition--Environmental
aspects--United States. Acid pollution of
rivers, lakes, etc.--United States. Aquatic
ecology--United States. Aquatic animals--
Effect of water pollution on--United States.
LC Classification: QH545.A17 A238 1991
Dewey Class No.: 574.5/2632/0973 20
Geographic Area Code: n-us---

Acidification of surface waters in eastern
Canada and its relationship to aquatic biota
/ J.R.M. Kelso ... [et al.].
Published/Created: Ottawa: Dept. of
Fisheries and Oceans, 1986. Related
Names: Kelso, J. R. M. Description: iv, 42
p.: ill.; 28 cm. ISBN: 0660120585 (pbk.):
Notes: Bibliography: p. 36-39. Subjects:
Acid pollution of rivers, lakes, etc.--
Canada, Eastern. Fishes--Effect of water
pollution on--Canada, Eastern. Water
acidification--Canada, Eastern. Series:
Canadian special publication of fisheries
and aquatic sciences, 0706-6481; 87 LC
Classification: QH545.A2 A26 1986
Dewey Class No.: 363.73/86/09713 20
Geographic Area Code: n-cn---

Acute toxicities of organic chemicals to fathead
minnows (Pimephales promelas) / by
Center for Lake Superior Environmental
Studies, University of Wisconsin-Superior;
L.T. Brooke ... [et al.], editors;
contributors, Cheryl Anderson ... [et al.].
Published/Created: Superior, Wis., U.S.A.:
Distributed by The Center, c1984- Related
Names: Brooke, L. T. University of
Wisconsin-Superior. Center for Lake

Superior Environmental Studies. Description: v. <1 >: ill.; 29 cm. Subjects: Fathead minnow--Effect of water pollution on. Organic compounds--Toxicology. Fishes--Effect of water pollution on. LC Classification: QL638.C94 A27 1984 Dewey Class No.: 591.2/4 19

Administrative structures and implementation of the Community directives on the dangerous substances discharged into the aquatic environment: final report/ European Commission. Published/Created: Luxembourg: Office for Official Publications of the European Communities, 1996. Related Names: European Commission. Description: 44, 62, [3] p.; 30 cm. ISBN: 9282755827 Notes: "CR-93-95-055-EN-C"--P. [4] of cover. Subjects: Water--Pollution--Government policy--European Union countries. LC Classification: TD255 .A36 1996 Dewey Class No.: 363.73/9456/094 21

Agricultural best management practices for protecting water quality in Georgia. Published/Created: [Atlanta]: State Soil & Water Conservation Commission, Georgia, [1994] Related Names: Georgia. State Soil & Water Conservation Commission. Description: v, 35 p.: col. ill.; 28 cm. Notes: Title from cover. "September 1994"--P. [3] of cover. Includes bibliographical references (p. 34). Subjects: Agricultural pollution--Georgia. Water quality management--Georgia. Best management practices (Pollution prevention)--Georgia. LC Classification: TD195.A34 A36 1994 Dewey Class No.: 363.73809758 22

Agricultural chemicals and groundwater protection: emerging management and policy: proceedings of a conference held October 22-23, 1987, St. Paul, Minnesota / sponsored by Freshwater Foundation ... [et al.]. Published/Created: Navarre, Minn., U.S.A. (Box 90, Navarre 55392-0090):

The Foundation, c1988. Related Names: Freshwater Foundation. Description: iii, 235 p.: ill.; 28 cm. Notes: Includes bibliographical references. Subjects: Agricultural chemicals--Environmental aspects--Congresses. Groundwater--Pollution--Congresses. Water quality management--Government policy--Congresses. Agricultural chemicals industry--Congresses. LC Classification: TD427.A35 A37 1988 Dewey Class No.: 363.73/8 20

Agricultural management and water quality / edited by Frank W. Schaller, George W. Bailey. Edition Information: 1st ed. Published/Created: Ames: Iowa State University Press, 1983. Related Names: Schaller, Frank W., 1914- Bailey, George W., 1933- Iowa State University. Environmental Research Laboratory (Athens, Ga.) National Conference on Agricultural Management and Water Quality (1981: Iowa State University) Description: xviii, 472 p.: ill.; 24 cm. ISBN: 081380082X Notes: Papers presented at the National Conference on Agricultural Management and Water Quality, held at Iowa State University, in May 1981, and sponsored by Iowa State University and U.S. Environmental Protection Agency's (EPA) Environmental Research Laboratory, Athens, Ga. Includes bibliographies and index. Subjects: Agricultural pollution--United States--Congresses. Water quality management--United States--Congresses. Farm management--United States--Congresses. LC Classification: TD428.A37 A36 1983 Dewey Class No.: 363.7/394 19 Geographic Area Code: n-us---

Agricultural nonpoint source pollution: watershed management and hydrology / edited by William F. Ritter, Adel Shirmohammadi. Published/Created: Boca Raton, Fla.: Lewis Publishers, c2001. Related Names: Ritter, William F.

Shirmohammadi, Adel, 1952- Description: 342 p.: ill., map; 24 cm. ISBN: 1566702224 (alk. paper) Notes: Includes bibliographical references and index. Subjects: Agricultural pollution--United States. Nonpoint source pollution--United States. Watershed management--United States. Water quality management--United States. LC Classification: TD428.A37 A362 2001 Dewey Class No.: 628.1/684 21 Geographic Area Code: n-us---

Agricultural practices and water quality. Edited by Ted L. Willrich and George E. Smith. Edition Information: [1st ed.] Published/Created: Ames, Iowa State University Press [1971, c1970] Related Names: Willrich, Ted L., ed. Smith, George Edward, 1913- ed. Iowa State University. Description: xxvii, 415 p. illus. 24 cm. ISBN: 0813817455 Notes: "Result of a conference, The role of agriculture in clean water, held at Iowa State University, Ames, Iowa, November 18-20, 1969." Includes bibliographies. Subjects: Water--Pollution. Agriculture. LC Classification: TD420 .A33 Dewey Class No.: 628/.1684

Agricultural salinity management in India / editors, N.K. Tyagi and P.S. Minhas. Published/Created: Karnal: Central Soil Salinity Research Institute, [1998] Related Names: Tyagi, N. K. Minhas, P. S. Central Soil Salinity Research Institute (Karnal,,l, India) Description: viii, 526 p.: ill.; 25 cm. Summary: Contributed articles. Notes: Includes bibliographical references and index. Subjects: Soils, Salts in--India. Irrigation water--Pollution--India. Water salinization--Control--India. Agriculture--Economic aspects--India. Water quality management--India. LC Classification: S599.6.I5 A47 1998 Overseas Acquisitions No.: I-E-2004-329684; 35-91

Agriculture and groundwater quality. Published/Created: Ames, Iowa: Council for Agricultural Science and Technology, [1985] Related Names: Council for

Agricultural Science and Technology. Description: 62 p.: ill.; 28 cm. Notes: "May 1985." Bibliography: p. 49-54. Subjects: Agricultural pollution. Groundwater--Pollution. Water quality. Series: Report / Council for Agricultural Science and Technology, 0194-4088; no. 103 Report (Council for Agricultural Science and Technology); no. 103. LC Classification: S589.75 .A37 1985 Dewey Class No.: 631.4/1 19

Agriculture and water quality: international perspectives / edited by John B. Braden and Stephen B. Lovejoy. Published/Created: Boulder, Colo.: L. Rienner, 1990. Related Names: Braden, John B. Lovejoy, Stephen B. Related Titles: Agriculture & water quality. Description: xii, 224 p.: ill.; 24 cm. ISBN: 1555871836 (alk. paper): Notes: Cover title: Agriculture & water quality. Includes bibliographical references. Subjects: Water quality management--Government policy. Water--Pollution--Prevention--Government policy. Agricultural pollution--Prevention--Government policy. LC Classification: HC79.W32 A37 1990 Dewey Class No.: 363.73/1 20

Agriculture and water quality: proceedings of a conference organized by the Agricultural Development and Advisory Service and the Agricultural Research Council, 17-19 December 1974. Published/Created: London: H.M.S.O., 1976. Related Names: Great Britain. Agricultural Development and Advisory Service. Agricultural Research Council (Great Britain) Description: xii, 469 p., [2] fold. p. of plates: ill., maps; 25 cm. ISBN: 0112408931: Notes: Includes bibliographical references. Subjects: Agricultural wastes--Environmental aspects--Great Britain --Congresses. Water--Pollution--Great Britain--Congresses. Series: Technical bulletin - Ministry of Agriculture, Fisheries and

Food; 32 Technical bulletin (Great Britain. Ministry of Agriculture, Fisheries and Food); 32. LC Classification: S217 .A6134 no. 32 TD930 Dewey Class No.: 630/.8 s 628.1/61 National Bibliography No.: . GB76-25843 Geographic Area Code: e-uk- --

Agriculture and water quality: proceedings of an interdisciplinary symposium, April 23 and 24, 1991 / editors, M.H. Miller ... [et al]. Published/Created: Guelph, Ont. Canada: Centre for Soil and Water Conservation, University of Guelph, c1992. Related Names: Miller, M. H. University of Guelph. Centre for Soil and Water Conservation. Description: ii, 213 p.: ill.; 24 cm. ISBN: 0889552983 Notes: "Sponsors: Centre for Soil and Water Conservation, University of Guelph, the Water Network, Waterloo Centre for Groundwater Research, University of Waterloo, Canadian Center for Toxicology, Ontario Chapter, Soil and Water Conservation Society"--T.p. verso. Includes bibliographical references. Subjects: Agriculture--Environmental aspects--Congresses. Water--Pollution-- Congresses. Water quality--Congresses. LC Classification: TD428.A37 A37 1992 Dewey Class No.: 363.73/1 20

Agriculture, hydrology, and water quality / edited by P.M. Haygarth and S.C. Jarvis. Published/Created: Wallingford, UK; New York: CABI Pub., c2002. Related Names: Haygarth, P. M. Jarvis, S. C. Description: xii, 502 p.: ill.; 25 cm. ISBN: 0851995454 (alk. paper) Contents: Machine generated contents note: Contributors -- Acknowledgements -- Note on Terminology and Abbreviations -- Introduction: an Interdisciplinary Approach for Agriculture, Hydrology and Water Quality -- PM.Haygarth and S.C.Jarvis --Section 1.Agriculture: Potential Sources of Water Pollution -- Introduction: Agriculture as a Potential

Source of Water Pollution -- A.N.Sharpley -- 1.Nitrogen -- D.Hatch, K.Goulding and D.Murphy -- 2.Phosphorus -- P Leinweber, B.L.Turner and R.Meissner -- 3.Manures -- D.R.Chadwick and S.Chen -- 4.Pesticides and Persistent Organic Pollutants -- B.Gevao and K.C.Jones -- 5.Heavy Metals -- W de Vries, PF.A.M.R6mkens, T.van Leeuwen and J.J.B.Bronswijk -- 6.Human Enteric Pathogens -- D.L.Jones, G.Campbell and C.W Kaspar -- 7.Sediment -- TR.Harrod and F.D.Theurer -- 8.Nutrient Balances -- D.B.Beegle, L.E.Lanyon and J.T.Sims --Section 2.Hydrology: the Carrier and Transport of Water Pollution -- Introduction: Modelling Hydrological and Nutrient Transport Processes -- K.Beven -- 9.Hydrological Source Management of Pollutants at the Soil Profile Scale -- R.W McDowell, A.N.Sharpley, P.J.A.Kleinman and W.J.Gburek -- 10.Hydrological Mobilization of Pollutants at the Field/Slope Scale -- D.Nash, D.Halliwell andJ.Cox -- 11.Modelling Hydrological Mobilization of Nutrient Pollutants at the Catchment Scale -- Y.J.P.Van Herpe, PA.Troch, PF.Quinn and S.Anthony -- 12.Pollutant-Sediment Interactions: Sorption, Reactivity and Transport of Phosphorus -- D.S.Baldwin, A.M, Mitchell and J.M.Olley -- 13.Quantifying Sediment and Nutrient Pathways within Danish Agricultural Catchments -- B.Kronvang, R.Grant, A.R.Laubel and M.L.Pedersen -- 14.Development of Geographical Information Systems for Assessing Hydrological Aspects of Diffuse Nutrient and Sediment Transfer from Agriculture -- A.I.Fraser -- 15.Wetlands as Regulators of Pollutant Transport -- M.S.A.Blackwell, D.V Hogan and E.Maltby --Section 3.Water Quality: Impacts and Case Studies from Around the World -- Introduction: Impacts of Agriculture on Water Quality Around the World -- G.Harris -- 16.Solutions to Nutrient Management Problems in the Chesapeake Bay -- Watershed, USA -- J.T Sims and F.J.Coale

-- 17.Nutrient and Pesticide Transfer from Agricultural Soils to Water in New Zealand -- K.C.Cameron, H.J.Di and L.M.Condron -- 18.Land, Water and People: Complex Interactions in the Murrumbidgee River Catchment, New South Wales, Australia -- S.Pengelly and G.Fishburn -- 19.Managing the Effects of Agriculture on Water Quality in Northern Ireland -- R.H.Foy and WC.K.O'Connor -- 20.Conflicts and Problems with Water Quality in the Upper Catchment of the Manyame River, Zimbabwe -- KE.Motsi, E.Mangwayana and K.E.Giller -- 21.Dryland Salinization: a Challenge for Land and Water Management in the Australian Landscape -- J.Williams, G.R.Walker and T.Hatton -- 22.Quantifying Nutrient Limiting Conditions in Temperate River Systems -- A.C.Edwards and PA.Chambers -- Index. Notes: Includes bibliographical references and index. Subjects: Agricultural pollution. Water--Pollution. Hydrology. Water quality. LC Classification: TD428.A37 A366 2002 Dewey Class No.: 628.1/684 21

Air & water pollution control. Portion of Title: Air and water pollution control Serial Key Title: Air & water pollution control Abbreviated Title: Air water pollut. control Published/Created: [Washington, D.C.]: Bureau of National Affairs, [1986- Related Names: Bureau of National Affairs (Arlington, Va.) Related Titles: Air pollution control. Water pollution control. Description: v.; 28 cm. Vol. 1, no. 1 (Oct. 8, 1986)- Ceased with Mar. 27, 1996 issue. Current Frequency: Biweekly Merger of: Air pollution control (Washington, D.C.) 0196-7150 (DLC) 87644823 (OCoLC)5877912 Water pollution control (Washington, D.C.: 1979) 0194-0147 (DLC) 87644825 (OCoLC)5319928 Continued by: BNA's environmental compliance bulletin 1073-5798 (DLC) 96643515 (OCoLC)29484008 ISSN: 0890-

0396 Cancelled/Invalid LCCN: sn 86012016 CODEN: AWPCE3 Notes: Title from caption. Latest issue consulted: Vol. 1, no. 5 (Dec. 3, 1986). Includes supplements. SERBIB/SERLOC merged record Merger of the bulletins from the loose-leaf services: Air pollution control; and: Water pollution control. Has separately published loose-leaf services: Air pollution control; and: Water pollution control. Subjects: Water--Pollution--Law and legislation--United States -- Periodicals. Air--Pollution--Law and legislation--United States --Periodicals. LC Classification: KF3786.A3 A37 NAL Class No.: KF3786.A3A37 Dewey Class No.: 344.73/04634/05 347.304463405 19 Postal Registration No.: 533920 USPS

Air & water pollution: a compliance manual. Published/Created: Neenah, Wis.: J.J. Keller & Associates, c1992- Related Names: J.J. Keller & Associates. Related Titles: Air and water pollution. Description: 1 v. (loose-leaf): ill.; 30 cm. ISBN: 1877798061 Subjects: Air--Pollution--Law and legislation--United States. Water--Pollution--Law and legislation--United States. LC Classification: KF3775 .A918 Dewey Class No.: 344.73/046342 347.30446342 20 Geographic Area Code: n-us---

Air & water pollution: a guide to federal regulations. Published/Created: Neenah, Wis.: J.J. Keller, c1992- Related Names: J.J. Keller & Associates. Related Titles: Air and water pollution. Description: 1 v. (loose-leaf): ill.; 30 cm. ISBN: 0934674973 Subjects: Air--Pollution--Law and legislation--United States. Water--Pollution--Law and legislation--United States. LC Classification: KF3812 .A947 1992 Dewey Class No.: 344.73/046342 347.30446342 20 Geographic Area Code: n-us---

Air and water analysis: new techniques and data / edited by R.W. Frei, J. Albaigeì s.

Published/Created: New York: Gordon and Breach Science Publishers, c1986. Related Names: Frei, R. W. (Roland W.) Albaigeì s, J. Symposium on the Analytical Chemistry of Pollutants (14th: 1984: Barcelona, Spain) International Congress on Analytical Techniques in Environmental Chemistry (3rd: 1984: Barcelona, Spain) Description: viii, 319 p.: ill.; 24 cm. ISBN: 2881241220 (France) Notes: Selection from papers presented at the 14th Annual Symposium on the Analytical Chemistry of Pollutants and the 3rd Congress on Analytical Techniques in Environmental Chemistry, held in Barcelona, Nov. 21-23, 1984. Includes bibliographies and indexes. Subjects: Environmental chemistry--Congresses. Air--Analysis--Congresses. Water--Analysis--Congresses. Air--Pollution--Measurement--Congresses. Water--Pollution--Measurement--Congresses. Series: Current topics in environmental and toxicological chemistry, 0275-2581; v. 9 LC Classification: TD193 .A39 1986 Dewey Class No.: 628.5 19

Air and water pollution control law, 1982: a comprehensive examination of the law pertaining to the control of air and water pollution with emphasis on recent developments / editors, Phillip D. Reed, Gregory S. Wetstone. Published/Created: Washington, DC (1346 Connecticut Ave., N.W., Washington 20036): Environmental Law Institute, c1982. Related Names: Reed, Phillip D. Wetstone, Gregory. Environmental Law Institute. Description: viii, 783 p.; 28 cm. Notes: "A preliminary version of this volume, Air and water pollution control law, 1981, was distributed at the Conference on Water and Air Pollution Law ... November 1981, Washington, D.C."--T.p. verso. Includes bibliographical references. Subjects: Air--Pollution--Law and legislation--United States. Water--Pollution--Law and legislation--United States. LC

Classification: KF3812 .A95 1982 Dewey Class No.: 344.73/04634 347.3044634 19 Geographic Area Code: n-us---

Air and water pollution control law: May 25-27, 1989, Washington, D.C.: ALI-ABA course of study materials / cosponsored by the Environmental Law Institute. Published/Created: Philadelphia, PA: American Law Institute-American Bar Association Committee on Continuing Professional Education, c1989. Related Names: Environmental Law Institute. American Law Institute--American Bar Association Committee on Continuing Professional Education. Description: xv, 534 p.; 28 cm. Notes: Spine title: Air and water pollution. "C432." Includes bibliographical references. Subjects: Air--Pollution--Law and legislation--United States. Water--Pollution--Law and legislation--United States. LC Classification: KF3812.Z9 A34 1989 Dewey Class No.: 344.73/04634 347.3044634 20 Geographic Area Code: n-us---

Air and water pollution. Published/Created: Oxford [etc.] New York, Pergamon Press. Related Titles: International journal of air pollution. International journal of air and water pollution. Description: p. cm. Current Frequency: Unknown Continued by: SERLOC succeeding entry Atmospheric environment and Water Research Notes: PREMARC/SERLOC merged record Subjects: Air--Pollution--Periodicals. [from old catalog] Water--Pollution--Periodicals. [from old catalog] LC Classification: RA576.A1 A35

Air and water pollution. Serial Key Title: Air and water pollution Abbreviated Title: Air water pollut. Published/Created: New York; London: Pergamon Press, 1961-1966. Description: 6 v.: ill.; 26 cm. Vol. 5, no. 1 (Nov. 1961)-v. 10, no. 11/12 (Nov./Dec. 1966). Current Frequency: Quarterly Continues: International journal of air and water pollution (OCoLC)2262569 Split into: Atmospheric environment 0004-6981 (DLC) 76012698 (OCoLC)1518568 Water research (Oxford) 0043-1354 (DLC) 76006152 (OCoLC)1769499 ISSN: 0568-3408 Notes: "An international journal." Articles chiefly in English; some in French or German. Split into: Atmospheric environment, and: Water research. Subjects: Air--Pollution--Periodicals. Water--Pollution--Periodicals. LC Classification: RA576.A1 I5 NLM Class No.: W1 AI702 Dewey Class No.: 363 11 Language Code: engfreger

Air toxics and water monitoring: 21 June, 1995, Munich, FRG / George M. Russwurm, chair/editor; sponsored by The Commission of the European Communities, Directorate General for Science, Research, and Development ... [et al.]. Published/Created: Bellingham, Wash., USA: SPIE--the International Society for Optical Engineering, c1995. Related Names: Russwurm, George M. Commission of the European Communities. Directorate General for Science, Research, and Development. Society of Photo-optical Instrumentation Engineers. Description: v, 178 p.: ill.; 28 cm. ISBN: 0819418617 Notes: Includes bibliographic references and index. Subjects: Air--Pollution--Measurement--Congresses. Water--Pollution--Measurement--Congresses. Stochastic processes--Congresses. Neural networks (Computer science)--Congresses. Series: Proceedings Europto series Proceedings / SPIE--the International Society for Optical Engineering; v. 2503 Proceedings EurOpt series. Proceedings of SPIE--the International Society for Optical Engineering; v. 2503. LC Classification: TD890 .A384 1995 Dewey Class No.: 628.5/028/7 20

Air/water pollution report. Serial Key Title: Air-water pollution report Abbreviated Title: Air-water pollut. rep. Published/Created: Silver Spring, MD: Business Publishers Description: 33 v.; 28 cm. Began in 1963 and ceased with May 22, 1995. Current Frequency: Weekly Merger of: Environment week 1041-8105 (DLC) 89645265 (OCoLC)18851780 Air/water pollution report's environment week 1082-8575 (DLC) 95648405 (OCoLC)32587940 Absorbed: Air and water news 0002-2187 (DLC)sc 79005508 (OCoLC)1581066 ISSN: 0002-2608 Cancelled/Invalid LCCN: sc 78000636 CODEN: AWPREE Notes: Description based on: Vol. 29, no. 3 (Jan. 21, 1991); title from caption. Latest issue consulted: Vol. 32, no. 50 (Dec. 19, 1994). Merged with: Environment week, to form: Air/water pollution report's environment week. Subjects: Air--Pollution--United States--Periodicals. Water--Pollution--United States--Periodicals. Air Pollution--Periodicals. Water Pollution--Periodicals. LC Classification: TD883.2 A645 NLM Class No.: W1 AI724 Dewey Class No.: 363.73/0973 20

Air/water pollution report. Variant Title: Air water pollution report Serial Key Title: Air/water pollution report (1998) Abbreviated Title: Air/water pollut. rep. (1998) Published/Created: Silver Spring, MD: Business Publishers, 1998- Description: v.; 28 cm. Vol. 36, no. 10 (Mar. 9, 1998)- Current Frequency: Weekly Continues: Air/water pollution report's environment week 1082-8575 (DLC) 95648405 (OCoLC)32587940 ISSN: 1098-8041 Incorrect ISSN: 1041-8105 Cancelled/Invalid LCCN: sn 98004337 Notes: Title from caption. Latest issue consulted: Vol. 36, no. 36 (Sept. 14, 1998). SERBIB/SERLOC merged record Subjects: Air--Pollution--United States--Periodicals. Water--Pollution--United States--Periodicals. Air--Pollution--Law

and legislation--United States --
Periodicals. Water--Pollution--Law and
legislation--United States --Periodicals. LC
Classification: TD883.2 A645 Dewey
Class No.: 363.73/0973 20

Air/water pollution report's environment week.
Portion of Title: Air water pollution
report's environment week Running Title:
Environment week Serial Key Title:
Air/water pollution report's environment
week Abbreviated Title: Air/water pollut.
rep. environ. week Published/Created:
Silver Spring, MD: Business Publishers,
Inc., 1995-1998. Description: 4 v.; 28 cm.
Vol. 33, no. 22 (May 29, 1995)-v. 36, no. 9
(Mar. 2, 1998). Current Frequency:
Weekly Merger of: Environment week
1041-8105 (DLC) 89645265
(OCoLC)18851780 Air/water pollution
report 0002-2608 (DLC)sc 78000636
(OCoLC)1780747 Absorbed: Environment
report 0013-9203 (DLC)sn 96041580
(OCoLC)1568070 Continued by: Air/water
pollution report (Silver Spring, Md.: 1998)
1098-8041 (DLC) 98646595
(OCoLC)38891849 ISSN: 1082-8575
Incorrect ISSN: 1041-8105
Cancelled/Invalid LCCN: sn 95039487
Notes: Title from caption.
SERBIB/SERLOC merged record Formed
by the union of: Environment week, and:
Air/water pollution report, and continues
the volume numbering of the latter;
Absorbed: Environment report. Subjects:
Air--Pollution--United States--Periodicals.
Water--Pollution--United States--
Periodicals. Air--Pollution--Law and
legislation--United States --Periodicals.
Water--Pollution--Law and legislation--
United States --Periodicals. LC
Classification: TD883.2 A645 Dewey
Class No.: 363.73/0973 20

Alaska's legacy of oil and hazardous substance
pollution: cleanup and management of
Alaska's contaminated sites / Alaska
Department of Environmental

Conservation, Division of Spill Prevention
and Response, Contaminated Sites
Program. Published/Created: [Juneau,
Alaska]: The Program, [2007] Related
Names: Alaska. Contaminated Sites
Program. Description: 39 p.: col. ill.; 28
cm. Computer File Information: This item
was harvested from the Dept. of
Environmental Conservation, Division of
Spill Prevention and Response,
Contaminated Sites Program web site:
http://www.dec.state.ak.us/spar/csp/docs/cs
story.pdf in June 2007 for the Alaska State
Publications Program; remote access
available via StaticURL. Notes: "January
2007 "Fiscal year 2006"--Cover. Includes
bibliographical references (p. 39).
Additional Formats: Also available in
electronic format via Internet. Subjects:
Alaska. Contaminated Sites Program. Oil
spills--Cleanup--Alaska. Oil pollution of
soils--Alaska. Oil pollution of water--
Alaska. Soil pollution--Alaska. Water--
Pollution--Alaska.

Alberta's Clean Water Act. Published/Created:
Edmonton, Alta.: Environment Council of
Alberta, [1985] Related Names: Lilley,
John. Environment Council of Alberta.
Alberta. Alberta Environment. Description:
iii, 101 p.; 28 cm. Contents: Conclusions
and recommendations of the review of the
Clean Water Act: report of the
Environment Council of Alberta to the
Minister of the Environment -- Review of
the Clean Water Act: staff report / prepared
by John Lilley. Notes: "March 1985."
Includes bibliographies. Subjects: Water--
Pollution--Law and legislation--Alberta.
LC Classification: KEA422.W3 A73 1985
Dewey Class No.: 344.7123/046343
347.1230446343 19 Geographic Area
Code: n-cn-ab

Algae abstracts: a guide to the literature.
Prepared from material supplied by Water
Resources Scientific Information Center,
Office of Water Resources Research,

Department of the Interior, Washington, D.C. Published/Created: New York, IFI/Plenum [1973- Related Names: Water Resources Scientific Information Center. Description: v. 29 cm. ISBN: 0306671816 (v. 1) Incomplete Contents: v. 1. To 1969.--v. 2. 1970-1972.--v. 3. 1972-1974. Subjects: Algology--Abstracts. Water--Pollution--Abstracts. Eutrophication--Abstracts. Algology--Indexes. Water--Pollution--Indexes. Eutrophication--Indexes. LC Classification: QK564.5 .A38 Dewey Class No.: 589/.3/08

Algal bloom detection, monitoring, and prediction: 3rd workshop "public health": Istituto superiore di sanitaÌ€, Rome, April 21-22, 1998: proceedings / edited by Giorgio Catena and Enzo Funari. Published/Created: Roma: Istituto superiore di sanitaÌ€, c1999. Related Names: Catena, Giorgio. Funari, E. Istituto superiore di sanitaÌ€ (Italy) Description: iii, 95 p.: ill. Notes: Includes bibliographical references. English, with summary in Italian. Subjects: Algal blooms--Mediterranean Region--Congresses. Algae--pathogenicity--Mediterranean Region--Congresses. Environmental Monitoring--Mediterranean Region--Congresses. Water Pollution, Chemical--prevention & control --Mediterranean Region--Congresses. Series: Rapporti ISTISAN, 1123-3117; 99/8 LC Classification: QK568.B55 A44 1999 NLM Class No.: W1 RA489J 1999 no.8 WA 689 A394 1999 Dewey Class No.: 579.8/17738 21 Language Code: eng ita

Almanac of enforceable state laws to control nonpoint source water pollution. Published/Created: Washington, D.C.: Environmental Law Institute, c1998. Related Names: Environmental Law Institute. Description: ii, 293 p.; 28 cm. ISBN: 0911937811 Notes: "ELI Project #970301"--T.p. verso. Subjects: Nonpoint source pollution--Law and legislation--

United States--States. Water--Pollution--Law and legislation--United States --States. Water quality management--United States--States. Series: Environmental Law Institute research report LC Classification: KF3790.Z95 A43 1998 Dewey Class No.: 344.73/046343 21

Alum sludge in the aquatic environment / prepared by Dennis B. George ... [et al.].; prepared for AWWA Research Foundation. Published/Created: Denver, CO: AWWA Research Foundation and American Water Works Association, c1991. Related Names: George, Dennis B. AWWA Research Foundation. Description: xix, 224 p.: ill.; 28 cm. ISBN: 0898675316 Notes: Includes bibliographical references (p. 165-175). Subjects: Water treatment plant residuals--Environmental aspects. Sewage disposal in rivers, lakes, etc.--Environmental aspects. Aquatic organisms--Effect of water pollution on. Aluminum--Toxicology. LC Classification: TD899.W3 A15 1991

Ambient toxicity testing in Chesapeake Bay: year 8 report. Published/Created: Annapolis, Md.: Chesapeake Bay Program, [2000] Related Names: Chesapeake Bay Program (U.S.) Description: 1 v. (various pagings): ill., maps; 28 cm. Notes: Includes bibliographical references. Subjects: Toxicity testing--Chesapeake Bay (Md. and Va.) Water--Pollution--Toxicology--Chesapeake Bay (Md. and Va.) Chesapeake Bay (Md. and Va.)--Environmental conditions. Series: CBP/TRS; 243-00 LC Classification: QH541.15.T68 A42 2000 Dewey Class No.: 577.27/09163/47 21

Ambient water quality criteria for dissolved oxygen, water clarity and chlorophyll a for Chesapeake Bay and its tidal tributaries: 2004 addendam / U.S. Environmental Protection Agency, Region III, Chesapeake Bay Program Office and Region III, Water Protection Division in coordination with the Office of Water, Office of Science and Technology. Published/Created: Anapolis, MD: U.S. Environmental Protection Agency, Region III, Chesapeake Bay Program Office, 2004. Related Names: United States. Environmental Protection Agency. Region III. Chesapeake Bay Program Office. United States. Environmental Protection Agency. Region III. Water Protection Division. United States. Environmental Protection Agency. Office of Water. United States. Environmental Protection Agency. Office of Science and Technology. Description: v, 103 p.: ill., maps; 28 cm. Notes: "October 2004." Includes bibliographical references. Subjects: Water--Dissolved oxygen--Chesapeake Bay (Md. and Va.) Water quality--Standards--Chesapeake Bay (Md. and Va.) Water--Pollution--Chesapeake Bay (Md. and Va.) Chlorophyll. LC Classification: TD225.C43 A473 2004 Dewey Class No.: 363.739/4620916347 22

Ambient water quality criteria for dissolved oxygen, water clarity and chlorophyll a for Chesapeake Bay and its tidal tributaries / U.S. Environmental Protection Agency, Region III, Chesapeake Bay Program Office and Region III, Water Protection Division in coordination with the Office of Water, Office of Science and Technology. Published/Created: [Anapolis, MD]: U.S. Environmental Protection Agency, Region III, Chesapeake Bay Program Office, [2003] Related Names: United States. Environmental Protection Agency. Region III. Chesapeake Bay Program Office. United States. Environmental Protection Agency. Region III. Water Protection Division. United States. Environmental Protection Agency. Office of Water. United States. Environmental Protection Agency. Office of Science and Technology. Description: 1 v. (various pagings): ill.; 28 cm. Notes: "April 2003." Includes bibliographical references. Subjects: Water--Dissolved oxygen--

Chesapeake Bay (Md. and Va.) Water quality--Standards--Chesapeake Bay (Md. and Va.) Water--Pollution--Chesapeake Bay (Md. and Va.) Chlorophyll. LC Classification: TD225.C43 A47 2003 Dewey Class No.: 363.739/4620916347 22

Ambio. Serial Key Title: Ambio Abbreviated Title: Ambio Published/Created: [Stockholm]: Royal Swedish Academy of Sciences; [Boston: Universitetsforlaget] [distributor], c1972- Related Names: Kungl. Svenska vetenskapsakademien. Description: v.: ill.; 30 cm. Vol. 1, no. 1 (Feb. 1972)- Current Frequency: Monthly (except January, April, July, October), <Feb. 1992-> ISSN: 0044-7447 CODEN: AMBOCX Notes: Distributor varies: Lawrence, KS: Allen Press Inc., <Feb. 1992-May 2003>. Latest issue consulted: Vol. 33, no. 3 (May 2003). In English. SERBIB/SERLOC merged record ACQN: aq 94014245 RECD: v. 1/1972 through v.26/1997, nr.1-3, 5-8 (7 RECD 3-98) Replaces 9025747 pub. moved from Sweden to Kansas Faxonc title no. 030902 Additional Formats: Also available to subscribers via the World Wide Web. Ambio (Online) (OCoLC)47720314 Subjects: Ecology--Periodicals. Pollution--Environmental aspects--Periodicals. Air Pollution--Periodicals. Environment--Periodicals. Environmental Health--Periodicals. Water Pollution--Periodicals. LC Classification: QH540 .A52 NLM Class No.: W1 AM103K Dewey Class No.: 301.31/05 Postal Registration No.: 008102 USPS

American justice. A civil action / A&E Television Network. Portion of Title: Civil action Published/Created: United States: A&E Television Network, 1998. United States. Related Names: Kurtis, Bill, host. Arts and Entertainment Network. Copyright Collection (Library of Congress) Description: 1 videocassette of 1 (VHS) (ca. 50 min.): sd., col.; 1/2 in. viewing copy. Summary: Documents the ground water contamination case in the book by the same title written by Jonathan Harr. Notes: Not viewed. Sources used: OCLC #40742944. Performer Notes: Host, Bill Kurtis. Acquisition Source: Received: 4/21/1999; viewing copy; copyright deposit--407; Copyright Collection. Subjects: Groundwater--Pollution--Law and legislation--Massachusetts --Woburn. Form/Genre: Documentary. migfg LC Classification: VAG 4920 (viewing copy)

America's defense monitor. Water, land, people, conflict / [Center for Defense Information]. Portion of Title: Water, land, people, conflict Published/Created: United States: [Center for Defense Information], 1998. United States. Related Names: Copyright Collection (Library of Congress) Description: 1 videocassette of 1 (ca. 29 min.): sd., col.; 1/2 in. viewing copy. Summary: Today, the greatest threats facing any nation's security may not be military threat. Increasingly, they are complex issues related to the environment such as: population growth, water scarcity, pollution, and economic stability. Notes: Copyright: no reg. Show number 1143. On CDI data base: Produced: July 5, 1998. Summary taken from CDI data base. Source used: MBRS accessioning materials; CDI data base. Acquisition Source: Received: 11-6-1998; viewing copy; copyright deposit--407; Copyright Collection. Form/Genre: Documentary--Television series. migfg LC Classification: VAG 1675 (viewing copy)

Ammonia in the aquatic environment / Environment Canada, Health Canada. Published/Created: Ottawa: Environment Canada, c2000. Related Names: Canada. Environment Canada. Canada. Health Canada. Description: viii, 96 p.: ill., maps; 28 cm. ISBN: 0662291921 Notes: Issued also in French under title: Ammoniac dans le milieu aquatique. At head of title:

Canadian Environmental Protection Act, 1999. Includes bibliographical references: p. 77-90. Co-published by: Health Canada. Additional Formats: Available also on the Internet. Subjects: Ammonia--Toxicology--Canada. Ammonia--Environmental aspects--Canada. Aquatic organisms--Effect of water pollution on--Canada. Plants--Effect of water pollution on--Canada. Environmental monitoring--Canada. Ammoniac--Toxicologie--Canada. Ammoniac--Aspect de l'environnement--Canada. Organismes aquatiques, Effets de la pollution de l'eau sur les--Canada. Plantes, Effets de la pollution de l'eau sur les--Canada. Environnement--Surveillance--Canada. Series: Priority substances list assessment report LC Classification: TD887.A66 A46 2000 Canadian Class No.: TD195* TD195 A44 A45 2000 COP.CA.2.2002-401 Dewey Class No.: 363.738/4 21 Government Document No.: En40-215/55E National Bibliographic Agency No.: 009803289

An Annual report of the surveillance program of North Carolina as it relates to the environmental factors of air and water / conducted by the North Carolina State Board of Health, Sanitary Engineering Division, [and] Laboratory Division in cooperation with participating municipalities and local county health departments. Cover Title: Environmental surveillance Published/Created: [Raleigh, N.C.: The Board Related Names: North Carolina. Sanitary Engineering Division. North Carolina. State Board of Health. Laboratory Division. Description: v.: maps; 28 cm. Current Frequency: Annual Continued by: Environmental radiation surveillance program (OCoLC)9201031 Notes: Description based on: 1963. Latest issue consulted: 1964. SERBIB/SERLOC merged record Subjects: Radioactive pollution--North Carolina--Periodicals. Air quality management--North Carolina--Periodicals. Water quality management--

North Carolina--Periodicals. Environmental monitoring--North Carolina--Periodicals. LC Classification: TD196.R3 A56 Dewey Class No.: 363.7/32/09756 19

An assessment of recovery and key processes affecting the response of surface waters to reduced levels of acid precipitation in the Adirondack and Catskill Mountains: final report / prepared for the New York State Energy Research and Development Authority; prepared by U.S. Geological Survey ... [et al.]. Portion of Title: Recovery and key processes affecting the response of surface waters to reduced levels of acid precipitation in the Adirondack and Catskill Mountains Published/Created: [Albany, N.Y.]: NYSERDA, [2005] Related Names: New York State Energy Research and Development Authority. Geological Survey (U.S.) Description: 1 v. (various pagings): col. ill.; 28 cm. Notes: "NYSERDA 6486; July 2005." Includes bibliographical references. Additional Formats: Also available in electronic format. Subjects: Acid pollution of rivers, lakes, etc.--New York (State) --Adirondack Mountains. Acid pollution of rivers, lakes, etc.--New York (State) --Catskill Mountains. Water quality--New York (State)--Adirondack Mountains. Water quality--New York (State)--Catskill Mountains. Series: NYSERDA report; 05-03 LC Classification: TD427.A27 A87 2005 Dewey Class No.: 363.738/609747 22 Government Document No.: ENE 800-4 ASSRK 205-8513 nydocs

An Assessment of regional ground-water contamination in Illinois / by John A. Helfrich ... [et al.]; prepared for the Illinois Hazardous Waste Research and Information Center. Published/Created: Savoy, Ill. (1808 Woodfield Dr., Savoy 61874): The Center, [1988] Related Names: Helfrich, John A. Illinois.

Hazardous Waste Research and Information Center. Illinois State Water Survey. Description: xii, 46 p.: ill., maps; 28 cm. Notes: "Illinois State Water Survey." "Printed May 1988." "HWRIC Project 85-005." "HWRIC RR 023." Includes bibliographical references (p. 43-46). Subjects: Groundwater--Pollution--Illinois. Water--Pollution--Illinois. LC Classification: MLCM 91/12952 (T) Dewey Class No.: 363.7394

An Economic analysis of phosphorus control and other aspects of R76-1 / prepared by James E. Ciecka ... [et al.]. Published/Created: [Chicago]: State of Illinois, Institute for Environmental Quality, 1978. Related Names: Ciecka, James E. Description: v, 120 p.: graphs; 28 cm. Notes: "Project no. 80.063." Includes bibliographies. Subjects: Water--Pollution--Illinois--Economic aspects. Sewage--Purification--Phosphate removal--Economic aspects. Series: IIEQ document; no. 78/16 LC Classification: TD224.I3 E25 Dewey Class No.: 614.7/72 Geographic Area Code: n-us-il

An Economic analysis of proposed amendments to water pollution regulations phosphorus discharges R87-6: final report / prepared by Blaser, Zeni, and Company; principal investigators, William L. Blaser ... [et al.]; prepared for Illinois Department of Energy and Natural Resources, Office of Research and Planning. Published/Created: Springfield, IL (325 W. Adams, Room 300, Springfield 62704-1892): The Office, [1989] Related Names: Blaser, William L. Blaser, Zeni, and Company. Illinois. Dept. of Energy and Natural Resources. Office of Research and Planning. Description: x, 132, [45] p.; 28 cm. Notes: "ILENR/RE-EA-89/03." Includes bibliographical references. Subjects: Water--Pollution--Law and legislation--Compliance costs --Illinois. Phosphorus compounds--Environmental aspects--Government policy--Illinois. LC Classification: HC107.I33 W3225 1989 Geographic Area Code: n-us-il

An Economic impact analysis of proposed amendments to public water supply regulations pending before the Illinois Pollution Control Board R85-14: final report / prepared by Blaser, Zeni and Company; principal investigator, William L. Blaser; prepared for Illinois Department of Energy and Natural Resources, Energy and Environmental Affairs Division. Published/Created: Springfield, IL: The Division, [1987] Related Names: Blaser, William L. Blaser, Zeni, and Company. Illinois. Division of Energy and Environmental Affairs. Description: 1 v. (various pagings); 28 cm. Notes: "ILENR/RE-EA-87/13." Includes bibliographical references. Subjects: Water--Pollution--Law and legislation--Economic aspects --Illinois. LC Classification: HC107.I33 W3227 1987 Dewey Class No.: 363.73/94/09773 20 Geographic Area Code: n-us-il

An Economic impact study of proposed amendments to Illinois administrative code 604.203 and 605.104 of subtitle F public water supplies (trihalomethanes), R84-12: final report / prepared by Blaser, Zeni and Company; principal investigators, William L. Blaser ... [et al.]; prepared for Illinois Department of Energy and Natural Resources, Office of Research and Planning. Published/Created: Springfield, IL (325 W. Adams, Room 300, Springfield 62704-1892): The Office, [1989] Related Names: Blaser, William L. Blaser, Zeni, and Company. Illinois. Dept. of Energy and Natural Resources. Office of Research and Planning. Description: ii, xiv, 105, [63] p.; 28 cm. Notes: "ILENR/RE-EA-89/02." Includes bibliographical references. Subjects: Water--Pollution--Law and legislation--Compliance costs --Illinois. Trihalomethanes--Environmental aspects--

Government policy --Illinois. LC Classification: HC107.I33 W3238 1989 Geographic Area Code: n-us-il

An Evaluation of stream liming effects on water quality and spawning of migratory fishes in Maryland coastal plain streams: 1988 results: final report / prepared by H.S. Greening ... [et al.]; prepared for Living Lakes, Inc. and Maryland Department of Natural Resources, Chesapeake Bay Research and Monitoring Division, Tidewater Administration. Published/Created: Annapolis, Md.: Chesapeake Bay Research and Monitoring, 1989. Related Names: Greening, Holly. Living Lakes, Inc. Maryland. Chesapeake Bay Research and Monitoring Division. Description: xix, 211 p.: ill., maps; 28 cm. Notes: "AD-89-5." Bibliography: p. 207-211. Subjects: Fishes--Effect of water quality on--Maryland. Fishes--Effect of water pollution on--Maryland. Acid pollution of rivers, lakes, etc.--Maryland. Aquatic liming--Maryland. Water quality bioassay--Maryland. LC Classification: SH174 .E93 1989

An Evaluation of the effects of the North Carolina phosphate detergent ban / North Carolina Department of Environment, Health, and Natural Resources, Division of Environmental Management, Water Quality Section. Published/Created: [Raleigh]: The Department, [1991] Related Names: North Carolina. Division of Environmental Management. Water Quality Section. Description: ii, 42 p.: ill.; 28 cm. Notes: "May 1991." Includes bibliographical references (p. 37-38). Subjects: Detergent pollution of rivers, lakes, etc.--North Carolina. Phosphates--Environmental aspects--North Carolina. Water quality--North Carolina. Series: Report / N.C. Dept. of Environment, Health, and Natural Resources, Division of Environmental Management; no. 91-04 Report (North Carolina. Division of

Environmental Management); no. 91-04. LC Classification: TD427.D4 E95 1991 Dewey Class No.: 363.73/946 20 Geographic Area Code: n-us-nc

Analysis of avian mortality at the North Shore restoration area of Lake Apopka in 1998-1999 / prepared for St. Johns River Water Management District; prepared by Exponent. Published/Created: Bellevue, WA (15375 SE 30th Pl., Bellevue 98007): Exponent, [2003] Related Names: St. Johns River Water Management District (Fla.) Exponent (Firm: Bellevue, Wash.) Description: 1 v. (various pagings): ill. (some col.), col. maps; 28 cm. Notes: Includes bibliographical references. Subjects: Birds--Mortality--Florida--Apopka, Lake. Birds--Effect of water pollution on--Florida--Apopka, Lake. Series: Special publication; SJ2004-SP1 Special publication (St. Johns River Water Management District (Fla.)); SJ2004-SP1. LC Classification: QL684.F6 A69 2003 Dewey Class No.: 598.1727/3/09759 22 Geographic Area Code: n-us-fl

Analysis of nutrient and ancillary water-quality data for surface and ground water of the Willamette Basin, Oregon, 1980-90 / by Bernadine A. Bonn ... [et al.]; prepared as part of the National Water-Quality Assessment Program. Published/Created: Portland, Or.: U.S. Geological Survey; Denver, Colo.: U.S. Geological Survey, Earth Science Information Center, Open-File Reports Section [distributor], 1995. Related Names: Bonn, Bernadine. National Water Quality Assessment Program (U.S.) Geological Survey (U.S.) Description: xi, 88 p.: ill., maps (some col.); 28 cm. Notes: "U.S. Department of the Interior, U.S. Geological Survey"--T.p. verso. Shipping list no.: 96-0080-P. One transparency in pocket. Includes bibliographical references (p. 83-84). Subjects: Nutrient pollution of water--Oregon--Willamette River Watershed. Nutrient pollution of water--

Oregon--Sandy River Watershed. Water quality--Oregon--Willamette River Watershed. Water quality--Oregon--Sandy River Watershed. Series: Water-resources investigations report; 95-4036 LC Classification: TD427.N87 A53 1995 NAL Class No.: GB701.W375 no.95-4036 Dewey Class No.: 553.7/0973 s 363.73/942/097953 21 Government Document No.: I 19.42/4:95-4036

Animal waste management team: Section 319, Nonpoint Source Control Program, project final report / developed by South Dakota Association of Conservation Districts in cooperation with Natural Resources Conservation Service. Published/Created: South Dakota: South Dakota Dept. of Environment and Natural Resources, [1998] Related Names: South Dakota Association of Conservation Districts. United States. Natural Resources Conservation Service. Description: 1 v. (various foliations): ill., maps; 28 cm. Notes: Cover title. "May 1, 1998." Includes bibliographical references. Subjects: Nonpoint Source Control Program (S.D.) Animal waste--Environmental aspects--South Dakota. Water--Pollution--South Dakota. Nonpoint source pollution--South Dakota. Water quality management--South Dakota. LC Classification: TD930.2 .A55 1998 Geographic Area Code: n-us-sd

Annotated bibliography of PLUARG reports. Published/Created: Windsor, Ont.: International Joint Commission, Great Lakes Regional Office, 1979. Related Names: International Reference Group on Great Lakes Pollution from Land Use Activities. International Joint Commission. Great Lakes Regional Office. Related Titles: Annotated bibliography of P.L.U.A.R.G. reports. Description: xii, 121 p.; 28 cm. Notes: At head of title: International Reference Group on Great Lakes Pollution from Land Use Activities (PLUARG). Subjects: International

Reference Group on Great Lakes Pollution from Land Use Activities--Bibliography--Catalogs. Water--Pollution--Great Lakes (North America)--Bibliography --Catalogs. LC Classification: Z5862.2.W3 A56 1979 TD223.3 Dewey Class No.: 016.3637/394 19 National Bibliography No.: C*** Geographic Area Code: nl-----

Annotated bibliography on water, sanitation, and diarrhoeal diseases: roles and relationships. Published/Created: Dhaka, Bangladesh: International Centre for Diarrhoeal Disease Research, Bangladesh, c1986. Related Names: Khan, Md. Shamsul Islam. International Centre for Diarrhoeal Disease Research, Bangladesh. Description: iii, 186 p.; 26 cm. Notes: "December 1986." Compilation and documentation: M. Shamsul Islam Khan ... et al. Subjects: Water--Pollution--Bibliography. Diarrhea--Etiology--Bibliography. Water-borne infections--Bibliography. Series: Specialized bibliography series; no. 12 Specialized bibliography series (Dhaka, Bangladesh) LC Classification: Z5862.2.W3 A57 1986 RA591 Dewey Class No.: 016.6163/427071 20 Overseas Acquisitions No.: B E 4736

Annual pollution report: air emissions and water discharges / Minnesota Pollution Control Agency. Portion of Title: Air emissions and water discharges. Published/Created: St. Paul, Minn.: Minnesota Pollution Control Agency, 2000- Related Names: Minnesota Pollution Control Agency. Description: v.: col. ill., maps; 28 cm. 2000- Current Frequency: Annual Subjects: Air--Pollution--Minnesota--Statistics--Periodicals. Water--Pollution--Minnesota--Statistics--Periodicals. LC Classification: TD883.5.M5 A685

Annual progress report. Published/Created: Springfield, Ill.: Illinois Dept. of Agriculture, Bureau of Soil and Water

Conservation, [1993- Related Names: Illinois. Bureau of Soil and Water Conservation. Description: v.: ill.; 28 cm. Vols. for 1993- called also 13th- 1993- Current Frequency: Annual Continues: Annual progress report (Illinois. Division of Natural Resources) 0884-433X (DLC) 85645432 (OCoLC)12351796 Notes: Title from cover. "Activities completed toward reducing water pollution resulting from agricultural sources of soil erosion." SERBIB/SERLOC merged record ACQN: aq 98005411 13th,1993-17th,1997; Subjects: Water--Pollution--Illinois--Periodicals. Soil erosion--Illinois--Periodicals. Agricultural pollution--Illinois--Periodicals. LC Classification: TD224.I3 A795 Dewey Class No.: 363.7/394 19

Annual progress report. Serial Key Title: Annual progress report - Illinois. Division of Natural Resources Abbreviated Title: Annu. prog. rep. - Ill., Div. Nat. Resour. Published/Created: [Springfield, Ill.]: Illinois Dept. of Agriculture, Division of Natural Resources, Related Names: Illinois. Division of Natural Resources. Description: v.: ill.; 28 cm. Began with issue for 1981. Current Frequency: Annual Continues: Progress report (OCoLC)8864382 Continued by: Annual progress report (Illinois. Bureau of Soil and Water Conservation) (DLC) 94640698 (OCoLC)28947368 ISSN: 0884-433X Notes: Description based on: 1984; title from cover. SERBIB/SERLOC merged record Acquisition Source: Illinois Dept. of Agriculture, Division of Natural Resources, Illinois State Fairgrounds, Springfield, Ill. 62702 Subjects: Water--Pollution--Illinois--Periodicals. Soil erosion--Illinois--Periodicals. Agricultural pollution--Illinois--Periodicals. LC Classification: TD224.I3 A795 Dewey Class No.: 363.7/394 19

Annual report / Long Island Sound Study.
Other Title: Long Island Sound Study ...
annual report Published/Created: [S.l.]:
Long Island Sound Study, [1986?]- Related
Names: Long Island Sound Study
(Organization) Description: v.: ill.; 28 cm.
1986- Current Frequency: Annual Notes:
Title from cover. Latest issue consulted:
1989/1990. Occasional accompanied by
errata sheet. SERBIB/SERLOC merged
record Subjects: Estuarine animals--Effect
of water pollution on--Long Island Sound
(N.Y. and Conn.)--Periodicals. Poisons--
Long Island Sound (N.Y. and Conn.)--
Periodicals. Toxicology--Long Island
Sound (N.Y. and Conn.) --Periodicals.
Anoxemia--Long Island Sound (N.Y. and
Conn.)--Periodicals. Estuarine area
conservation--Long Island Sound (N.Y.
and Conn.)--Periodicals. LC Classification:
QL195 .A56 Dewey Class No.: 577.7/3461

Annual report on the Maryland Water Pollution
Control Fund. Published/Created:
Baltimore, Md.: Maryland Dept. of the
Environment, Related Names: Maryland.
Dept. of the Environment. Description: v.;
28 cm. Current Frequency: Annual Notes:
Description based on: Jan. 1988; title from
cover. SERBIB/SERLOC merged record
Subjects: Maryland Water Pollution
Control Fund. Water--Pollution--Maryland.
Water pollution control industry--
Maryland. LC Classification: HC107.M33
W322

Annual report on the Virginia Water Quality
Improvement Fund, Nonpoint Source
Program. Portion of Title: Virginia Water
Quality Improvement Fund, Nonpoint
Source Program Other Title: Annual report
on the Virginia WQIA fund Cover Title:
Report of the Department of Conservation
and Recreation, annual report on the
Virginia Water Quality Improvement Fund
Nonpoint Source Program to the Governor
and the General Assembly of Virginia
Published/Created: Richmond, Va.: Dept.

of Conservation and Recreation, 1998-
Related Names: Virginia Water Quality
Improvement Fund. Nonpoint Source
Program. Virginia. Dept. of Conservation
and Recreation. Description: v.: forms,
maps; 28 cm. 1998- Current Frequency:
Annual Continued by: Annual report on the
Virginia Water Quality Improvement
Fund. Cooperative Point Source Pollution
Control Program (DLC) 2003201413
(OCoLC)53832083 Notes: "Report by ...
director of the Virginia Department of
Conservation and Recreation to the
Governor, Commonwealth of Virginia, and
the Virginia General Assembly."
SERBIB/SERLOC merged record ACQN:
aq 98012613 1998; Commonwealth of
Virginia and the Virginia General
Assembly; Senate Document No. 21; NST
- 7/22/98. Subjects: Virginia Water Quality
Improvement Fund. Nonpoint Source
Program--Periodicals. Water quality
management--Virginia--Periodicals.
Nonpoint source pollution--Virginia--
Periodicals. Series: Senate document
Senate document (Virginia. General
Assembly. Senate) LC Classification: J87
.V9 date b subser TD224.V8

Annual report on the Virginia Water Quality
Improvement Fund. Cooperative Point
Source Pollution Control Program. Portion
of Title: Cooperative Point Source
Pollution Control Program Point Source
Pollution Control Other Title: Water
Quality Improvement Fund annual report
Cover Title: Annual report on the Virginia
Water Quality Improvement Fund. Point
Source Pollution Control
Published/Created: [Richmond]: Dept. of
Environmental Quality Related Names:
Virginia Water Quality Improvement
Fund. Cooperative Point Source Pollution
Control Program. Virginia. Dept. of
Environmental Quality. Description: v.; 28
cm. Current Frequency: Annual Continues:
Annual report on the Virginia Water
Quality Improvement Fund, Nonpoint

Source Program (DLC) 98641004
(OCoLC)38501540 Notes: "Report by ...
director, Virginia Department of
Environmental Quality to the Governor,
Commonwealth of Virginia, and the
Virginia General Assembly." Description
based on: 2001. Subjects: Virginia Water
Quality Improvement Fund. Cooperative
Point Source Pollution Control Program--
Periodicals. Water quality management--
Virginia--Periodicals. Nonpoint Source
Pollution--Virginia--Periodicals. LC
Classification: J87 .V9

Annual water quality data report for the Waste
Isolation Pilot Plant. Cover Title: Annual
water quality data report
Published/Created: Carlsbad, NM: U.S.
Dept. of Energy Related Names: United
States. Dept. of Energy. Waste Isolation
Pilot Plant (N.M.) Description: v.: ill.,
maps; 28 cm. Vol. for Mar. 1988 called
also May 1988 on cover. Began in 1986.
Current Frequency: Annual
Cancelled/Invalid LCCN: sn 88040040
Notes: Description based on: 1987. Latest
issue consulted: Mar. 1988.
SERBIB/SERLOC merged record
Subjects: Waste Isolation Pilot Plant
(N.M.)--Periodicals. Radioactive waste
disposal--Environmental aspects--New
Mexico--Eddy County--Measurement--
Periodicals. Groundwater--Quality--New
Mexico--Eddy County--Measurement --
Periodicals. Radioactive pollution of
water--New Mexico--Measurement --
Periodicals. Water quality--New Mexico--
Measurement--Periodicals. LC
Classification: TD427.R3 A56 Dewey
Class No.: 363.73/942/0978942 20
Government Document No.: E 1.28/15:

Application of PCR technologies for virus
detection in groundwater / prepared by
Morteza Abbaszadegan ... [et al.].
Published/Created: Denver, CO: AWWA
Research Foundation: American Water
Works Association, c1998. Related Names:

Abbaszadegan, Morteza. AWWA Research
Foundation. Description: xxi, 60 p.: ill.
(some col.); 28 cm. ISBN: 0898679346
(recycled paper) Notes: "Sponsored by
AWWA Research Foundation." Includes
bibliographical references (p. 55-57).
Subjects: Viral pollution of water.
Enteroviruses--Analysis. Polymerase chain
reaction. Groundwater--Analysis. LC
Classification: TD427.V55 A66 1998
Dewey Class No.: 628.1/61 21

Applied isotope hydrogeology: a case study in
Northern Switzerland / F.J. Pearson, Jr. ...
[et al.] with contributions by J.N. Andrews
... [et al.]. Published/Created: Amsterdam;
New York: Elsevier; New York, NY,
U.S.A.: Distributors for the U.S. and
Canada, Elsevier Science Pub. Co., 1991.
Related Names: Pearson, F. J. Nationale
Genossenschaft fuîr die Lagerung
Radioaktiver Abfaîlle (Switzerland)
Description: xxiv, 439, A19 p.: ill., maps;
25 cm. ISBN: 0444889833 (acid-free
paper) Notes: Three folded maps in pocket.
Report on isotope data gathered by a
program sponsored by the Nationale
Genossenschaft fuîr die Lagerung
Radioaktiver Abfaîlle. "May 1991"--Ser.
t.p. Includes bibliographical references (p.
421-439). Subjects: Radioactive waste
disposal in the ground--Environmental
aspects--Switzerland. Radioactive tracers
in water pollution research --Switzerland.
Groundwater flow--Switzerland--
Measurement. Series: Studies in
environmental science; 43 Technical
report; 88-01 Technischer Bericht NTB;
88-01. LC Classification: TD898.13.S9
A66 1991 Dewey Class No.: 628.1/685 20
Geographic Area Code: e-sz---

Appropriate technology in water supply and
waste disposal: proceedings of a session
sponsored by the Research Council on
Environmental Impact Analysis of the
ASCE Technical Council on Research at
the ASCE National Convention in

Chicago, Illinois, October 16-20, 1978 / Charles G. Gunnerson, and John M. Kalbermatten, editors. Published/Created: New York: American Society of Civil Engineers, c1979. Related Names: Gunnerson, Charles G. Kalbermatten, John M. ASCE Technical Council on Research. Environmental Impact AnalysisResearch Council. Description: 270 p., [1] leaf of plates: ill.; 22 cm. Notes: "Originally published as ASCE preprint 3453." Includes bibliographical references and indexes. Subjects: Water-supply engineering--Congresses. Sewage disposal--Congresses. Sanitary engineering--Congresses. Pollution--Congresses. Water-supply--Developing countries--Congresses. Sewage disposal--Developing countries--Congresses. Sanitary engineering--Developing countries--Congresses. LC Classification: TD201 .A68 Dewey Class No.: 628/.09172/4 Geographic Area Code: d------

Aquatic and surface photochemistry / edited by George R. Helz, Richard G. Zepp, Donald G. Crosby. Published/Created: Boca Raton: Lewis Publishers, c1994. Related Names: Helz, G. R. Zepp, Richard G. Crosby, Donald G. Description: 552 p.: ill.; 25 cm. ISBN: 0873718712 (acid-free paper) Notes: Includes bibliographical references and index. Subjects: Water--Pollution. Photochemistry. Water--Purification--Photocatalysis. LC Classification: TD423 .H45 1994 Dewey Class No.: 628.1/68 20 Electronic File Information: Publisher description http://www.loc.gov/catdir/enhancements/fy 0744/93022832-d .html

Aquatic ecosystems / [edited] by Arvind Kumar. Published/Created: New Delhi: A.P.H. Pub. Corp., 2003. Related Names: Kumar, Arvind, 1953- Description: xii, 437 p.: ill.; 25 cm. ISBN: 8176484547 Summary: In the Indian context; contributed articles. Notes: Includes

bibliographical references and index. Subjects: Aquatic ecology--India. Water--Pollution--India. Water quality management--India. LC Classification: QH541.5.W3 A677 2003 Overseas Acquisitions No.: I-E-2003-305900; 49-90

Aquatic ecotoxicology: fundamental concepts and methodologies / editors, Alain Boudou, Francis Ribeyre. Published/Created: Boca Raton, Fla.: CRC Press, c1989. Related Names: Boudou, Alain, 1949- Ribeyre, Francis. Description: 2 v.: ill.; 27 cm. ISBN: 0849348285 (v. 1) 0849348293 (v. 2) Notes: Includes bibliographies and indexes. Subjects: Aquatic ecology. Water--Pollution--Toxicology. Water--Pollution--Environmental aspects. Environmental monitoring. LC Classification: QH541.5.W3 A68 1989 Dewey Class No.: 574.5/263 19

Aquatic environment / [editor], Ashutosh Gautam. Published/Created: New Delhi: Ashish Pub. House, 1992. Related Names: Gautam, Ashutosh, 1965- Description: 144 p.: ill., maps; 23 cm. ISBN: 8170244870: Summary: Contributed articles covering various aspects of ecology, pollution, and conservation of aquatic environment and resources in India. Notes: Includes index. Includes bibliographical references. Subjects: Aquatic ecology--India. Water--Pollution--Environmental aspects--India. Aquatic resources conservation--India. Water conservation--India. Stream conservation--India. LC Classification: QH183 .A684 1992 Dewey Class No.: 574.5/263/0954 20 Overseas Acquisitions No.: I-E-70050 Geographic Area Code: a-ii---

Aquatic mesocosm studies in ecological risk assessment / edited by Robert L. Graney, James H. Kennedy, John H. Rodgers. Published/Created: Boca Raton: Lewis Publishers, c1994. Related Names: Graney, Robert L. Kennedy, James H., 1947-

Rodgers, John H. (John Hasford), 1950-
Description: 723 p.: ill.; 25 cm. ISBN:
0873715926 Notes: "A symposium ... held
at the 11th Annual Meeting of the Society
of Environmental Toxicology and
Chemistry, in Arlington, Virginia in
November of 1990"--Chapter 1, introd.
Includes bibliographical references and
index. Subjects: Ecological risk
assessment--Congresses. Aquatic ecology--
Congresses. Pesticides--Risk assessment--
Congresses. Water--Pollution--Congresses.
Series: SETAC special publications series
LC Classification: QH541.15.R57 A68
1994 Dewey Class No.: 574.5/263 20
Electronic File Information: Publisher
description
http://www.loc.gov/catdir/enhancements/fy
0744/93003144-d .html

Aquatic pollutants and biologic effects, with
emphasis on neoplasia / edited by H. F.
Kraybill ... [et al.]. Published/Created: New
York: New York Academy of Sciences,
1977. Related Names: Kraybill, H. F.
(Herman Fink), 1914- New York Academy
of Sciences. Description: 604 p.: ill.; 23
cm. ISBN: 0890720444 Notes: "This series
of papers is the result of a conference
entitled aquatic pollutants and biologic
effects with emphasis on neoplasia, held by
the New York Academy of Sciences on
September 27, 28, and 29, 1976." Includes
bibliographical references. Subjects:
Aquatic animals--Effect of water pollution
on--Congresses. Veterinary oncology--
Congresses. Carcinogens--Congresses.
Water--Pollution--Toxicology--
Congresses. Water pollutants--Congresses.
Carcinogens, Environmental--Congresses.
Neoplasms--Veterinary--Congresses.
Neoplasms--Etiology--Congresses. Series:
Annals of the New York Academy of
Sciences; v. 298 Annals of the New York
Academy of Sciences; v. 298. LC
Classification: Q11 .N5 vol. 298
QP82.2.W36 NLM Class No.: W1

AN626YL v. 298 WA689 A656 1976
Dewey Class No.: 508/.1 s 591.2/4

Aquatic pollution and toxicology / [editor],
R.K. Trivedy. Edition Information: 1st ed.
Published/Created: Jaipur: ABD
Publishers: Distribution, Oxford Book Co.,
2001. Related Names: Trivedy, R. K.
Description: 240 p.: ill., maps; 23 cm.
ISBN: 8185771189 Summary: With
reference to India; contributed articles.
Notes: Includes bibliographical references.
Subjects: Water--Pollution--Environmental
aspects--India. Water--Pollution--
Toxicology--India. Aquatic organisms--
Effect of water pollution on. Fishes--Effect
of water pollution on. LC Classification:
QH545.W3 A634 2001 Overseas
Acquisitions No.: I-E-2001-416255; 35-90

Aquatic pollution studies, 1902-1966:
commemorating the fiftieth anniversary of
the National Research Council of Canada.
Published/Created: Ottawa: Fisheries
Research Board of Canada, 1966. Related
Names: National Research Council of
Canada. Fisheries Research Board of
Canada. Description: ca. 400 p.: ill.; 26 cm.
Notes: Includes bibliographical references.
Subjects: National Research Council of
Canada. Fishes--Effect of water pollution
on. Water--Pollution--Environmental
aspects--Canada. LC Classification: SH174
.A68 Dewey Class No.: 597/.024 19
Geographic Area Code: n-cn---

Aquatic sciences and fisheries abstracts. ASFA
3, Aquatic pollution and environmental
quality. Variant Title: Aquatic sciences and
fisheries abstracts from CSA. ASFA 3,
Aquatic pollution and environmental
quality <Feb. 2005-> Portion of Title:
Aquatic pollution and environmental
quality Serial Key Title: Aquatic sciences
and fisheries abstracts. ASFA 3, Aquatic
pollution and environmental quality
Abbreviated Title: Aquat. sci. fish. abstr.,
ASFA 3 Aquat. pollut. environ. qual.
Published/Created: Bethesda, MD:

Cambridge Scientific Abstracts, 1990-
Related Names: United Nations. Office for
Ocean Affairs and the Law of the Sea.
Description: v.; 23 cm. Vol. 20, no. 1 (Feb.
1990)- Current Frequency: Bimonthly
Continues in part: Aquatic sciences &
fisheries abstracts 0044-8516 (DLC)
74648034 (OCoLC)1796271 ISSN: 1045-
6031 Notes: Title from cover. Latest issue
consulted: Vol. 34, no. 1 (Feb. 2005)
(surrogate). Compiled by: United Nations
Office for Ocean Affairs and the Law of
the Sea and other United Nations agencies
with the collaboration of national research
centers worldwide. Complements: Aquatic
sciences and fisheries abstracts. ASFA 1,
Biological sciences and living resources,
and: Aquatic sciences and fisheries
abstracts. ASFA 2, Ocean technology,
policy and non-living resources. Additional
Formats: Also published online. Aquatic
sciences and fisheries abstracts. ASFA 3,
Aquatic pollution and environmental
quality (Online) 1555-6247 Related Item:
Aquatic sciences and fisheries abstracts.
ASFA 1, Biological sciences and living
resources 0140-5373 (DLC)sn 80002939
(OCoLC)3924687 Subjects: Aquatic
organisms--Effect of water pollution on--
Abstracts --Periodicals. Water--Pollution--
Abstracts--Periodicals. Conservation of
Natural Resources--Abstracts.
Conservation of Natural Resources--
Periodicals. Environmental Monitoring--
Abstracts. Environmental Monitoring--
Periodicals. Environmental Pollution--
Abstracts. Environmental Pollution--
Periodicals. Fisheries--Abstracts. Fisheries-
-Periodicals. Marine Biology--Abstracts.
Marine Biology--Periodicals. Water
Pollution--Abstracts. Water Pollution--
Periodicals. LC Classification: QH545.W3
A636 NLM Class No.: Z 5322.M3 A6567a
NAL Class No.: Z5322.M3A65 Dewey
Class No.: 551 11 Postal Registration No.:
008461 USPS National Bibliographic
Agency No.: 9208454 DNLM SR0066738
DNLM

Aquatic toxicity from in-situ oil burning /
prepared by EVS Consultants for
Emergencies Science Division,
Environmental Technology Centre,
Environment Canada. Variant Title:
Newfoundland offshore burn experiment
Cover Title: NOBE, Newfoundland
offshore burn experiment
Published/Created: [S.l.]: Marine Spill
Response Corporation, 1995. Related
Names: E.V.S. Consultants. Environmental
Technology Centre (Canada). Emergencies
Science Division. Description: xii, 51 p.;
28 cm. ISBN: 066222891X Notes: "En40-
487/1995-E". Includes bibliographical
references (p. 50-51). Summary in French.
Subjects: Oil spills--Newfoundland and
Labrador--Management. In situ
remediation--Newfoundland and Labrador.
Aquatic organisms--Effect of oil spills on--
Newfoundland and Labrador. Oil pollution
of water. Oil spills--Environmental aspects.
LC Classification: TD427.P4 A68 1995
Language Code: eng fre

Aquatic toxicology / edited by Jerome O.
Nriagu. Published/Created: New York:
Wiley, c1983. Related Names: Nriagu,
Jerome O. Description: xiii, 525 p.: ill.; 24
cm. ISBN: 0471889016: Notes: "A Wiley-
Interscience publication." Includes
bibliographies and index. Subjects: Water--
Pollution--Environmental aspects. Water--
Pollution--Toxicology. Series: Advances in
environmental science and technology; v.
13 LC Classification: TD180 .A38 vol. 13
QH545.W3 Dewey Class No.: 628 s
574.5/263 19

Aquatic toxicology / editor, Lavern J. Weber.
Published/Created: New York: Raven
Press, c1982-<c1984 > Related Names:
Weber, Lavern J. Description: v. <1-2 >:
ill.; 25 cm. ISBN: 0890044392 (v. 1)
Notes: Includes bibliographies and
indexes. Subjects: Aquatic organisms--
Effect of water pollution on. Water--
Pollution--Toxicology. Water quality

bioassay. Toxicology--Period. Water
pollutants--Toxicity--Period. LC
Classification: QH545.W3 A65 1982 NLM
Class No.: W1 AQ927 Dewey Class No.:
574.2/4 19

Aquatic toxicology [microform].
Published/Created: Amsterdam,
Netherlands: Elsevier/North Holland
Biomedical Press, c1981- Description: v.:
ill.; 24 cm. Vol. 1, no. 1 (Apr. 1981)-
Current Frequency: 8 issues a year, <1983-
> Former Frequency: Six no. a year ISSN:
0166-445X Notes: ACQN: aq 94008562
Additional Formats: Microfilm. Lausanne,
Switzerland: Elsevier Sequoia, S.A.:
University Microfilms International
[distributor]. microfilm reels; 35 mm.
Subjects: Aquatic animals--Effect of water
pollution on--Periodicals. Water--
Pollution--Environmental aspects--
Periodicals. Toxicology--Periodicals.
Water Pollutants--toxicity--Periodicals. LC
Classification: Microfilm (o) 89/2737

Aquatic toxicology [microform].
Published/Created: Amsterdam,
Netherlands: Elsevier/North Holland
Biomedical Press, c1981- Description: v.:
ill.; 24 cm. Vol. 1, no. 1 (Apr. 1981)-
Current Frequency: 8 issues a year, <1983-
> Former Frequency: Six no. a year ISSN:
0166-445X Notes: SERBIB/SERLOC
merged record Additional Formats:
Microfilm. Ann Arbor, Mich.: University
Microfilms International. microfilm reels;
35 mm. Subjects: Aquatic animals--Effect
of water pollution on--Periodicals. Water--
Pollution--Environmental aspects--
Periodicals. Toxicology--Periodicals.
Water Pollutants--toxicity--Periodicals. LC
Classification: Microfilm (o) 89/2737

Aquatic toxicology and environmental fate: a
symposium / sponsored by ASTM
Committee E-47 on Biological Effects and
Environmental Fate. Serial Key Title:
Aquatic toxicology and environmental fate
Abbreviated Title: Aquat. toxicol. environ.

fate Published/Created: Philadelphia, PA:
ASTM, c1986-c1988. Related Names:
ASTM Committee E-47 on Biological
Effects and Environmental Fate.
Symposium on Aquatic Toxicology.
Symposium on Aquatic Toxicology and
Hazard Assessment. Description: 3 v.: ill.;
24 cm. 9th v. (14-16 Apr. 1985)-11th v.
(10-12 May 1987). Continues: Symposium
on Aquatic Toxicology. Aquatic
toxicology and hazard assessment 1040-
306X (DLC) 88659953
(OCoLC)10268729 Continued by: Aquatic
toxicology and hazard assessment
(Philadelphia, Pa.: 1989) (DLC)sn
91024152 ISSN: 1045-4713 Notes:
Contains papers presented at: the
Symposium on Aquatic Toxicology, 9th v.;
Symposium on Aquatic Toxicology and
Hazard Assessment, 10th v.-11th v.
SERBIB/SERLOC merged record
Subjects: Aquatic animals--Effect of water
pollution on--Congresses. Aquatic plants--
Effect of water pollution on--Congresses.
Water--Pollution--Toxicology--
Congresses. Environmental impact
analysis--Congresses. Water quality
bioassay--Congresses. Series: ASTM
special technical publication LC
Classification: QH545.W3 S95a NAL
Class No.: QH545.W3S95 Dewey Class
No.: 574.5/263 20

Aquatic toxicology and risk assessment. Serial
Key Title: Aquatic toxicology and risk
assessment Abbreviated Title: Aquat.
toxicol. risk assess. Published/Created:
Philadelphia, PA: ASTM, 1990-c1991.
Related Names: ASTM Committee E-47
on Biological Effects and Environmental
Fate. ASTM Committee E-47 on
Biological Effects and Environmental Fate.
Subcommittee E47.01 on Aquatic
Toxicology. Symposium on Aquatic
Toxicology and Risk Assessment.
Description: 2 v.: ill.; 24 cm. 13th v. (16-
18 Apr. 1989)-14th v. (22-24 Apr. 1990).
Current Frequency: Annual Continues:

Aquatic toxicology and hazard assessment (Philadelphia, Pa.: 1989) (DLC)sn 91024152 ISSN: 1056-6864 Incorrect ISSN: 1040-306X Cancelled/Invalid LCCN: sn 90035477 Notes: Contains papers from the Symposium on Aquatic Toxicology and Risk Assessment, sponsored by ASTM Committee E-47 on Biological Effects and Environmental Fate and its Subcommittee E47.01 on Aquatic Toxicology. SERBIB/SERLOC merged record Subjects: Aquatic animals--Effect of water pollution on--Congresses. Aquatic plants--Effect of water pollution on--Congresses. Water--Pollution--Toxicology--Congresses. Water quality bioassay--Congresses. Series: STP ASTM special technical publication. LC Classification: QH545.W3 S95a NAL Class No.: QH545.W3S95 Dewey Class No.: 574.5/263 20

Aquatic toxicology and water quality management / edited by Jerome O. Nriagu, J.S.S. Lakshminarayana. Published/Created: New York: Wiley, c1989. Related Names: Nriagu, Jerome O. Lakshminarayana, J. S. S. Workshop on Aquatic Toxicology (13th: 1987?: Moncton, N.B.) Description: xiv, 292 p.: ill.; 25 cm. ISBN: 047161551X Notes: "This volume is based on refereed contributions from the Thirteenth Annual Aquatic Toxicology Workshop held in Moncton, New Brunswick, Canada"--P. "A Wiley-Interscience publication." Subjects: Water quality management--Congresses. Organic water pollutants--Toxicology--Congresses. Water chemistry--Congresses. Fishes--Effect of water pollution on--Congresses. Series: Wiley series in Advances in environmental science and technology; v. 22 Advances in environmental science and technology; v. 22. LC Classification: TD180 .A38 vol. 22 TD365 Dewey Class No.: 628 s 363.7/394 19 Electronic File Information: Publisher description

http://www.loc.gov/catdir/description/wiley032/88007282. html Table of Contents http://www.loc.gov/catdir/toc/onix03/88007282.html

Aquatic toxicology research focus / Elias P. Svensson (editor). Published/Created: New York: Nova Science, c2007. Projected Publication Date: 0804 Related Names: Svensson, Elias P. Description: p. cm. ISBN: 9781604561920 (hardcover) Subjects: Water--Pollution--Toxicology. LC Classification: QH90.8.T68 A675 2007 Dewey Class No.: 571.9/5 22

Aquatic toxicology. Serial Key Title: Aquatic toxicology Abbreviated Title: Aquat. toxicol. Published/Created: Amsterdam, Netherlands: Elsevier/North Holland Biomedical Press, c1981- Description: v.: ill.; 24 cm. Vol. 1, no. 1 (Apr. 1981)- Current Frequency: Irregular, <1991- > Former Frequency: Six no. a year 8 issues a year, <1983- > ISSN: 0166-445X CODEN: AQTODG Notes: Latest issue consulted: Vol. 59, no. 3/4 (Sept. 24, 2002). SERBIB/SERLOC merged record Indexed selectively by: Chemical abstracts 0009-2258 Additional Formats: Also available online. Aquatic toxicology (Amsterdam, Netherlands: Online) (DLC) 2003233137 (OCoLC)38524852 Subjects: Aquatic animals--Effect of water pollution on--Periodicals. Water--Pollution--Environmental aspects--Periodicals. Toxicology--Periodicals. Water Pollutants--toxicity--Periodicals. LC Classification: QH545.W3 A66 NLM Class No.: W1 AQ926 Dewey Class No.: 574.5/263

Aquatic toxicology: molecular, biochemical, and cellular perspectives / edited by Donald C. Malins, Gary K. Ostrander. Published/Created: Boca Raton, FL: Lewis Publishers, c1994. Related Names: Malins, Donald C. Ostrander, Gary Kent. Description: 539 p.: ill.; 25 cm. ISBN: 0873715454 (acid-free paper) Notes: Includes bibliographical references and index. Subjects: Aquatic organisms--Effect of water pollution on. Fishes--Effect of water pollution on. Molecular toxicology. LC Classification: QH90.8.T68 A67 1994 Dewey Class No.: 574.2/8 20 Electronic File Information: Publisher description http://www.loc.gov/catdir/enhancements/fy 0744/93006117-d .html

Aquifer susceptibility in Virginia, 1998-2000 / by David L. Nelms ... [et al.]. Published/Created: Richmond, Va. (1730 East Parham Rd., Richmond, 23228): U.S. Dept. of the Interior, U.S. Geological Survey; Denver, CO: U.S. Geological Survey, Branch of Information Services [distributor], 2003. Related Names: Nelms, David L. Virginia. Office of Drinking Water. Geological Survey (U.S.) Description: vii, 58 p.: ill. (some col.), maps (1 col.); 28 cm. Cancelled/Invalid LCCN: 2004368639 Notes: "Prepared in cooperation with Virginia Department of Health, Office of Drinking Water." Includes bibliographical references (p. 39-44). Subjects: Groundwater--Pollution--Virginia. Water--Pollution potential--Virginia. Aquifers--Virginia. Series: Water-resources investigations report; 03-4278 LC Classification: GB701 .W375 no. 03-4278 TD224.V8 Geographic Area Code: n-us-va

Areal distribution and concentrations of contaminants of concern in surficial streambed and lakebed sediments, Lake Erie-Lake Saint Clair drainages, 1990-97 / by S.J. Rheaume ... [et al.] Published/Created: Lansing, Mich.: U.S. Dept. of the Interior, U.S. Geological Survey; Denver, CO: Branch of Information Services [distributor], 2001. Related Names: Rheaume, S. J. Geological Survey (U.S.) Description: viii, 60 p.: ill. (some col.), col. maps; 28 cm. Notes: Includes bibliographical references (p. 38-40). Subjects: Contaminated sediments--Erie, Lake. Contaminated sediments--Saint

Clair, Lake (Mich. and Ont.) Water--Pollution--Erie, Lake. Water--Pollution--Saint Clair, Lake (Mich. and Ont.) Lake sediments--Erie, Lake--Analysis. Lake sediments--Saint Clair, Lake (Mich. and Ont.) --Analysis. Series: Water-resources investigations report; 00-4200 LC Classification: GB701 .W375 no. 00-4200 TD878.3.E75 Geographic Area Code: n-us-mi n-cn-on

Arizona mining BADCT guidance manual: aquifer protection program. Published/Created: Phoenix, Ariz.: Arizona Dept. of Environmental Quality, [1998] Related Names: Arizona. Dept. of Environmental Quality. Description: 1 v. (various pagings): ill.; 28 cm. Notes: Includes bibliographical references and index. Subjects: Mineral industries--Environmental aspects--Arizona. Groundwater--Pollution--Arizona. Water quality management--Arizona. LC Classification: TD428.M56 A75 1998 Dewey Class No.: 363.739/4 21 Government Document No.: ENQ 1.2:B 12 A 68 azdocs

Arkansas nonpoint source pollution management program. Published/Created: [Little Rock, Ark.: Arkansas Soil and Water Conservation Commission, 1994] Related Names: Arkansas Soil and Water Conservation Commission. Description: 1 v. (various pagings): maps; 28 cm. Notes: Cover title. "March 1994." Subjects: Water quality management--Arkansas. Nonpoint source pollution--Arkansas. LC Classification: TD224.A8 A76 1994 Dewey Class No.: 363.73/946/09767 20 Government Document No.: YA.S 683/8:A 6/994/[final] ardocs

Arsenic contamination in east-central Illinois ground waters: final report / prepared by Illinois State Water Survey, Aquatic Chemistry Section; principal investigators, Thomas R. Holm, Charles D. Curtiss III; prepared for Illinois Department of Energy and Natural Resources, Energy and Environmental Affairs Division. Published/Created: Springfield, IL: The Division: [Available through ENR Clearinghouse]; Springfield, VA: Available from National Technical Information Service, [1988] Related Names: Holm, Thomas R. Curtiss, Charles D. Illinois State Water Survey. Illinois. Dept. of Energy and Natural Resources. Illinois. Division of Energy and Environmental Affairs. Description: viii, 63 leaves: ill.; 28 cm. Notes: "ILENR/RE-WR-88/16." "Printed: July 1988." "Contract: WR 6." "Project: 87/2002." Includes bibliographical references (leaves 60-63). Subjects: Groundwater--Pollution--Illinois. Water-supply--Illinois. Arsenic--Toxicology. LC Classification: MLCM 91/12951 (T) Dewey Class No.: 363.7394

Arsenic in Bangladesh: report on the 500-village rapid assessment project / editor[s], Quazi Quamruzzaman, Mahmudur Rahman, Allison Quazi; analysis and text, Shafiul Azam Ahmed, Golam Morshed. Published/Created: Dhaka: Dhaka Community Hospital, in association with Bangladesh Arsenic Victims Rehabilitaion Trust, 2000. Related Names: Quamruzzaman, Quazi. Rahman, Mahmudur. Quazi, Allison. Dhaka Community Hospital. Bangladesh Arsenic Vitims Rehabilitation Trust. Description: 48 p.: ill. (some col.), map; 25 cm. Notes: "A Ministry of Health and Family Welfare-UNDP financed survey." Includes statistical tables. Includes bibliographical references (p. 30). Subjects: Water--Pollution--Bangladesh. Drinking water--Arsenic content--Bangladesh. Water--Purification--Arsenic removal--Bangladesh. LC Classification: TD420 .A77 2000 Overseas Acquisitions No.: B-E-00-410878; 26

Arsenic management modelling, a pilot study in Narayanganj-Narsingdi area: progress

report. Published/Created: Dhaka: Surface Water Modelling Centre, [1998] Related Names: Surface Water Modelling Centre (Bangladesh) Dansk hydraulisk institut. Vandkvalitetsinstituttet. Bangladesh University of Engineering and Technology. Description: ii, 13, [12] leaves: ill., maps; 30 cm. Summary: With refernce to two regions in Bangladesh. Notes: Cover title. "April 1998." "In association with Danish Hydraulic Institute, Water Quality Institute, Bangladesh, University of Engineering and Technology." Subjects: Drinking water--Arsenic content--Bangladesh. Water--Purification--Arsenic removal--Bangladesh. Groundwater--Pollution--Bangladesh--Management. LC Classification: TD427.A77 A765 1998 Overseas Acquisitions No.: B-E-2002-290648; 18

Ashio koì„doku jiken kenkyuì„. Published/Created: 1974. Related Names: Kano, Masanao, 1931- [from old catalog] ed. Description: p. cm. Notes: Romanized. Subjects: Environmental policy--Watarase River watershed, Japan. [from old catalog] Water--Pollution--Wararase River Watershed, Japan. [from old catalog] Copper--Toxicology. LC Classification: HC463.W35 A8 Geographic Area Code: a-ja---

Assessing contaminant sensitivity of endangered and threatened species: effluent toxicity tests / by F. James Dwyer ... [et al.]. Published/Created: Washington, D.C.: United States Environmental Protection Agency, Office of Research and Development, [1999] Related Names: Dwyer, F. James. Description: v, 9 p.; 28 cm. Notes: Includes bibliographical references (p. 6-7). Subjects: Fishes--Effect of water pollution on. Sewage--Environmental aspects. Toxicity testing. LC Classification: SH174 .A873 1999

Assessing contaminant sensitivity of endangered and threatened species:

toxicant classes / by F. James Dwyer ... [et al.]. Published/Created: Washington, DC: United States Environmental Protection Agency, Office of Research and Development, [1999] Related Names: Dwyer, F. James. United States. Environmental Protection Agency. Office of Research and Development. Description: v, 15 p.; 28 cm. Notes: Includes bibliographical references (p. 9). Subjects: Fishes--Effect of water pollution on. Toxicity testing. LC Classification: SH174 .A875 1999 Dewey Class No.: 571.9/517 21

Assessing the sensitivity of high altitude New Mexican wilderness lakes to acidic precipitation and trace metal contamination / by Thomas R. Lynch ... [et al.]. Published/Created: Las Cruces, N.M.: New Mexico Water Resources Research Institute, [1988] Related Names: Lynch, Thomas R. New Mexico Water Resources Research Institute. Geological Survey (U.S.) United States. Forest Service. Description: x, 177 p.: ill., maps; 28 cm. Notes: "Technical completion report." Financially supported by the U.S. Forest Service. Financed in part by the U.S. Dept. of the Interior, Geological Survey. "November 1988." Bibliography: p. 74-81. Subjects: Acid pollution of rivers, lakes, etc.--New Mexico. Lakes--New Mexico. Water quality--New Mexico. Acid precipitation (Meteorology)--Environmental aspects --New Mexico. Water--Pollution--New Mexico. Metals--Environmental aspects--New Mexico. Series: WRRI report; no. 234 LC Classification: GB705.N6 N64 no. 234

Assessment and control of nonpoint source pollution of aquatic ecosystems: a practical approach / edited by Jeffrey A. Thornton ... [et al.]. Published/Created: Paris: UNESCO; New York, USA: Parthenon Pub. Group, c1999. Related Names: Thornton Jeffrey A. Description: xi, 466

p.: ill.; 24 cm. ISBN: 1850703841 Notes: Includes bibliographical references (p. 427-544) and index. Subjects: Nonpoint source pollution. Water--Pollution. Series: Man and the biosphere series; v. 23 LC Classification: TD424.8 .A87 1998 Dewey Class No.: 628.1/68 21 Electronic File Information: Publisher description http://www.loc.gov/catdir/enhancements/fy0646/97009771-d .html

Assessment of freshwater quality: report on the results of the WHO/UNEP programme on health-related environmental monitoring. Published/Created: [S.l.]: United Nations Environment Programme: World Health Organization, c1988. Related Names: Global Environmental Monitoring System. United Nations Environment Programme. World Health Organization. Description: 80 p.: ill., maps; 29 cm. Notes: At head of title: Global Environment Monitoring System. Bibliography: p. 75-80. Subjects: Water quality. Water--Pollution. LC Classification: TD370 .A87 1988

Assessment of interstate streams in the Susquehanna River Basin, monitoring report ... Published/Created: Harrisburg, Pa: Water Quality and Monitoring Program, Susquehanna River Basin Commission, 1999- Related Names: Susquehanna River Basin Commission. Susquehanna River Basin Commission. Division of Water Quality and Monitoring Programs. Susquehanna River Basin Commission. Division of Watershed Assessment and Protection. Susquehanna River Basin Commission. Water Quality and Monitoring Program. Susquehanna River Basin Commission. Watershed Assessment and Protection Program. Description: v.: ill., maps; 29 cm. No. 12 (July 1, 1997 to June 30, 1998)- Current Frequency: Annual Continues: Water quality of interstate streams in the Susquehanna River Basin, monitoring report for ... and ... water years (DLC)sf

90093008 (OCoLC)22334428 Notes: Latest issue consulted: No. 19 (July 1, 2004 to June 30, 2005). Issued by: Susquehanna River Basin Commission. Water Quality and Mentoring Program, 1997/1998-1999/2000; by: Watershed Assessment and Protection Program, 2000/2001-2003/2004; by: Watershed Assessment and Protection Division, 2004/2005- Additional Formats: Also issued in electronic format. Subjects: Water quality--Susquehanna River Watershed--Statistics --Periodicals. Water--Pollution--Susquehanna River Watershed--Statistics --Periodicals. Series: Publication / Susquehanna River Basin Commission Publication (Susquehanna River Basin Commission) LC Classification: TD225.S895 A835 Government Document No.: RIV 128-3 WATQI 89-12367 nydocs

Assessment of nonpoint source impacts on Illinois water quality. Published/Created: Springfield, Ill. (2200 Churchill Rd., Springfield 62794-9276): Illinois Environmental Protection Agency, Division of Water Pollution Control, 1988. Related Names: Illinois. Environmental Protection Agency. Illinois. Division of Water Pollution Control. Description: iv, 113 p.: ill., maps; 28 cm. Notes: Cover title: Assessment of nonpoint source impacts on Illinois water resources. "April, 1988." "IEPA/WPC/88-020"--Cover. Includes bibliographical references (p. 112-113). Subjects: Water--Pollution--Illinois. Groundwater--Pollution--Illinois. Water quality management--Illinois. LC Classification: MLCM 91/12932 (T) Dewey Class No.: 363.7394

Assessment of nonpoint source pollution, state of South Carolina / prepared by Bureau of Water Pollution Control, Division of Water Quality and Shellfish Sanitation, Water Quality Planning and Standards Section. Edition Information: Rev. Published/Created: Columbia, S.C.: South

Carolina Dept. of Health and Environmental Control, Bureau of Water Pollution Control, [1989] Related Names: South Carolina. Bureau of Water Pollution Control. South Carolina. Division of Water Quality and Shellfish Sanitation. Water Quality Planning and Standards Section. Related Titles: Assessment of nonpoint source pollution for the state of South Carolina. Description: iv, 83, [6], iii, 66 p.: ill., maps; 28 cm. Notes: Cover title: Assessment of nonpoint source pollution for the state of South Carolina. "June 1988." "October, 1989"--Cover. South Carolina state doc. no.: H3496Wat 2.N55-2. Includes bibliographical references (p. 65-66). Subjects: Pollution--South Carolina. Water--Pollution--South Carolina. LC Classification: MLCM 92/12288 (T)

Assessment of nonpoint sources of pollution in urbanized watersheds: a guidance document for municipal officials / [prepared by Betsy Shreve-Gibb and William Boucher of Metcalf & Eddy, Inc.]. Published/Created: Hartford, Conn.: State of Connecticut, Dept. of Environmental Protection, Bureau of Water Management, Planning and Standards Division, [1995] Related Names: Shreve-Gibb, Betsy. Boucher, William. Connecticut. Dept. of Environmental Protection. Connecticut. Bureau of Water Management. Planning and Standards Division. Metcalf & Eddy. Description: 1 v. (various pagings): ill.; 28 cm. Notes: "April 1995." Includes bibliographical references. Subjects: Water quality--Evaluation. Nonpoint source pollution. Urbanization--Environmental aspects. Urban runoff--Environmental aspects. Watershed management. Series: DEP bulletin; no. 22 LC Classification: TD367 .A87 1995 Dewey Class No.: 628.1/68 20

Assessment of the impact of the emission of certain organochlorine compounds:

chlorophenols, chloropropenes, and epichlorohydrin on the aquatic environment (toxicity, persistence, bioaccumulation, and other ecotoxicological data): final report / prepared by, Klaas R. Krijgsheld. Published/Created: [Strasbourg]: European Science Foundation, [1984] Related Names: Krijgsheld, Klaas R. European Science Foundation. European Economic Community. Description: ii, ii, 3-163 p.: ill.; 30 cm. Notes: A study carried out under the responsibility of the European Science Foundation (ESF) for the European Economic Community under contract No U (83) 637. Bibliography: p. 139-163. Subjects: Organochlorine compounds--Environmental aspects. Organochlorine compounds--Bioaccumulation. Water--Pollution--Environmental aspects. LC Classification: QH545.O72 A87 1984 Dewey Class No.: 574.2/4 19

Assessment of water pollution / [editor], S.R. Mishra. Published/Created: New Delhi: APH Pub. Corp., 1996. Related Names: Mishra, S. R. Description: xvii, 485 p.: ill.; 23 cm. ISBN: 8170247101 Summary: With reference to India; contributed articles. Notes: Includes bibliographical references and index. Subjects: Limnology--India. Water--Pollution--Environmental aspects--India. LC Classification: QH96.57.A1 A87 1996 Dewey Class No.: 574.5/2632 20 Overseas Acquisitions No.: I-E-95911313 Geographic Area Code: a-ii---

Assessment of water pollution in Calicut city with special reference to Cannoly canal [microform]: final report. Published/Created: Kozhikode: Water Quality and Environment Division, Centre for Water Resources Development and Management, 1996. Related Names: Centre for Water Resources Development and Management (Calicut, India) Description: 128 leaves; 29 cm. Notes:

"Funded by: Department of Science, Technology, and Environment, Government of Kerala." Includes bibliographical references (p. 127). Additional Formats: Microfiche. New Delhi: Library of Congress Office; Washington, D.C.: Library of Congress Photoduplication Service, 2001. 3 microfiches. Master microform held by: DLC. Subjects: Water--Pollution--India--Calicut. LC Classification: Microfiche 2001/60219 (T) Overseas Acquisitions No.: I-E-00-370467; 35-91; 49-24

ASTM standards on Aquatic toxicology and hazard evaluation / sponsored by ASTM Committee E-47 on Biological Effects and Environmental Fate. Portion of Title: Standards on Aquatic toxicology and hazard evaluation Aquatic toxicology and hazard evaluation Published/Created: Philadelphia, PA: ASTM, 1993. Related Names: ASTM Committee E-47 on Biological Effects and Environmental Fate. Description: vii, 538 p.: ill.; 28 cm. ISBN: 0803117787 Notes: "ASTM publication code number (PCN): 03-547093-16." Includes bibliographical references and index. Subjects: Water--Pollution--Toxicology--Standards--United States. LC Classification: RA591.5 .A87 1993 Dewey Class No.: 615.9/02 20 Geographic Area Code: n-us---

Atlas of America's polluted waters. Published/Created: Washington, DC (1200 Pennsylvania Ave., N.W., Washington 20460): Assessment and Watershed Protection Division, Office of Wetlands, Oceans, and Watersheds, U.S. Environmental Protection Agency, 2000. Related Names: United States. Environmental Protection Agency. Description: 1 atlas (53 p.): col. ill., col. maps; 22 x 28 cm. Scale Information: Scales differ. Notes: Shows streams, rivers, coastal shorelines, lakes, estuaries, and wetlands that do not meet state water quality standards. "May 2000." "EPA 840-B-002. Also available via Internet from the EPA web site. Address as of 7/26/00: http://www.epa.gov/OWOW/tmdl/; current access is available via PURL. Subjects: Water--Pollution--United States--Maps. LC Classification: G1201.C35 A8 2000 Geographic Class No.: 3701 Dewey Class No.: 363.739/42/0973022 21

Atmospheric deposition of contaminants to the Great Lakes and coastal waters: proceedings from a session at SETAC's 15th annual Meeting, 30 October-3 November 1994, Denver, Colorado / edited by Joel E. Baker. Published/Created: Pensacola, Fla.: SETAC Press, c1997. Related Names: Baker, Joel E., 1959- SETAC Foundation for Environmental Education. SETAC (Society). Meeting (15th: 1994: Denver, Colo.) Description: xxvi, 451 p.: ill., maps; 26 cm. ISBN: 1880611104 (alk. paper) Notes: "Publication sponsored by the Society of Enivonmental Toxicology and Chemistry (SETAC) and the SETAC Foundation for Environmental Education." Includes bibliographical references and index. Subjects: Atmospheric deposition--Environmental aspects--United States--Congresses. Water--Pollution--United States--Congresses. Series: SETAC technical publications series LC Classification: TD427.A84 A86 1997 Dewey Class No.: 628.1/68 21 Geographic Area Code: n-us---

Atmospheric nutrient input to coastal areas: reducing the uncertainties / Richard A. Valigura ... [et al.] Published/Created: [Silver Spring, MD]: U. S. Department of Commerce, National Oceanic and Atmospheric Administration, Ocean Office, [1996] Related Names: Valigura, Richard A. United States. NOAA Coastal Ocean Program Office. Description: 1 v. (various pagings): ill., maps (some col.); 28 cm. Notes: "Science for solution." "June

1996." Includes bibliographical references. Subjects: Nutrient pollution of water--United States. Nitrogen cycle--Analysis. Water--Nitrogen content--United States. Coastal ecology--United States. Coastal zone management--Environmental aspects--United States. Marine eutrophication--United States. Series: NOAA Coastal Ocean Program decision analysis series; no. 9 LC Classification: TD427.N87 A88 1996 Dewey Class No.: 363.739/4 21

Atmospheric pollutants in natural waters / edited by Steven J. Eisenreich. Published/Created: Ann Arbor, Mich.: Ann Arbor Science Publishers, c1981. Related Names: Eisenreich, S. J. Description: xi, 512 p.: ill.; 24 cm. ISBN: 0250403692 Notes: Includes bibliographical references and index. Subjects: Water--Pollution. Air--Pollution. LC Classification: TD425 .A85 Dewey Class No.: 628.1/68 19

Atrazine in North American surface waters: a probabilistic aquatic ecological risk assessment / Jeffrey M. Giddings ... [et al.]. Published/Created: Pensacola, FL: Society of Environmental Toxicology and Chemistry, c2005. Related Names: Giddings, J. M. (Jeffrey M.) Description: xxxvii, 392 p.: ill.; 23 cm. ISBN: 1880611783 (pbk.: alk. paper) Notes: Includes bibliographical references and index. Subjects: Atrazine--Toxicology--North America. Aquatic organisms--Effect of water pollution on--North America. Atrazine--Environmental aspects--North America. Ecological risk assessment--North America. LC Classification: SB952.A82 A87 2005 Dewey Class No.: 363.17/92/097 22 Geographic Area Code: n------

Bacteria TMDL for Birch Creek Watershed, Virginia / submitted by Virginia Department of Environmental Quality; prepared by George Mason University and the Louis Berger Group, Inc. Portion of Title: Bacteria total maximum daily load for Birch Creek Watershed, Virginia Published/Created: [Richmond, Va.]: Dept. of Environmental Quality, [2004] Related Names: Louis Berger Group. George Mason University. Virginia. Dept. of Environmental Quality. Description: 1 v. (various pagings): col. ill., col. maps; 28 cm. Notes: Title from cover. "April, 2004"--cover. Includes bibliographical references. c. 1-2, December 2004, dep., KFI Additional Formats: Report also available in PDF format at Virginia DEQ web site. Subjects: Water quality biological assessment--Virginia--Birch Creek Watershed (Pittsylvania County-Halifax County) Escherichia coli--Virginia--Birch Creek Watershed (Pittsylvania County-Halifax County) Water--Pollution--Total maximum daily load--Virginia--Birch Creek Watershed (Pittsylvania County-Halifax County) LC Classification: TD224.V8 B29 2004

Bacteria TMDL for Carter Run, Fauquier County, Virginia / submitted by Virginia Department of Environmental Quality. Portion of Title: Bacteria total maximum daily load for Carter Run, Fauquier County, Virginia Running Title: Carter Run bacteria TMDL Edition Information: Rev. Published/Created: [Richmond, Va.]: Dept. of Environmental Quality, [2005] Related Names: Virginia. Dept. of Environmental Quality. Description: v, 51, [2] p.: ill. (some col.), maps (some col.); 28 cm. Notes: Title from cover. "January, 2005 (Revised)"--cover. Includes bibliographical references. c. 1-2, July 2005, dep., KFI Additional Formats: Report also available in PDF format at Virginia DEQ web site. Subjects: Water--Pollution--Total maximum daily load--Virginia --Carter Run Watershed (Fauquier County) Water quality biological assessment--Virginia--Carter Run Watershed (Fauquier County) Escherichia coli--Virginia--Carter Run Watershed (Fauquier County) LC Classification:

TD224.V8 B32 2005 Dewey Class No.: 363.739/4209755275 22

Bacteria TMDL for Deep Run, Stafford and Fauquier County, Virginia / submitted by Virginia Department of Environmental Quality; prepared by Rappahannock-Rapidan Regional Commission [and] Enginering Concepts, Inc. Portion of Title: Bacteria total maximum daily load for Deep Run, Stafford and Fauquier County, Virginia Running Title: Bacteria TMDL for Deep Run Published/Created: [Richmond, Va.]: Virginia Dept. of Environmental Quality, [2004] Related Names: Engineering Concepts, Inc. Rappahannock Rapidan Regional Commission. Virginia. Dept. of Environmental Quality. Description: 1 v. (various pagings): ill. (some col.), maps (some col.); 28 cm. Notes: Title from cover. "April 2004"--cover. Includes bibliographical references. Engineering Concepts, Inc. of Fincastle, Virginia supported this study as a contractor to Rappahannock Rapidan Regional Commission through funding provided by the Virginia Department of Environmental Quality Contract #8679. c. 1-2, July 2005, dep., KFI Additional Formats: Report also available in PDF format at Virginia DEQ web site. Subjects: Water--Pollution--Total maximum daily load--Virginia--Deep Run Watershed (Stafford County-Fauquier County) Water quality biological assessment--Virginia--Deep Run Watershed (Stafford County-Fauquier County) Escherichia coli--Virginia--Deep Run Watershed (Stafford County-Fauquier County) LC Classification: TD224.V8 B33 2004

Bacteria TMDL for Matadequin Creek, Hanover County, Virginia / submitted by Virginia Department of Environmental Quality. Variant Title: Bacteria total maximum daily load for Matadequin Creek, Hanover County, Virginia Running

Title: Matadequin Creek bacteria TMDL Published/Created: [Richmond, Va.]: Dept. of Environmental Quality, [2004] Related Names: Virginia. Dept. of Environmental Quality. Description: 1 v. (various pagings): ill. (some col.), maps (some col.); 28 cm. Notes: Title from cover. "October, 2004 (Revised)"--cover. Includes bibliographical references. c. 1-2, July 2005, dep., KFI Additional Formats: Report also available in PDF format at Virginia DEQ web site. Subjects: Water--Pollution--Total maximum daily load--Virginia --Matadequin Creek Watershed (Hanover County) Water quality biological assessment--Virginia--Matadequin Creek Watershed (Hanover County) Escherichia coli--Virginia--Matadequin Creek Watershed (Hanover County) LC Classification: TD224.V8 B335 2004 Dewey Class No.: 363.739/46209755462 22

Bacteria TMDL for Roses Creek Watershed, Virginia / submitted by Virginia Department of Environmental Quality; prepared by George Mason University and the Louis Berger Group, Inc. Portion of Title: Bacteria total maximum daily load for Roses Creek Watershed, Virginia Published/Created: [Richmond, Va.]: Dept. of Environmental Quality, [2004] Related Names: Louis Berger Group. George Mason University. Virginia. Dept. of Environmental Quality. Description: 1 v. (various pagings): col. ill., col. maps; 28 cm. Notes: Title from cover. "April, 2004"--cover. Includes bibliographical references. c. 1-2, December 2004, dep., KFI Additional Formats: Report also available in PDF format at Virginia DEQ web site. Subjects: Water quality biological assessment--Virginia--Roses Creek Watershed (Brunswick County) Escherichia coli--Virginia--Roses Creek Watershed (Brunswick County) Water--Pollution--Total maximum daily load--Virginia--Roses Creek Watershed

(Brunswick County) LC Classification: TD224.V8 B336 2004

Bacteria TMDL for South Mayo River, Patrick County, Virginia / submitted by Virginia Department of Environmental Quality. Portion of Title: Bacteria total maximum daily load for South Mayo River, Patrick County, Virginia Edition Information: Rev. Published/Created: [Richmond, Va.]: Dept. of Environmental Quality, [2004] Related Names: Virginia. Dept. of Environmental Quality. Description: v, 52, [2] p.: ill. (some col.), col. maps; 28 cm. Notes: Cover title. "January, 2004 (Revised February, 2004)"--Cover. Includes bibliographical references. c. 1-2, December 2004, dep., KFI Additional Formats: Report also available in PDF format at Virginia DEQ web site. Subjects: Water quality biological assessment--Virginia--South Mayo River (Patrick County) Escherichia coli--Virginia--South Mayo River (Patrick County)) Water--Pollution--Total maximum daily load--Virginia--South Mayo River (Patrick County)

Bacteria TMDL for Tuckahoe Creek, Little Tuckahoe Creek, Anderson, Broad, Georges and Readers Branches, and Deep Run, Henrico, Goochland and Hanover Counties, Virginia / submitted by Virginia Department of Environmental Quality. Running Title: Tuckahoe Creek watershed bacteria TMDL Published/Created: [Richmond, Va.]: Dept. of Environmental Quality, [2004] Related Names: Virginia. Dept. of Environmental Quality. Description: 1 v. (various pagings): ill. (some col.), col. maps; 28 cm. Notes: Title from cover. "July, 2004"--cover. Includes bibliographical references. c. 1-2, June 2005, dep., KFI Additional Formats: Report also available in PDF format at Virginia DEQ web site. Subjects: Water--Pollution--Total maximum daily load--Virginia --Tuckahoe Creek Watershed

(Henrico County) Water--Pollution--Total maximum daily load--Virginia --Tuckahoe Creek Watershed (Goochland County) Water--Pollution--Total maximum daily load--Virginia --Tuckahoe Creek Watershed (Hanover County) Water quality biological assessment--Virginia--Tuckahoe Creek Watershed (Henrico County) Escherichia coli--Virginia--Tuckahoe Creek Watershed (Henrico County) LC Classification: TD224.V8 B337 2004

Bacteria TMDL for White Oak Swamp, Henrico County, Virginia / submitted by Virginia Department of Environmental Quality. Variant Title: Bacteria total maximum daily load for White Oak Swamp, Henrico County, Virginia Running Title: White Oak Swamp bacteria TMDL Published/Created: [Richmond, Va.]: Dept. of Environmental Quality, [2004] Related Names: Virginia. Dept. of Environmental Quality. Description: 1 v. (various pagings): ill. (some col.), maps (some col.); 28 cm. Notes: Title from cover. "July, 2004"--cover. Includes bibliographical references. c. 1-2, July 2005, dep., KFI Additional Formats: Report also available in PDF format at Virginia DEQ web site. Subjects: Water--Pollution--Total maximum daily load--Virginia--White Oak Swamp Watershed (Henrico County) Water quality biological assessment--Virginia--White Oak Swamp Watershed (Henrico County) Escherichia coli--Virginia--White Oak Swamp Watershed (Henrico County) LC Classification: TD224.V8 B34 2004

Bacteria TMDLs for Abrams Creek and Upper and Lower Opequon Creek located in Frederick and Clarke County, Virginia / submitted by Virginia Department of Environmental Quality, Virginia Department of Conservation and Recreation; prepared by Department of Biological Systems Engineering, Virginia

Tech. Portion of Title: Bacteria total maximum daily load for Abrams Creek and Upper and Lower Opequon Creek, Virginia Running Title: Final bacteria TMDLs for Abrams and Opequon Creeks Published/Created: [Richmond, Va.]: Dept. of Environmental Quality, [2004] Related Names: Virginia. Dept. of Conservation and Recreation. Virginia Polytechnic Institute and State University. Dept. of Biological Systems Engineering. Virginia. Dept. of Environmental Quality. Description: [1], 250, [2] p.: col. ill., col. maps; 28 cm. Notes: Cover title. "October 2003 Revised January 2004"--Cover. Includes bibliographical references. c. 1-2, January 2005, dep., KFI Additional Formats: Report also available in PDF format at Virginia DEQ web site. Subjects: Water quality biological assessment--Virginia--Abrams Creek Watershed (Frederick County) Water quality biological assessment--Virginia--Opequon Creek Watershed (Frederick County-Clarke County) Escherichia coli--Virginia--Abrams Creek Watershed (Frederick County) Escherichia coli--Virginia--Opequon Creek Watershed (Frederick County-Clarke County) Water--Pollution--Total maximum daily load--Virginia --Abrams Creek Watershed (Frederick County) Water--Pollution--Total maximum daily load--Virginia --Opequon Creek Watershed (Frederick County-Clarke County) LC Classification: TD224.V8 B288 2004

Bacteria TMDLs for Cedar Run and Licking Run, Virginia / submitted by Virginia Department of Environmental Quality. Portion of Title: Bacteria total maximum daily load for Cedar Run and Licking Run, Virginia Published/Created: [Richmond, Va.]: Dept. of Environmental Quality, [2004] Related Names: Virginia. Dept. of Environmental Quality. Description: 1 v. (various pagings): ill. (some col.), maps (some col.); 28 cm. Notes: Cover title.

"June, 2004"--Cover. Includes bibliographical references. c. 1-2, January 2005, dep., KFI Additional Formats: Report also available in PDF format at Virginia DEQ web site. Subjects: Water quality biological assessment--Virginia--Cedar Run Watershed (Fauquier County-Prince William County) Water quality biological assessment--Virginia--Licking Run Watershed (Fauquier County-Prince William County) Escherichia coli--Virginia--Cedar Run Watershed (Fauquier County-Prince William County) Escherichia coli--Virginia--Licking Run Watershed (Fauquier County-Prince William County) Water--Pollution--Total maximum daily load--Virginia--Cedar Run Watershed (Fauquier County-Prince William County) Water--Pollution--Total maximum daily load--Virginia --Licking Run Watershed (Fauquier County-Prince William County) LC Classification: TD224.V8 B325 2004

Bacteria TMDLs for the Goose Creek Watershed / submitted by Virginia Department of Environmental Quality, Virginia Department of Conservation and Recreation; prepared by Interstate Commission on the Potomac River Basin. Variant Title: Bacteria total maximum daily loads for the Goose Creek Watershed Published/Created: [Richmond, Va.]: Dept. of Environmental Quality, [2003] Related Names: Virginia. Dept. of Environmental Quality. Virginia. Dept. of Conservation and Recreation. Interstate Commission on the Potomac River Basin. Description: 1 v. (various pagings): ill. (some col.), maps (some col.); 28 cm. Notes: Title from cover. "February 5, 2003"--cover. Includes bibliographical references. c. 1-2, July 2005, dep., KFI Additional Formats: Report also available in PDF format at Virginia DEQ web site. Subjects: Water--Pollution--Total maximum daily load--Virginia--Goose Creek Watershed (Loudoun County-Fauquier County) Water

quality biological assessment--Virginia--Goose Creek Watershed (Loudoun County-Fauquier County) Escherichia coli--Virginia--Goose Creek Watershed (Loudoun County-Fauquier County) LC Classification: TD224.V8 B345 2003 Dewey Class No.: 363.739/4620975528 22

Bacterial indicators of pollution / editor, Wesley O. Pipes. Published/Created: Boca Raton, Fla.: CRC Press, c1982. Related Names: Pipes, Wesley O. Workshop on Water Quality and Health Significance of Bacterial Indicators of Pollution (1978: Drexel University). Description: 174 p.: ill.; 27 cm. ISBN: 0849359708 Notes: Includes edited papers from the Workshop on Water Quality and Health Significance of Bacterial Indicators of Pollution, held at Drexel University, Philadelphia, 1978. Includes bibliographical references and index. Subjects: Sanitary microbiology--Congresses. Escherichia--Identification--Congresses. Indicators (Biology)--Congresses. Bacteria--Analysis. Water pollution--Analysis. LC Classification: QR48 .B33 1982 NLM Class No.: WA 689 B131 Dewey Class No.: 628.1/61 19

Bacteriological map of India in relation to rural drinking water / compiled by J.W. Bhattacherjee ... [et al.]. Published/Created: Lucknow: Industrial Toxicology Research Centre, [c1989] Related Names: Bhattacherjee, J. W. Industrial Toxicology Research Centre (India) National Drinking Water Mission (India) Description: 1 atlas (36 p.): maps; 29 cm. Scale Information: Scale not given. Notes: "August 1989." "National Drinking Water Mission, Dept. of Rural Development, Govt. of India"--T.p. Subjects: Drinking water--Contamination--India--Maps. Water-supply, Rural--India--Maps. Water--Pollution--India. Water-supply--India. LC Classification: G2281.N856 B3 1989 Geographic Class No.: 7651 NLM Class No.: KK4794 WA

689 Dewey Class No.: 363.6/1/0954022 20
Overseas Acquisitions No.: I-E-65382

Bacteriological, and chemical pollutants.
Papers by Charles W. Hendricks ... et al.
Published/Created: New York, MSS
Information Corp. [1973] Related Names:
Hendricks, Charles W. Description: 151 p.
illus. 24 cm. ISBN: 0842270760 Notes:
Includes bibliographical references.
Subjects: Water--Pollution. Pathogenic
bacteria. Chemicals. Water pollution--
Collected works. Water pollution,
Radioactive--Collected works. Series:
Fresh water pollution,1 MSS topics in
ecology series LC Classification: TD420
.F75 vol. 1 NLM Class No.: WA689 F887
Dewey Class No.: 628.1/68/08 s 628.1/68

Baseline loadings of nitrogen, phosphorus, and
sediments from Illinois watersheds:
October 1980-September 1996 / prepared
by Matthew B. Short. Published/Created:
[Springfield, IL]: Illinois Environmental
Protection Agency, 1999. Related Names:
Short, Matthew B. Illinois. Bureau of
Water. Description: 1 v. (various pagings):
maps (some col.); 28 cm. Notes:
"November 1999." "Bureau of Water,
Division of Water Pollution Control,
Surface Water Section, Central Monitorig
Unit." "IEPA/BOW/99-020"-- Cover.
Includes bibliographical references.
Subjects: Nutrient pollution of water--
Mississippi River Watershed. Nitrogen--
Environmental aspects--Illinois.
Phosphorus--Environmental aspects--
Illinois. Suspended sediments--Illinois.
Watersheds--Illinois. LC Classification:
TD427.N87 B37 1999

Baseline water quality of Minnesota's principal
aquifers / prepared by Don Jakes ... [et al.].
Published/Created: St. Paul, Minn.:
Minnesota Pollution Control Agency,
Ground Water and Solid Waste Division,
Program Development Section, Ground
Water Unit, [1998] Related Names: Jakes,
Don. Minnesota Pollution Control Agency.

Description: 1 v. (various pagings): ill.; 28
cm. Notes: Errata slip laid in. "March
1998." Includes bibliographical references.
Subjects: Water quality--Minnesota.
Groundwater--Pollution--Minnesota.
Aquifers--Minnesota. LC Classification:
TD224.M6 B37 1998 Dewey Class No.:
363.739/42/09776 21

Baseline water quality of Minnesota's principal
aquifers, Region 1: Northeastern
Minnesota / prepared by Ground Water
Monitoring and Assessment Program.
Published/Created: St. Paul, Minn.:
Minnesota Pollution Control Agency,
Environmental Outcomes Division,
Environmental Monitoring and Analysis
Section, Ground Water and Toxics
Monitoring Unit, [1999] Related Names:
Ground Water Monitoring and Assessment
Program (Minn.) Description: viii, 71 p.:
maps; 28 cm. Notes: "February, 1999."
Includes bibliographical references (p. 29-
31). Subjects: Water quality--Minnesota.
Groundwater--Pollution--Minnesota.
Aquifers--Minnesota. LC Classification:
TD224.M6 B3723 1999 Dewey Class No.:
363.739/42/09776 21

Baseline water quality of Minnesota's principal
aquifers, Region 3: Northwest Minnesota /
prepared by Ground Water Monitoring and
Assessment Program. Published/Created:
St. Paul, Minn.: Minnesota Pollution
Control Agency, Environmental Outcomes
Division, Environmental Monitoring and
Analysis Section, Ground Water and
Toxics Monitoring Unit, [1999] Related
Names: Ground Water Monitoring and
Assessment Program (Minn.) Description:
vi, 58 p.: ill., maps; 28 cm. Notes:
"February, 1999." Includes bibliographical
references (p. 29-31). Subjects: Water
quality--Minnesota. Groundwater--
Pollution--Minnesota. Aquifers--
Minnesota. LC Classification: TD224.M6
B3725 1999 Dewey Class No.:
363.739/42/09776 21

Baseline water quality of Minnesota's principal aquifers, Region 5: Southeast Minnesota / prepared by Ground Water Monitoring and Assessment Program. Published/Created: St. Paul, Minn.: Minnesota Pollution Control Agency, Environmental Outcomes Division, Environmental Monitoring and Analysis Section, Ground Water and Toxics Monitoring Unit, [1999] Related Names: Ground Water Monitoring and Assessment Program (Minn.) Minnesota Pollution Control Agency. Environmental Outcomes Division. Description: vi, 80 p.: ill., maps; 28 cm. Notes: "April, 1999." Includes bibliographical references (p. 33-35). Subjects: Water quality--Minnesota. Groundwater--Pollution--Minnesota. Aquifers--Minnesota. LC Classification: TD224.M6 B3727 1999 Dewey Class No.: 363.739/4 21

Baseline water quality of Minnesota's principal aquifers, Region 6: Twin Cities metropolitan area / prepared by Ground Water Monitoring and Assessment Program. Portion of Title: Twin Cities metropolitan area Published/Created: St. Paul, Minn.: Minnesota Pollution Control Agency, Environmental Outcomes Division, Environmental Outcomes Monitoring and Analysis Section, Ground Water and Toxics Monitoring Unit, [1999] Related Names: Ground Water Monitoring and Assessment Program (Minn.) Minnesota Pollution Control Agency. Environmental Outcomes Division. Description: viii, 67 p.: maps; 28 cm. Notes: "January, 1999." Includes bibliographical references (p. 26-28). Subjects: Water quality--Minnesota--Minneapolis Metropolitan Area. Water quality--Minnesota--Saint Paul Metropolitan Area. Groundwater--Pollution--Minnesota--Minneapolis Metropolitan Area. Groundwater--Pollution--Minnesota--Saint Paul Metropolitan Area. Aquifers--Minnesota--Minneapolis Metropolitan Area. Aquifers--

Minnesota--Saint Paul Metropolitan Area. LC Classification: TD224.M6 B37 1999 Dewey Class No.: 363.739/4 21

BASINS [electronic resource] / U.S. Environmental Protection Agency, Office of Water, Office of Science and Technology. Edition Information: Version 3.1. Published/Created: Washington, D.C.: U.S. Environmental Protection Agency, Office of Water, Office of Science and Technology, [2004] Related Names: United States. Environmental Protection Agency. Office of Water. United States. Environmental Protection Agency. Office of Science and Technology. Description: 1 CD-ROM: col.; 4 3/4 in. Computer File Information: System requirements: Pentium PC, 300-Mhz IBM compatible; Windows NT, 2000, or XP; 512 MB RAM; 300 MB hard drive space; ArcView 3.1, 3.2, or 3.3 if using grid data; Spatial Analyst. Scale Information: Scale not given. Summary: A multipurpose environmental analysis system for use by regional, state, and local agencies in performing watershed and water-quality-based studies, to address three objectives: to facilitate examination of environmental information, to provide an integrated watershed and modeling framework, and to support analysis of point and nonpoint source management alternatives. Notes: Title from container. "August 2004." "EPA-823-C-04-004." Subjects: Watersheds--United States--Databases. Water quality management--United States--Databases. Watershed management--United States--Databases. Water--Pollution--United States--Point source identification--Databases. Nonpoint source pollution--United States--Databases. Water--Pollution--United States--Total maximum daily load --Databases. Water Supply--United States--Databases. Water Pollution--United States--Databases. Subject Keywords: Maps--Digital LC

Classification: G3701.C3 2004 .B3 CD
Geographic Class No.: 3701

Benthic TMDL for Quail Run, Rockingham
County, Virginia / submitted by Virginia
Dept. of Environmental Quality; prepared
by Dept. of Biological Systems
Engineering, Virginia Tech. Portion of
Title: Benthic total maximum daily load
for Quail Run, Rockingham County,
Virginia Published/Created: [Richmond,
VA]: Virginia Dept. of Environmental
Quality, [2003] Related Names: Virginia.
Dept. of Environmental Quality. Virginia
Polytechnic Institute and State University.
Dept. of Biological Systems Engineering.
Description: iii, 68 p.: ill. (some col.), col.
maps; 28 cm. Notes: Title from cover.
"February 6, 2003"--cover. Includes
bibliographical references. c. 1-2,
December 2004, dep., KFI Additional
Formats: Report also available in PDF
format at Virginia DEQ web site. Subjects:
Water quality biological assessment--
Virginia--Quail Run (Rockingham County)
Benthos--Virginia--Quail Run
(Rockingham County) Water--Pollution--
Total maximum daily load--Virginia--
Quail Run (Rockingham County) LC
Classification: TD224.V8 B458 2003

Benthic TMDL for Toms Brook in Shenandoah
County, Virginia / submitted by Virginia
Department of Environmental Quality,
Virginia Department of Conservation and
Recreation; prepared by Department of
Biological Systems Engineering, Virginia
Tech. Published/Created: [Richmond,
Va.?]: Virginia Dept. of Environmental
Quality, [2004] Related Names: Virginia.
Dept. of Environmental Quality. Virginia.
Dept. of Conservation and Recreation.
Virginia Polytechnic Institute and State
University. Dept. of Biological Systems
Engineering. Description: iv, 93 p.: ill.
(some col.), charts, maps (some col.); 28
cm. Notes: Title from cover. "January 9,
2004." Includes bibliographical references.

c. 1-2, June 2005, dep., KFI Additional
Formats: Report also available in PDF
format at Virginia DEQ web site. Subjects:
Water quality biological assessment--
Virginia--Toms Brook (Shenandoah
County) Benthos--Virginia--Toms Brook
(Shenandoah County) Water--Pollution--
Total maximum daily load--Virginia--
Toms Brook (Shenandoah County) LC
Classification: TD224.V8 B46 2004
Dewey Class No.: 363.739/4620975595 22

Bibliography of water quality research reports.
Serial Key Title: Bibliography of water
quality research reports Abbreviated Title:
Bibliogr. water qual. res. rep.
Published/Created: Washington, Office of
Research and Monitoring, U.S.
Environmental Protection Agency. Related
Names: United States. Environmental
Protection Agency. Office of Research and
Monitoring. Description: 27 cm. ISSN:
0090-2055 Notes: SERBIB/SERLOC
merged record Acquisition Source:
(Environmental Protection Agency,
Publications Distribution Section) Route 8,
Box 166, Hwy. 70, West, Raleigh, N.C.,
27612 Subjects: Water--Pollution--
Bibliography. Water quality management--
Bibliography. Series: Water pollution
control research series Water pollution
control research series. LC Classification:
Z5862.2.W3 B53 Dewey Class No.:
016.3339/1

Bibliography: scientific journal articles based
on MMS Environmental research / [edited]
by Michele Tetley and Kenyon Wells.
Published/Created: [Herndon, Va.] (381
Elden St., Herndon 22070): U.S. Dept. of
the Interior, Minerals Management
Service, Environmental Studies Branch,
Environmental Studies Program, 1993.
Related Names: Tetley, Michele. Wells,
Kenyon. Environmental Studies Program
(U.S.) United States. Minerals
Management Service. Branch of
Environmental Studies. Related Titles:

Bibliography of scientific journal articles based on MMS Environmental research. Scientific journal articles based on MMS environmental research. Description: ii, 312 p.; 28 cm. Notes: "MMS 93-0069." "December 1993." The bibliography covers articles published in scientific journals that were based upon MMS funded research. "U.S. G.P.O.: 1994-301-078:80423." Subjects: United States. Dept. of the Interior. Minerals Management Service. Gulf of Mexico OCS Region.--Catalogs. United States. Dept. of the Interior. Minerals Management Service. Gulf of Mexico OCS Region.--Bibliography -- Catalogs. Continental shelf--United States--Bibliography. Offshore oil well drilling--Environmental aspects--United States--Bibliography. Oceanography--United States--Bibliography. Ecology--United States--Bibliography. Water--Pollution--Bibliography. Petroleum industry and trade--Environmental aspects--United States--Bibliography. Offshore oil industry--Environmental aspects--United States --Bibliography. Offshore gas industry--Environmental aspects--Bibliography. Series: OCS statistical report LC Classification: Z6004.P6 B56 1993 GC85 Dewey Class No.: 016.3338/2314/0973 20

Big South Fork/Bear Creek Nonpoint Source Interstate Demonstration Project: final report. Published/Created: Frankfort, Ky.: Kentucky Dept. for Environmental Protection, Division of Water, Water Quality Branch, Nonpoint Source Section, [1999] Related Names: Kentucky. Water Quality Branch. Nonpoint Source Section. Description: ix, 74 p.: ill., maps 28 cm. Notes: Includes bibliographical references (p. 40-42). Subjects: Nonpoint source pollution--Bear Creek (Tenn. and Ky.) Nonpoint source pollution--Rock Creek (Tenn. and Ky.) Water quality biological assessment--Bear Creek (Tenn. and Ky.) Water quality biological assessment--Rock Creek (Tenn. and Ky.) Acid mine drainage--Environmental aspects--Cumberland River, Big South Fork (Tenn. and Ky.) Series: Technical report; no. 3 Technical report (Kentucky. Water Quality Branch. Nonpoint Source Section); no. 3. LC Classification: TD223.2 .B54 1999 Dewey Class No.: 363.739/4/09769135 21 Geographic Area Code: n-us-tn n-us-ky

Binational study regarding the presence of toxic substances in the Rio Grande/Rio Bravo and its tributaries along the boundary portion between the United States and Mexico: final report, September 1994 = Estudio binacional sobre la presencia de sustancias toì xicas en el Rio Bravo/Rio Grande y sus afluentes, en su porcioì n fronteriza entre Meì xico y Estados Unidos: informe final, septiembre de 1994. Parallel Title: Estudio binacional sobre la presencia de sustancias toì xicas en el Rio Bravo/Rio Grande y sus afluentes, en su porcioì n fronteriza entre Meì xico y Estados Unidos Published/Created: [El Paso, Tex.]: International Boundary and Water Commission; [Mexico]: National Water Commission of Mexico; [Washington, D.C.]: U.S. Environmental Protection Agency, [1994] Related Names: United States. Environmental Protection Agency. Description: 246 p.; 28 cm. Notes: Cover title. Subjects: Water--Pollution--Rio Grande Watershed. Fishes--Effect of water pollution on--Rio Grande Watershed. LC Classification: TD223 .B464 1994 Dewey Class No.: 363.73/84/097644 20 Geographic Area Code: n-us--- n-mx---

Binding, uptake, and toxicity of alum sludge. Published/Created: Vegreville, AB: Alberta Environmental Centre, [1986] Related Names: Alberta Environmental Centre. Description: viii, 117 p.: ill.; 28 cm. Notes: "February 5, 1987," "AECV86-R10"--Cover. Includes bibliographies. Subjects: Rainbow trout--Effect of water pollution on--Alberta --Edmonton.

Rainbow trout--Effect of water pollution on--Alberta --Calgary Region. Sewage sludge--Environmental aspects--Alberta--Edmonton Region. Sewage sludge--Environmental aspects--Alberta--Calgary Region. Aluminum--Toxicology. LC Classification: QL638.S2 B47 1986 Dewey Class No.: 597/.55 20 Geographic Area Code: n-cn-ab

Bioaccumulation in aquatic systems: contributions to the assessment: proceedings of an international workshop, Berlin, 1990 / edited by R. Nagel and R. Loskill. Edition Information: 1st ed. Published/Created: Weinheim, Federal Republic of Germany; New York: VCH, 1991. Related Names: Nagel, R. (Roland), 1944- Loskill, R. (Renate) Description: xv, 239 p.: ill.; 25 cm. ISBN: 3527283951 (VCH, Weinheim) 156081201X (VCH, New York) Notes: Includes bibliographical references and index. Subjects: Water--Pollution--Congresses. Organic compounds--Bioaccumulation--Congresses. LC Classification: QH545.W3 B53 1991 Dewey Class No.: 574.5/263 20

Bioassay techniques and environmental chemistry. Gary E. Glass, editor. Published/Created: [Ann Arbor, Mich.] Ann Arbor Science Publishers [1973] Related Names: Glass, Gary E., ed. American Chemical Society. Description: xi, 499 p. illus. 24 cm. ISBN: 0250400170 Notes: Papers presented at a symposium held at the national American Chemical Society meeting, Washington, D.C., 1971. Includes bibliographical references. Subjects: Biological assay. Indicators (Biology) Water--Pollution--Measurement. Aquatic animals--Effect of water pollution on. LC Classification: QH90.57.B5 B56 Dewey Class No.: 574/.028

Bioavailability: physical, chemical, and biological interactions / edited by Jerry L. Hamelink ... [et al.]. Published/Created: Boca Raton: Lewis Publishers, c1994.

Related Names: Hamelink, J. L. SETAC (Society) Description: 239 p.: ill.; 27 cm. ISBN: 1566700868 Notes: Proceedings of a workshop held in Pellston, Michigan, in 1992, sponsored by the Society of Environmental Toxicology and Chemistry. Includes bibliographical references and index. Subjects: Water--Pollution--Toxicology--Congresses. Water--Pollution--Environmental aspects--Congresses. Water chemistry--Environmental aspects--Congresses. Aquatic organisms--Effect of water pollution on --Congresses. Series: SETAC special publications series LC Classification: QH545.W3 B545 1994 Dewey Class No.: 574.5/263 20 Electronic File Information: Publisher description http://www.loc.gov/catdir/enhancements/fy0744/93049061-d .html

Biogeochemistry and distribution of suspended matter in the North Sea and implications to fisheries biology / [editors, Stephan Kempe ... et al.]. Published/Created: Hamburg: Im Selbstverlag des Geologisch-Palaìˆontologischen Institutes der Universitaìˆt, Hamburg, 1988. Related Names: Kempe, Stephan. Description: xxiv, 547 p.: ill., maps; 25 cm. Notes: Translation of: Biogeochemie und Verteilung von Schwebstoffen in der Nordsee und ihr Bezug zur Fischereibiologie. Includes bibliographical references. Subjects: Sediment, Suspended--North Sea. Marine pollution--North Sea. Fishes--Effect of sediments on. Fishes--Effect of water pollution on. Series: Mitteilungen aus dem Geologisch-Palaìˆontologischen Institut der Universitaìˆt Hamburg, 0072-1115; Heft 65 LC Classification: QE1 .H3 Heft 65

Biological and photobiological action of pollutants on aquatic microorganisms, by W. C. Neely [and others] Published/Created: Auburn, Ala., Water Resources Research Institute, Auburn University, 1973. Related Names: Neely,

W. C. (William Charles), 1931-
Description: vi, 121 p. illus. 28 cm. Notes:
"OWRR Project A-017-ALA." Includes
bibliographical references. Subjects:
Water--Microbiology. Water--Pollution--
Physiological effect. Microbial ecology.
Series: WRRI bulletin, 9 Auburn
University. Water Resources Research
Institute. WRRI bulletin, 9. LC
Classification: TC1 .A85 no. 9 QR105
Dewey Class No.: 333.9/1/008 s 576/.15

Biological aspects of freshwater pollution:
proceedings of the course held at the Joint
Research Centre of the Commission of the
European Communities. Ispra, Italy, 5-9
June 1978 / edited by O. Ravera. Edition
Information: 1st ed. Published/Created:
Oxford; New York: Published for the
Commission of the European Communities
by Pergamon Press, 1979. Related Names:
Ravera, O. Commission of the European
Communities. Joint Research Centre.
Description: ix, 214 p.: ill.; 25 cm. ISBN:
0080234429: Notes: "ISPRA Courses."
Includes bibliographies. Subjects: Water--
Pollution--Environmental aspects--
Congresses. Aquatic animals--Effect of
water pollution on--Congresses.
Freshwater ecology--Congresses. LC
Classification: QH545.W3 B56 1979
Dewey Class No.: 574.5/2632 National
Bibliography No.: GB79

Biological control of water pollution / edited by
Joachim Tourbier and Robert W. Pierson,
Jr.; introd. by Edward W. Furia.
Published/Created: [Philadelphia]:
University of Pennsylvania Press, 1976.
Related Names: Tourbier, J. Toby
(Joachim Toby) Pierson, Robert W.
Description: ix, 340 p.: ill.; 26 cm. Notes:
"Based, in part, upon concepts and
techniques presented at the International
Conference on Biological Water Quality
Improvement Alternatives, held on March
3, 4, and 5, 1975, at the University of
Pennsylvania." Includes bibliographical

references and index. Subjects: Sewage--
Purification--Biological treatment--
Congresses. Water--Pollution--Congresses.
LC Classification: TD745 .B53 Dewey
Class No.: 628.4

Biological monitoring in four tributaries to
Lake Tuscaloosa, Tuscaloosa County,
Alabama, 1986-88 / by Maurice F. Mettee
... [et al.]. Published/Created: Tuscaloosa,
Ala.: Geological Survey of Alabama, 1990.
Related Names: Mettee, Maurice F.
Geological Survey of Alabama.
Description: 43 p.: ill., map; 28 cm. Notes:
Includes bibliographical references.
Subjects: Coal mines and mining--
Environmental aspects--Alabama --
Tuscaloosa County. Water--Pollution--
Alabama--Tuscaloosa County. Acid mine
drainage--Environmental aspects--
Alabama --Tuscaloosa County. Sulfates--
Environmental aspects--Alabama--
Tuscaloosa County. Series: Circular; 148
Circular (Geological Survey of Alabama);
148. LC Classification: MLCM 92/05366
(T)

Biological monitoring of fish / edited by
Charles H. Hocutt, Jay R. Stauffer, Jr.
Published/Created: Lexington, Mass.:
Lexington Books, 1980. Related Names:
Hocutt, Charles H. Stauffer, Jay R.
Description: xi, 416 p.: ill.; 24 cm. ISBN:
066903309X Notes: Includes
bibliographies and indexes. Subjects:
Fishes--Effect of water pollution on.
Fishes--Monitoring. LC Classification:
SH174 .B55 Dewey Class No.:
628.1/68/0287 19

Biological monitoring of inland fisheries /
edited by John S. Alabaster.
Published/Created: London: Applied
Science Publishers, 1977. Related Names:
Alabaster, John S. European Inland
Fisheries Advisory Commission.
Description: xvi, 226 p.: ill., maps; 23 cm.
ISBN: 0853347190: Notes: "Proceedings
of a symposium of the European Inland

Fisheries Advisory Commission, Helsinki, Finland, 7-8 June 1976." Summary also in French. Includes bibliographies and index. Subjects: Fishes--Effect of water pollution on--Europe. Fishery management--Europe. Water quality management--Europe. LC Classification: SH174 .B56 Dewey Class No.: 639/.97/709294 Language Code: eng fre National Bibliography No.: GB77-21245 Geographic Area Code: e------

Biological Stream Characterization (BSC): a biological assessment of Illinois stream quality: a report of the Illinois Biological Stream Characterization Work Group. Published/Created: Springfield, Ill. (P.O. Box 19276, Springfield 62794-9276): Illinois Environmental Protection Agency, Division of Water Pollution Control, [1989] Related Names: Illinois. Division of Water Pollution Control. Related Titles: Biological assessment of Illinois stream quality. Illinois biological stream characterization work group report. Description: iv, 42 p.: ill., map; 28 cm. Notes: "July 1989." "September 1989"--Cover. "IEPA/WPC/89-275"--Cover. Includes bibliographical references (p. 29-31). Subjects: Water quality--Illinois--Measurement. Aquatic biology--Illinois. Water--Pollution--Illinois--Measurement. Stream measurements--Illinois. Fish populations--Illinois. Series: Special report no. 13 of the Illinois State Water Plan Task Force State Water Plan Task Force special report; 13. LC Classification: QH541.5.S7 B54 1989 Dewey Class No.: 351.82325

Biological test method. Acute lethality test using Daphnia spp / Environmental Proctection, Conservation and Protection, Environment Canada. Published/Created: Ottawa, Ont. Canada: Environment Canada, 1990. Related Names: Canada. Environmental Protection Directorate. Related Titles: Acute lethality test using Daphnia spp. Description: xxi, 57 p.: ill.; 28 cm. ISBN: 0662180763 (alk. paper)

Notes: Abstract in French. "Report EPS 1/RM/11." Includes bibliographical references (p. 37-40). Subjects: Water quality bioassay. Daphnia--Effect of water pollution on. Environmental monitoring--Canada. Series: Environmental protection series LC Classification: QH90.57.B5 B585 1990 Dewey Class No.: 574.92 20 Language Code: eng fre Geographic Area Code: n-cn---

Biological test method. Acute lethality test using rainbow trout. Published/Created: Ottawa, Ont. Canada: Environment Canada, c1990. Related Names: Canada. Environmental Protection Service. Description: xix, 51 p.; 28 cm. ISBN: 0662180747 Notes: Summary in English and French. "Report EPS 1/RM/9." Includes bibliographical references (p. 33-37). Subjects: Water quality bioassay. Rainbow trout--Effect of water pollution on. Water--Pollution. Environmental monitoring--Canada. Series: Environmental protections series LC Classification: QH90.57.B5 B586 1990 Dewey Class No.: 574.92 20 Language Code: eng fre Geographic Area Code: n-cn---

Biological test method. Fertilization assay using echinoids (sea urchins and sand dollars). Published/Created: Ottawa, Ont., Canada: Environmental Protection, Conservation and Protection, Environment Canada, [1992] Related Names: Canada. Conservation and Protection. Description: xxiii, 97 p.: ill.; 28 cm. ISBN: 0662204301 Notes: "December 1992." "Report EPS 1/RM/27." Includes bibliographical references. Subjects: Water quality bioassay. Sea urchins--Effect of water pollution on--Testing. Sand dollars--Effect of water pollution on--Testing. Sea urchins--Reproduction. Sand dollars--Reproduction. Marine pollution--Testing. Series: Environmental protection series LC

Classification: QH90.57.B5 B587 1992
Dewey Class No.: 593.9/5 20

Biological test method. Test of larval growth
and survival using fathead minnows /
Environmental Protection, Conservation
and Protection, Environment Canada.
Published/Created: Ottawa, Ont., Canada:
Environment Canada, [1992] Related
Names: Canada. Environmental Protection
Directorate. Description: xxiii, 70 p.: ill.;
28 cm. ISBN: 0662193970 Notes:
Abstracts in English and French. "February
1992." "Report EPS 1/RM/22." Includes
bibliographical references (p. 51-57).
Subjects: Water quality bioassay. Fathead
minnow--Larvae--Effect of water pollution
on --Testing. Series: Environmental
protection series LC Classification:
QH90.57.B5 B588 1992 Dewey Class No.:
597/.52 20 Language Code: eng fre

Biological test method. Toxicity tests using
early life stages of salmonid fish (rainbow
trout, coho salmon, or Atlantic salmon) /
Environmental Protection, Conservation
and Protection, Environment Canada.
Published/Created: Ottawa, Ont.:
Environment Canada, [1992] Related
Names: Canada. Conservation and
Protection. Description: xxv, 81 p.: ill.; 28
cm. ISBN: 0662205553 Notes: "December
1992." "Report EPS 1/RM/28." Includes
bibliographical references (p. 51-57).
Subjects: Water quality bioassay. Rainbow
trout--Effect of water pollution on--
Testing. Coho salmon--Effect of water
pollution on--Testing. Atlantic salmon--
Effect of water pollution on--Testing.
Series: Environmental protection series LC
Classification: QH90.57.B5 B589 1992
Dewey Class No.: 597/.55 20

Biomarkers in marine organisms: a practical
approach / edited by Philippe Garrigues ...
[et al.]. Edition Information: 1st ed.
Published/Created: Amsterdam; New
York: Elsevier, 2001. Related Names:
Garrigues, Philippe, 1953- Description:

550 p.: ill.; 25 cm. ISBN: 044482913X
(hardcover) Cancelled ISBN: 044492913X
(hardcover) Notes: Includes
bibliographical references. Subjects:
Marine organisms--Effect of water
pollution on. Water quality biological
assessment. Biochemical markers. LC
Classification: QH545.W3 B5665 2001
Dewey Class No.: 591.77 21

Biomonitoring of coastal waters and estuaries /
edited by Kees J.M. Kramer.
Published/Created: Boca Raton, Fla.: CRC
Press, c1994. Description: xxiv, 327 p.: ill.;
27 cm. ISBN: 0849348951 Notes: Includes
bibliographical references and indexes.
Subjects: Kramer, Kees J. M. Marine
organisms--Effect of water pollution on.
Environmental monitoring. Indicators
(Biology) Coastal ecology. LC
Classification: QH545.W3 B567 1994
Dewey Class No.: 574.5/2636 20
Electronic File Information: Publisher
description
http://www.loc.gov/catdir/enhancements/fy
0744/94010065-d .html

Biomonitoring of Environmental Status and
Trends (BEST) Program: environmental
contaminants and their effects on fish in
the Columbia River Basin / by Jo Ellen
Hinck ... [et al.]. Portion of Title:
Environmental contaminants and their
effects on fish in the Columbia River Basin
Published/Created: Reston, Va.: U.S. Dept.
of the Interior, U.S. Geological Survey,
2004. Related Names: Hinck, Jo Ellen.
Biomonitoring of Environmental Status
and Trends Program (Geological Survey)
Geological Survey (U.S.) Description: x,
126 p.: col. ill., col. maps; 28 cm. Notes:
Includes bibliographical references (p.
104-118). Subjects: Water quality
biological assessment--Columbia River
Watershed. Fishes--Effect of pollution on--
Columbia River Watershed. Series:
Scientific investigations report; 2004-5154
LC Classification: QH96.8.B5 B578 2004

Dewey Class No.: 577.6/427/09797 22
Geographic Area Code: n-cn-bc n-us-or n-us-wa

Biomonitoring of Environmental Status and
Trends (BEST) Program: environmental
contaminants and their effects on fish in
the Mississippi River basin / edited by
Christopher J. Schmitt. Published/Created:
[Reston, VA]: U.S. Dept. of the Interior,
U.S. Geological Survey, [2002] Related
Names: Schmitt, Christopher J. Geological
Survey (U.S.) Description: xiii, 241 p.: col.
ill., col. maps; 28 cm. Notes: Includes
bibliographical references (p. 199-218).
Subjects: Water quality biological
assessment--Mississippi River Watershed.
Fishes--Effect of pollution on--Mississippi
River Watershed. Series: Biological
science report, 1081-292X;
USGS/BRD/BSR--2002-0004 LC
Classification: QH96.8.B5 B58 2002
Dewey Class No.: 571.9/517/0977 21
Geographic Area Code: n-us---

Biomonitoring of Environmental Status and
Trends (BEST) Program: environmental
contaminants and their effects on fish in
the Rio Grande basin / by Christopher J.
Schmitt ... [et al.]. Portion of Title:
Environmental contaminants and their
effects on fish in the Rio Grande basin
Published/Created: [Reston, Va.]: U.S.
Geological Survey, 2004. Related Names:
Schmitt, Christopher J. Geological Survey
(U.S.) Biomonitoring of Environmental
Status and Trends Program (Geological
Survey) Description: x, 118 p.: col. ill., col.
maps; 28 cm. Notes: Includes
bibliographical references (p. 89-100).
Subjects: Water quality biological
assessment--Rio Grande Watershed.
Fishes--Effect of pollution on--Rio Grande
Watershed. Series: Scientific investigations
report; 2004-5108 LC Classification:
QH96.8.B5 B585 2004 Dewey Class No.:
597.1727/09764/4 22

Biomonitoring of Environmental Status and
Trends (BEST) Program: environmental
contaminants, health indicators, and
reproductive biomarkers in fish from the
Colorado River Basin / by Jo Ellen Hinck
... [et al]. Published/Created: Reston, VA:
U.S. Department of the Interior, U.S.
Geological Survey, 2006. Related Names:
Hinck, Jo Ellen. Biomonitoring of
Environmental Status and Trends Program
(Geological Survey) Geological Survey
(U.S.) Description: x, 119 p.: ill. (some
col.), col. maps; 28 cm. Notes: Includes
bibliographical references (p. 91-106).
Subjects: Water quality biological
assessment--Colorado River Watershed
(Colo.-Mexico) Fishes--Effect of water
pollution on--Colorado River Watershed
(Colo.-Mexico) Series: Scientific
investigations report; 2006-5163 LC
Classification: QH96.8.B5 B588 2006
Dewey Class No.: 363.739/42097913 22
Geographic Area Code: n-us-co n-mx---
Electronic File Information: Table of
contents only
http://www.loc.gov/catdir/toc/fy0704/2006
475735.html

Biomonitoring of polluted water: reviews on
actual topics / editor, A. Gerhardt.
Published/Created: Switzerland: Trans
Tech, c2000. Related Names: Gerhardt, A.
Description: x, 301 p.: ill., maps; 25 cm.
ISBN: 0878498451 Contents: Introduction:
Biomonitoring for the 21st century / A.
Gerhardt. Design of a national programme
for monitoring and assessing the health of
aquatic ecosystems, with specific reference
to the South African River Health
Programme / D.J. Roux -- Selected
methods: Cellular, histological and
biochemical biomarkers / M. Schramm ...
[et al.]. Biomonitoring with morphological
deformities in aquatic organisms / L.
Janssens de Bisthoven. Recent trends in
online biomonitoring for water quality
control / A. Gerhardt -- Aquatic
ecosystems: Groundwater biomonitoring /

F. MoÌˆsslacher and J. Notenboom --
Selected bioindicators: Protozoa in
polluted water biomonitoring / J.R. Pratt
and N.J. Bowers. Crustaceans as
bioindicators / M. Rinderhagen, J.
Ritterhoff and G.P. Zauke. Biomonitoring
and ecotoxicology: fish as indicators of
pollution-induced stress in aquatic systems
/ L. Cleveland, J.F. Fairchild and E.E.
Little. Submerged bryophytes in running
waters, ecological characteristics and their
use in biomonitoring / H. Tremp.
Biomonitoring using aquatic vegetation /
M.A. Lewis and W. Wang. Use of aquatic
macrophytes in monitoring and in
assessment of biological integrity / P.M.
Stewart, R.W. Scribailo and T.P. Simon.
Notes: Includes bibliographical references.
Subjects: Water quality biological
assessment. Environmental monitoring.
Water--Pollution--Measurement. Series:
Environmental research forum, 1421-0274;
v. 9 LC Classification: QH90.57.B5 B5897
2000 Dewey Class No.: 577.6/27/0287 21

Biotransformation and fate of chemicals in the
aquatic environment: proceedings of a
workshop held at the University of
Michigan Biological Station, Pellston,
Michigan, 14-18 August 1979 / editors,
Alan W. Maki, Kenneth L. Dickson, John
Cairns, Jr. Published/Created: Washington,
D.C.: American Society for Microbiology,
c1980. Related Names: Maki, Alan W.,
1947- Dickson, Kenneth L. Cairns, John,
1923- Description: x, 150 p.: ill.; 24 cm.
ISBN: 091482628X: Notes: Includes
bibliographies and index. Subjects:
Biodegradation--Congresses. Water--
Microbiology--Congresses. Aquatic
ecology--Congresses. Water--Pollution--
Measurement--Congresses. LC
Classification: QH530.5 .B56 Dewey Class
No.: 574.5/263

Blue Nile River from the Ethiopian border to
Khartoum: final report / edited by M.D. El
Khalifa. Published/Created: [Khartoum],

Sudan: Institute of Environmental Studies,
University of Khartoum, [1985] Related
Names: El Khalifa, M. D. JaÌ„miÊ¾at al-
KhartÌ£uÌ„m. MaÊ¾had al-DiraÌ„saÌ„t al-
BiÌ„Ê¾iÌ„yah. Environmental
Management in the Sudan (Program)
United States. Agency for International
Development. Description: v, 69 p., [3]
leaves of plates: ill.; 29 cm. Notes: At head
of title: Environmental Training &
Management in Africa (ETMA),
Environmental Management in the Sudan.
"Prepared for the United States Agency for
International Development, project no.
698-0427." "September 1985."
Bibliography: 67-69. Subjects: Water--
Pollution--Environmental aspects--Blue
Nile River Watershed (Ethiopia and
Sudan) LC Classification: TD319.N6 B58
1985 Dewey Class No.:
363.7/3942/096264 19 Geographic Area
Code: f-et--- fi----- fn-----

BNA's environmental compliance series on CD
[electronic resource]. Variant Title: Title in
installation instructions: CSCD Portion of
Title: Environmental compliance series on
CD Serial Key Title: BNA's environmental
compliance series on CD Abbreviated
Title: BNA's environ. compliance ser. CD
Published/Created: Washington, DC:
Bureau of National Affairs, c1996- Related
Names: Bureau of National Affairs
(Arlington, Va.) Related Titles: BNA's
environmental compliance bulletin. Waste
management guide. BNA's air pollution
control guide. BNA's water pollution
control guide. BNA's waste management
guide. Description: CD-ROMs; 4 3/4 in.
Dec. 1996- Current Frequency: Monthly
ISSN: 1094-4923 Cancelled/Invalid
LCCN: sn 97004274 Computer File
Information: System requirements: IBM-
compatible MS-DOS computer, with an
80486 processor or better; 8MB RAM,
hard disk drive with at least 6MB free
space; MS-DOS 5.0 or higher and
Windows 3.1 or higher; CD-ROM drive

(double speed or better) with driver software; color monitor (VGA or better). Summary: Contains: BNA's air pollution control guide; BNA's water pollution guide; and: BNA's waste management guide. Notes: Title from disc label. Latest issue consulted: Feb. 1998. Includes a user's guide called: Handbook, each monthly update also includes a printed newsletter called: "User's tips." SERBIB/SERLOC merged record Information contained in this publication is also available within the related publications: BNA's environment library on CD; and: BNA's environment library on compact disc. Provides archival coverage of prior issues of the publications: BNA's environmental compliance bulletin. Forms available on the CD-ROM: BNA's environmental & safety library interactive forms. Related Item: BNA's environment library on CD 1094-771X (DLC) 98645080 (OCoLC)34877185 BNA's environment library on compact disc 1094-7728 (DLC) 98645082 (OCoLC)32283423 BNA's environmental compliance bulletin 1073-5798 (DLC) 96643515 (OCoLC)29484008 BNA's environmental & safety library interactive forms 1530-3071 (DLC) 00213487 (OCoLC)42044255 Subjects: Environmental law--United States. Air--Pollution--Law and legislation--United States. Water--Pollution--Law and legislation--United States. Hazardous wastes--Law and legislation--United States. Refuse and refuse disposal--Law and legislation--United States. LC Classification: KF3775.A3 Dewey Class No.: 333 12

Boston Harbor and Massachusetts Bay: issues, resources, status, and management: proceedings of a seminar held June 13, 1985, Washington, D.C. Published/Created: [Washington, D.C.]: U.S. Dept. of Commerce, National Oceanic and Atmospheric Administration, NOAA Estuarine Programs Office, [1987] Related Names: United States. NOAA Estuarine Programs Office. Description: i, 131 p.: ill.; 28 cm. Notes: "February 1987"--Cover. Bibliography: p. 125-131. Subjects: Water--Pollution--Massachusetts--Boston--Congresses. Water--Pollution--Massachusetts--Massachusetts Bay --Congresses. Fishes--Effect of water pollution on--Massachusetts--Boston --Congresses. Massachusetts Bay (Mass.)--Congresses. Series: NOAA estuary-of-the month seminar series; no. 4 LC Classification: QH545.W3 B66 1987 Dewey Class No.: 333.91/64 20 Geographic Area Code: n-us-ma

Boston Harbor wastewater discharge survey, 1978-1981: Boston Harbor, Mystic River, Neponset River, Weymouth River / prepared by Massachusetts Department of Environmental Quality Engineering, Division of Water Pollution Control, Technical Services Branch. Published/Created: Westborough, Mass.: The Division, [1981] Related Names: Massachusetts. Division of Water Pollution Control. Technical Services Branch. Related Titles: Boston Harbor, including Mystic, Neponset, and Weymouth River basins. Description: 42 p.: ill.; 28 cm. Notes: Cover title: Boston Harbor, including Mystic, Neponset, and Weymouth River basins. "December 1981." "Part B, Wastewater discharge data, 1978-1981"--Cover. "Publication no.: #12950-44-50-9-82-CR." Subjects: Water--Pollution--Massachusetts--Boston Harbor Watershed. Sewage disposal in rivers, lakes, etc.--Massachusetts --Boston Harbor Watershed. Sewage disposal in the ocean--Massachusetts--Boston. Boston Harbor (Mass.) LC Classification: TD225.B7 B63 1981 Dewey Class No.: 363.7/3942/097446 19 Geographic Area Code: n-us-ma

Brevard County near shore ocean nutrification analysis / project director, Dr. John R.

Proni... [et al.]. Published/Created: Miami, Fla.: Atlantic Oceanographic and Meteorological Laboratory, [2005] Related Names: Proni, John R. Atlantic Oceanographic and Meteorological Laboratories. Description: viii, 84 p.: 1 col. ill., 1 col. map., charts (some col.); 28 cm. Summary: "The expert panel was charged with reviewing scientific data and literature to answer a series of questions aimed at determining the existence of elevated nutrients along Brevard County Beaches and the various impacts of elevated nutrients to near shore ecology and human health. The panel was also tasked with evaluating methods for detecting nutrient sources and the impact of cruise and gaming vessels. Lastly, the panel was asked to make recommendations for future monitoring and research."--p.1. Notes: "July 2005." Includes bibliographical references (p. 70-84). Additional Formats: Abstract available on the World Wide Web. Subjects: Nutrient pollution of water--Florida--Brevard County --Analysis. Organic wastes--Health aspects--Florida--Brevard County. Water quality management--Florida--Brevard County. Atlantic Coast (Fla.)--Environmental conditions. Series: NOAA technical report, OAR . AOML; 37 LC Classification: TD427.N87 B74 2005

Butyltin and copper monitoring in a northern Chesapeake Bay marina and river system in 1989: an assessment of tributyltin legislation / Lenwood W. Hall, Jr. ... [et al.]. Published/Created: Queenstown, Md.: University of Maryland, Agricultural Experiment Station, Wye Research and Education Center, [1990] Related Names: Hall, Lenwood W. Maryland. Chesapeake Bay Research and Monitoring Division. Description: ii, 18 p.: ill.; 28 cm. Notes: "Chesapeake Bay Research and Monitoring Division"--Cover. "CBRM-TR-92-1"--Cover. Includes bibliographical references (p. 17-18). Subjects: Water--Pollution--Chesapeake Bay (Md. and Va.) Water--Pollution--Maryland. Tributyltin--Environmental aspects--Chesapeake Bay (Md. and Va.) Dibutyltin--Environmental aspects--Chesapeake Bay (Md. and Va.) Copper--Environmental aspects--Chesapeake Bay (Md. and Va.) LC Classification: TD225.C43 B88 1990 Dewey Class No.: 363.738 21

Buzzards Bay, 1978-1979 wastewater discharge survey data / prepared by Water Quality and Research Section, Massachusetts Division of Water Pollution Control. Published/Created: Westborough, Mass.: The Division, [1979] Related Names: Massachusetts. Water Quality and Research Section. Description: 33 p.: maps; 28 cm. Notes: "Publication #11, 676-33-60-12-79-CR." "November 1979." "Part B"--Cover. Subjects: Water--Pollution--Massachusetts--Buzzards Bay Watershed --Statistics. Sewage disposal in rivers, lakes, etc.--Massachusetts --Buzzards Bay Watershed--Statistics. LC Classification: TD225.B98 B89 1979 Dewey Class No.: 363.7/3942/097448 19 Geographic Area Code: n-us-ma

Cadmium in the aquatic environment / edited by Jerome O. Nriagu, John B. Sprague. Published/Created: New York: Wiley, c1987. Related Names: Nriagu, Jerome O. Sprague, John B. Description: xii, 272 p.; 24 cm. ISBN: 0471858846 Notes: "A Wiley-Interscience publication." Includes bibliographies and index. Subjects: Cadmium--Environmental aspects. Water--Pollution--Environmental aspects. Aquatic organisms--Effect of water pollution on. Series: Advances in environmental science and technology, 0065-2563; v. 19 LC Classification: TD180 .A38 vol. 19 QH545.C37 Dewey Class No.: 628 s 574.5/263 19 Electronic File Information: Publisher description http://www.loc.gov/catdir/description/wiley037/87002198. html Table of Contents

http://www.loc.gov/catdir/toc/onix02/8700
2198.html

California 305(b) report on water quality.
Published/Created: [Sacramento, Calif.]:
State Water Resources Control Board
Related Names: California Environmental
Protection Agency. State Water Resources
Control Board. Description: v.: maps; 28
cm. Began in 1994- Current Frequency:
Biennial Continues: California report on
water quality (OCoLC)42324256 Notes:
"Prepared as required by Federal Clean
Water Act Section 305(b)." Title from
cover. Latest issue consulted: 2000.
Subjects: United States. Federal Water
Pollution Control Act --Periodicals. Water
quality--California--Periodicals. Water
quality management--California--
Periodicals. LC Classification: TD224.C3
C215

California beach closure report.
Published/Created: Sacramento: Division
of Water Quality, State Water Resources
Control Board, California Environmental
Protection Agency, -2003. Related Names:
California Environmental Protection
Agency. Division of Water Quality.
Related Titles: Beach closure report.
Description: -2002. Current Frequency:
Annual Continued by: California beach
advisory report (OCoLC)57253713 Notes:
Description based on: 2000; title from
cover. Additional Formats: Also issued
online with additional title: Beach closure
report. Subjects: Bathing beaches--Health
aspects--California--Periodicals. Beach
closures--California--Periodicals. Water--
Pollution--California--Periodicals. Water
quality management--California--
Periodicals. LC Classification: RA606
.C35 Government Document No.:
W745.B42 cadocs

California. Central Valley Regional Water
Pollution Study of water uses and pollution
in Stanislaus River Basin, San Joaquin
River watershed. Published/Created:

Sacramento, 1953. Description: 72 l. maps
(part fold.) 28 cm. Subjects: Water--
Pollution--Stanislaus River watershed.
[from old catalog] LC Classification:
TD224.C3 A516

California's rivers and streams: working toward
solutions / [prepared by Division of Water
Quality; design and illustrations by Sharon
Perrin]. Published/Created: [Sacramento,
Calif.]: State Water Resources Control
Board: California Environmental
Protection Agency, [1995] Related Names:
California. State Water Resources Control
Board. Division of Water Quality.
California. State Water Resources Control
Board. California Environmental
Protection Agency. Description: iv, 51 p.,
[1] folded leaf of plates: ill., col. maps; 28
cm. Notes: "January 1995"--Cover.
Includes index. Subjects: Water quality
management--California. Rivers--
California. Stream conservation--
California. Watershed management--
California. Water--Pollution--California.
LC Classification: TD224.C3 C357 1995

Canadian aquatic resources / edited by M.C.
Healey and R.R. Wallace.
Published/Created: Ottawa: Dept. of
Fisheries and Oceans: Canadian Govt. Pub.
Centre [distributor], 1987. Related Names:
Healey, M. C. (Michael Charles), 1942-
Wallace, R. R. Rawson Academy of
Aquatic Science. Description: viii, 533 p.:
ill.; 25 cm. ISBN: 0660124874: Notes:
"The Rawson Academy of Aquatic
Science." Includes bibliographical
references. Subjects: Water-supply--
Canada. Aquatic ecology--Canada. Water--
Pollution--Canada. Water resources
development--Canada. Series: Canadian
bulletin of fisheries and aquatic sciences,
0706-6503; 215 Canadian bulletin of
fisheries and aquatic sciences; bulletin 215.
LC Classification: TD226 .C348 1987
Dewey Class No.: 333.91/00971 20
Geographic Area Code: n-cn---

Carcinogenic, mutagenic, and teratogenic marine pollutants: impact on human health and the environment. Published/Created: Woodlands, Tex.: Published on behalf of World Health Organization Regional Office for Europe and United Nations Environment Programme, Portfolio Pub. Co.; Houston: Gulf Pub. Co., c1990. Related Names: World Health Organization. United Nations Environment Programme. Description: xix, 284 p.: ill.; 24 cm. ISBN: 0943255104 Notes: Includes bibliographical references. Subjects: Marine pollution. Marine pollution--Health aspects. Aquatic organisms--Effect of water pollution on. Carcinogens. Mutagens. Teratogenic agents. Series: Advances in applied biotechnology series; v. 5 LC Classification: QH545.W3 C37 1990 Dewey Class No.: 363.73/942 20

Carrying capacity of public water supply watersheds: a literature review of impacts on water quality from residential development / by James M. Doenges ... [et al.]. Published/Created: Hartford, CT: [Dept. of Environmental Protection, 1990] Related Names: Doenges, James M. Litchfield Hills Council of Elected Officials. Description: 1 v. (various pagings); 28 cm. ISBN: 0942085000 (alk. paper) Notes: "Prepared for the Litchfield Hills Council of Elected Officials." "March 1990." Includes bibliographical references. Subjects: Housing development--Environmental aspects. Water quality. Water--Pollution. Series: DEP bulletin; 11 DEP bulletin; no. 11. LC Classification: TD428.C64 C37 1990 Dewey Class No.: 333.33/714 20

Case law under the Federal Water Pollution Control Act amendments of 1972 / prepared for the Committee on Public Works and Transportation, U.S. House of Representatives, by the Congressional Research Service of the Library of Congress. Published/Created: Washington: U.S. G.P.O.: for sale by the Supt. of Docs., U.S. G.P.O., 1977. Related Names: Richardson, Roy. Costello, George. Library of Congress. Congressional Research Service. United States. Congress. House. Committee on Public Works and Transportation. Description: xvii, 126 p.; 24 cm. Notes: "By Roy Richardson... [and others] updated and edited by George Costello"--P. xvii. At head of title: 95th Congress, 1st session. Committee print. "95-35." Subjects: Water--Pollution--Law and legislation--United States --Digests. LC Classification: KF3786.A55 C37 Dewey Class No.: 344.73/046343/02648 347.3044634302648 19 Geographic Area Code: n-us---

Challenges to international waters: regional assessments in a global perspective / Global International Waters Assessment. Published/Created: Nairobi, Kenya: United Nations Environment Programme, 2006. Related Names: United Nations Environment Programme. Global International Waters Assessment. Description: 120 p.: col. ill., col. maps; 28 cm. ISBN: 9189584473 9789189584471 Notes: "The GIWA final report." "Published by the United Nations Environment Programme in collaboration with GEF, the University of Kalmar, and the Municipality of Kalmar, Sweden, and the Governments of Sweden, Finland and Norway"--T.p. verso. Includes bibliographical references (p. 92). Additional Formats: Also issued online. Subjects: Marine pollution. Water--Pollution. Environmental protection--International cooperation. Water-supply--International cooperation. Eutrophication. Global environmental change. Fishes--Conservation. Transboundary pollution. LC Classification: GC1085 .C47 2006

Changes in streamflow and water quality in selected nontidal basins in the Chesapeake Bay Watershed, 1985-2004 / by Michael J.

Langland ... [et al.]; in cooperation with the U.S. Environmental Protection Agency, Chesapeake Bay Program, Maryland Department of Natural Resources, and the Virginia Department of Environmental Quality. Published/Created: Reston, VA: U.S. Dept. of the Interior, U.S. Geological Survey, 2006. Related Names: Langland, Michael J. Description: vii, 75 p.: col. ill., col. maps; 28 cm. + 1 computer optical disc (4 3/4 in.) Notes: Includes bibliographical references (p. 74-75). Subjects: Streamflow--Chesapeake Bay Watershed. Water quality--Chesapeake Bay Watershed. Water--Pollution--Chesapeake Bay Watershed. Series: Scientific investigations report; 2006-5178 LC Classification: GB1216.6 C44 2006 Geographic Area Code: n-us-md

Changes in streamflow and water quality in selected nontidal sites in the Chesapeake Bay basin, 1985-2003 / by Michael J. Langland ... [et al.]; in cooperation with the U.S. Environmental Protection Agency, Chesapeake Bay Program, the Maryland Department of Natural Resources, and the Virginia Department of Environmental Quality. Published/Created: Reston, Va.: U.S. Geological Survey; Denver, CO: U.S. Geological Survey, Information Services [distributor], 2004. Related Names: Langland, Michael J. Description: v, 50 p.: col. ill., col. maps; 28 cm. Notes: Includes bibliographical references (p. 28). Additional Formats: Also available via Internet. Subjects: Streamflow--Chesapeake Bay Watershed. Water quality--Chesapeake Bay Watershed. Water--Pollution--Chesapeake Bay Watershed. Series: Scientific investigations report; 2004-5259 LC Classification: GB1216.6 .C43 2004

Characterization of hypoxia: topic 1, report for the integrated assessment on hypoxia in the Gulf of Mexico / Nancy N. Rabalais ... [et al.]. Portion of Title: Report for the integrated assessment on hypoxia in the Gulf of Mexico Published/Created: Silver Spring, Md.: U.S. Dept. of Commerce, National Oceanic and Atmospheric Administration, Coastal Ocean Program, [1999] Related Names: Rabalais, Nancy N., 1950- Description: xviii, 167 p.: ill., maps; 28 cm. Notes: "May 1999." Includes bibliographical references (p. 152-167). Subjects: Nutrient pollution of water--Mexico, Gulf of. Water--Dissolved oxygen--Mexico, Gulf of. Coastal ecology--Mexico, Gulf of. Hypoxia (Water) Series: NOAA Coastal Ocean Program decision analysis series; no. 15 LC Classification: TD427.N87 C43 1999 Dewey Class No.: 363.739/4/0916364 21 Geographic Area Code: nm-----

Charles River, 1980-1981 wastewater discharge data / prepared by Water Quality and Research Section, Massachusetts Division of Water Pollution Control. Published/Created: Westborough, Mass.: The Division, [1983] Related Names: Massachusetts. Water Quality and Research Section. Related Titles: Charles River wastewater discharge data, 1980-1981. Description: 35 p.: ill.; 28 cm. Notes: Cover title: Charles River wastewater discharge data, 1980-1981. "May 1983." Subjects: Water--Pollution--Massachusetts--Charles River Watershed. Sewage--Analysis. Sewage disposal plants--Massachusetts. LC Classification: TD224.M4 C4 1983 Dewey Class No.: 363.7/3942/097444 19 Geographic Area Code: n-us-ma

Charleston Harbor: issues, resources, status, and management: proceedings of a seminar held April 4, 1989, Washington, D.C. Published/Created: [Washington, D.C.?]: U.S. Dept. of Commerce, National Oceanic and Atmospheric Administration, NOAA Estuarine Programs Office, [1989] Related Names: United States. NOAA Estuarine Programs Office. Description: vii, 62 p.:

ill.; 28 cm. Notes: Item 250-E-32 Shipping
list no.: 89-445-P. "July 1989." Includes
bibliographical references. Subjects:
Water--Pollution--Environmental aspects--
South Carolina --Charleston--Congresses.
Watershed management--South Carolina--
Charleston --Congresses. Estuaries--South
Carolina--Charleston. Series: NOAA
estuary-of-the-month seminar series; no.
16 LC Classification: TD424.35.S6 C46
1989 Dewey Class No.: 333.91/64 20
Government Document No.: C 55.49:16
Geographic Area Code: n-us-sc

Chemical and radiochemical constituents in
water from wells in the vicinity of the
Naval Reactors Facility, Idaho National
Engineering and Environmental
Laboratory, Idaho, 1996 / by LeRoy L.
Knobel ... [et al.]; prepared in cooperation
with the U.S. Department of Energy.
Published/Created: Idaho Falls, Idaho: U.S.
Dept. of the Interior, U.S. Geological
Survey; Denver, CO: U.S. Geological
Survey, Information Services [distributor],
[1999] Related Names: Knobel, LeRoy L.
United States. Dept. of Energy. Geological
Survey (U.S.) Description: iv, 58 p.: ill.,
maps; 28 cm. Notes: "October 1999"
Chiefly tables. Includes bibliographical
references (p. 14-17). Subjects: Water
quality--Snake River Plain (Idaho and Or.)
Groundwater--Pollution--Snake River
Plain (Idaho and Or.) Radioactive pollution
of water--Snake River Plain (Idaho and
Or.) Groundwater--Pollution--Idaho--Idaho
National Engineering and Environmental
Laboratory Region. Series: U.S.
Geological Survey open-file report; 99-272
LC Classification: TD225.S47 C44 1999
Dewey Class No.: 363.739/42/097961 21

Chemical and radiochemical constituents in
water from wells in the vicinity of the
Naval Reactors Facility, Idaho National
Engineering and Environmental
Laboratory, Idaho, 1997-98 / by Roy C.
Bartholomay ... [et al.]; prepared in

cooperation with the U.S. Department of
Energy. Published/Created: Idaho Falls,
Idaho: U.S. Dept. of the Interior, U.S.
Geological Survey; Denver, CO: U.S.
Geological Survey, Information Services
[distributor], 2000. Related Names:
Bartholomay, Roy C. United States. Dept.
of Energy. Geological Survey (U.S.)
Description: iv, 52 p.: ill., maps; 28 cm.
Notes: "June 2000." Chiefly tables.
Includes bibliographical references (p. 13-
15). Subjects: Water quality--Snake River
Plain (Idaho and Or.) Groundwater--
Pollution--Snake River Plain (Idaho and
Or.) Radioactive pollution of water--Snake
River Plain (Idaho and Or.) Groundwater--
Pollution--Idaho--Idaho National
Engineering and Environmental
Laboratory Region. Series: U.S.
Geological Survey open-file report; 00-236
LC Classification: TD225.S47 C4423 2000
Dewey Class No.: 363.739/42/097961 21
Geographic Area Code: n-us-id n-us-or

Chemical and radiochemical constituents in
water from wells in the vicinity of the
Naval Reactors Facility, Idaho National
Engineering and Environmental
Laboratory, Idaho, 1999 / by Roy C.
Bartholomay ... [et al.]; prepared in
cooperation with the U.S. Department of
Energy. Published/Created: Idaho Falls,
Idaho: U.S. Dept. of the Interior, U.S.
Geological Survey; Denver, CO: U.S.
Geological Survey, Information Services
[distributor], [2001] Related Names:
Bartholomay, Roy C. United States. Dept.
of Energy. Geological Survey (U.S.)
Description: iv, 37 p.: ill., maps; 28 cm.
Notes: "January 2001." Chiefly tables.
Includes bibliographical references (p. 11-
14). Subjects: Water quality--Snake River
Plain (Idaho and Or.) Groundwater--
Pollution--Snake River Plain (Idaho and
Or.) Radioactive pollution of water--Snake
River Plain (Idaho and Or.) Groundwater--
Pollution--Idaho--Idaho National
Engineering and Environmental

Laboratory Region. Series: U.S. Geological Survey open-file report; 01-27 LC Classification: TD225.S47 C44 2001 Dewey Class No.: 363.739/42/097961 21 Geographic Area Code: n-us-id n-us-or

Chemical and radiochemical constituents in water from wells in the vicinity of the Naval Reactors Facility, Idaho National Engineering and Environmental Laboratory, Idaho, 2000 / by Roy C. Bartholomay ... [et al.]; prepared in cooperation with the U.S. Department of Energy. Published/Created: Idaho Falls, Idaho: U.S. Dept. of the Interior, U.S. Geological Survey; Denver, CO: U.S. Geological Survey, Information Services [distributor], [2001] Related Names: Bartholomay, Roy C. United States. Dept. of Energy. Geological Survey (U.S.) Description: iv, 34 p.: maps; 28 cm. Notes: "April 2001." Chiefly tables. Includes bibliographical references (p. 10-13). Subjects: Water quality--Snake River Plain (Idaho and Or.) Groundwater--Pollution--Snake River Plain (Idaho and Or.) Radioactive pollution of water--Snake River Plain (Idaho and Or.) Groundwater--Pollution--Idaho National Engineering and Environmental Laboratory Region. Series: U.S. Geological Survey open-file report; 02-148 LC Classification: TD225.S47 C44 2001b

Chemical constituents in ground water from 39 selected sites with an evaluation of associated quality assurance data, Idaho National Engineering and Environmental Laboratory and vicinity, Idaho / by LeRoy L. Knobel ... [et al.]; prepared in cooperation with the U.S. Department of Energy. Published/Created: Idaho Falls, Idaho: U.S. Dept. of the Interior, U.S. Geological Survey; Denver, CO: Information Services [distributor], [1999] Related Names: Knobel, LeRoy L. United States. Dept. of Energy. Geological Survey (U.S.) Description: v, 58 p.: maps; 28 cm.

Notes: "August 1999" "DOE/ID-22159"--Cover. Includes bibliographical references (p. 16-19). Subjects: Groundwater--Pollution--Idaho--Idaho National Engineering and Environmental Laboratory Region. Radioactive waste disposal--Environmental aspects--Idaho --Idaho National Engineering and Environmental Laboratory Region. Water quality--Idaho--Measurement. Series: U.S. Geological Survey open-file report; 99-246 LC Classification: TD224.I2 C487 1999 Dewey Class No.: 363.17/992/09796 21

Chemical quality of water, sediment, and fish in Mountain Creek Lake, Dallas, Texas, 1994-97 / by P.C. Van Metre ... [et al.]; in cooperation with the Southern Division Naval Facilities Engineering Command. Published/Created: Austin, Tex.: U.S. Department of the Interior, U.S. Geological Survey; Denver, CO: U.S. Geological Survey, Information Services [distributor], 2003. Related Names: Van Metre, Peter. United States. Naval Facilities Engineering Command. Southern Division. Geological Survey (U.S.) Description: v, 69 p.: ill. (some col.), col. maps; 28 cm. Notes: Includes bibliographical references (p. 60-63). Subjects: Water--Pollution--Texas--Mountain Creek Lake. Water--Pollution--Texas--Dallas Region. Series: Water-resources investigations report; 03-4082 LC Classification: GB701 .W375 no. 03-4082 TD224.T4 Geographic Area Code: n-us-tx

Chemicals in the aquatic environment: advanced hazard assessment / Lars Landner, ed. Published/Created: Berlin; New York: Springer-Verlag, c1989. Related Names: Landner, Lars. Description: xxii, 415 p.: ill.; 24 cm. ISBN: 0387508635 (U.S.: alk. paper) Notes: Includes bibliographies and indexes. Subjects: Water--Pollution--Environmental aspects. Water quality--Environmental aspects. Water chemistry--Environmental aspects. Environmental impact analysis. Series: Springer series on environmental management LC Classification: QH545.W3 C438 1989 Dewey Class No.: 363.73/84 20

Chemung River drainage basin survey series. Published/Created: [Albany] New York State Dept. of Health, Water Pollution Control Board [1955- Related Names: New York (State) Water Pollution Control Board. [from old catalog] Description: no. in v. illus., maps. 29 cm. Subjects: Water--Pollution--Chemung valley--Collected works. [from old catalog] LC Classification: TD225.C42 C44

Chernobyl -- what have we learned?: the successes and failures to mitigate water contamination over 20 years / edited by Yasuo Onishi, Oleg V. Voitsekhovich and Mark J. Zheleznyak. Published/Created: Dordrecht: Springer, c2007. Related Names: Onishi, Yasuo. Voiĭ†tï̦ sï̦jekhovich, O. V. (Oleg V.) Zheleznyak, Mark J. Description: x, 289 p.: ill. (some col.), maps; 25 cm. ISBN: 9781402053481 (hbk.) 1402053487 (hbk.) 9781402053498 (e-book) 1402053495 (e-book) Notes: Includes bibliographical references and index. Subjects: Radioactive pollution of water--Ukraine--ChornobylÊ¹. Radiation--Safety measures. Chernobyl Nuclear Accident, ChornobylÊ¹, Ukraine, 1986 --Environmental aspects. Series: Environmental pollution; v. 12 LC Classification: TD427.R3 C46 2007 National Bibliography No.: GBA695690 bnb National Bibliographic Agency No.: 013599771 Uk 980472806 GyFmDB

Chernobyl disaster and groundwater / editor, Vyacheslav Shestopalov. Published/Created: Lisse [Netherlands]; Exton, PA: A.A. Balkema, c2002. Related Names: Shestopalov, Vïï̦ aï̦jcheslav Mikhaiĭ†lovich. Description: 289 p.: ill., maps; 25 cm. ISBN: 9058092313 Contents: Characterization of the

Chernobyl disaster / V. Shestopalov and V. Gudzenko -- Radioactive contamination of groundwater within the Chernobyl exclusion zone / SP Dzhepo and AS Skal'skii -- Radioactive contamination of groundwater around the "shelter" object / V. Shestopalov ... [et al.] -- Radioactive contamination of groundwater within Kiev conurbation / V. Shestopalov and Yu Rudenko -- Distribution and migration of radionuclides in meliorated areas / A. Shevchenko -- Application of decision models to assessment of strontium-90 migration to water wells at the Chernobyl nuclear power plant / L. Smith ... [et al.] -- Anomalous zones of radionuclide migration in geological environment / V. Shestopalov, V. Bublias and A. Bohuslavsky -- Forecasting radionuclide migration in groundwater / V. Shestopalov, A. Bohuslavsky, and V. Bublias. Notes: Includes bibliographical references. Subjects: Groundwater--Pollution--Ukraine. Radioactive pollution of water--Ukraine. LC Classification: TD427.R3 C45 2002 Dewey Class No.: 628.1/685 22

Chesapeake Bay atmospheric deposition study: phase I--July 1990-June 1991 / principal investigators, Joel E. Baker ... [et al.]. Published/Created: Annapolis, MD: State of Maryland, Dept. of Natural Resources, Tidewater Administration, Chesapeake Bay Research and Monitoring Division, [1992] Related Names: Baker, Joel E., 1959- Maryland. Chesapeake Bay Research and Monitoring Division. Description: 1 v.: ill.; 28 cm. Notes: "December 1992." "Prepared for State of Maryland, Department of Natural Resources, Tidewater Administration, Chesapeake Bay Research and Monitoring Division, Annapolis, MD." "CBRM-AD-93-5"--Cover. Includes bibliographical references. Subjects: Atmospheric deposition--Environmental aspects--Chesapeake Bay Watershed. Water--Pollution--Chesapeake Bay Watershed. LC

Classification: TD427.A84 C48 1992
Dewey Class No.: 628.1/68/0916347 20
Geographic Area Code: n-us-md n-us-va

Chesapeake Bay basinwide nutrient reduction
strategy: Virginia point source programs:
second progress report / prepared by
Virginia Water Control Board, Chesapeake
Bay Office. Published/Created: Richmond,
VA (P.O. Box 11143, Richmond 23230):
The Board, [1992] Related Names:
Virginia. State Water Control Board.
Chesapeake Bay Office. Description: iv, 28
p.: ill.; 28 cm. Notes: "February 1992."
Subjects: Water--Pollution--Chesapeake
Bay (Md. and Va.) Water quality
management--Virginia. Nutrient pollution
of water--Chesapeake Bay (Md. and Va.)
Water--Pollution--Virginia--Point source
identification. Series: Information bulletin;
587 Information bulletin (Virginia. State
Water Control Board); 587. LC
Classification: TD225.C43 C443 1992
Dewey Class No.: 363.73/946/0916347 20
Geographic Area Code: n-us-va n-us-md

Chesapeake Bay basinwide toxics reduction
strategy: Virginia point source toxic
loading inventory phase I: summary report
/ prepared by Virginia Water Control
Board, Chesapeake Bay Office.
Published/Created: Richmond, VA (P.O.
Box 11143, Richmond 23230): The Board,
[1992] Related Names: Virginia. State
Water Control Board. Chesapeake Bay
Office. Related Titles: Point source toxic
loading inventory Phase I Description: iv,
18 p.; 28 cm. Notes: "August 1992."
Includes bibliographical references (p. 16).
Subjects: Water--Pollution--Virginia.
Water--Pollution--Chesapeake Bay
Watershed. Series: Information bulletin;
592 Information bulletin (Virginia. State
Water Control Board); 587. LC
Classification: TD224.V8 C49 1992
Dewey Class No.: 363.73/946/09755 20
Geographic Area Code: n-us-va

Chesapeake Bay nutrient and sediment
reduction tributary strategy for the James
River, Lynnhaven and Poquoson coastal
basins, March 2005. Variant Title: At head
of title: Commonwealth of Virginia
Published/Created: [Richmond, Va.]:
Commonwealth of Virginia, [2005]
Related Names: Virginia. Secretary of
Natural Resources. Virginia. Chesapeake
Bay Local Assistance Dept. Virginia. Dept.
of Conservation and Recreation. Virginia.
Dept. of Environmental Quality. Related
Titles: Tributary strategy goals for nutrient
and sediment reduction in the James River.
Description: x, 143 p.: ill., maps; 28 cm.
Notes: Title from cover. Rev. ed. of:
Tributary strategy goals for nutrient and
sediment reduction in the James River.
2000. Includes bibliographical references.
c. 1-2, June 2005, dep., KFI Additional
Formats: Also available as pdf document at
Virginia Secretary of Natural Resources
web site. Subjects: Rivers--Virginia.
Water--Pollution--Virginia--James River
Watershed. Nutrient pollution of water--
Virginia--James River Watershed.
Suspended sediments--Environmental
aspects--Virginia--James River Watershed.
Water quality management--Chesapeake
Bay (Md. and Va.) James River Watershed
(Va.) Series: Virginia's tributary strategies
LC Classification: TD224.V8 T75 2005
Dewey Class No.: 363.739/4609755 22

Chesapeake Bay nutrient and sediment
reduction tributary strategy, January 2005.
Variant Title: At head of title:
Commonwealth of Virginia
Published/Created: [Richmond, Va.]:
Commonwealth of Virginia, [2005]
Related Names: Virginia. Secretary of
Natural Resources. Virginia. Chesapeake
Bay Local Assistance Dept. Virginia. Dept.
of Conservation and Recreation. Virginia.
Dept. of Environmental Quality.
Description: 85 p.: ill.; 28 cm. Notes: Title
from cover. Includes bibliographical
references. c. 1-2, June 2005, dep., KFI

Additional Formats: Also available as pdf document at Virginia Secretary of Natural Resources web site. Subjects: Nutrient pollution of water--Chesapeake Bay Watershed. Rivers--Virginia. Water--Pollution--Chesapeake Bay Watershed. Suspended sediments--Environmental aspects--Chesapeake Bay Watershed. Water quality management--Chesapeake Bay (Md. and Va.) Chesapeake Bay Watershed. Series: Virginia's tributary strategies LC Classification: TD427.N87 C45 2005 Dewey Class No.: 363.739/460975518 22

Chesapeake Bay Water Quality Monitoring Program. Long-term benthic monitoring and assessment component, Level I comprehensive report / prepared for the Maryland Department of Natural Resources, Resource Assessment Service, Tidewater Ecosystem Assessments. Portion of Title: Long term benthic monitoring and assessment component, Level I comprehensive report Published/Created: Columbia, Md.: Versar Inc. Related Names: Chesapeake Bay Water Quality Monitoring Program (Md.) Maryland. Tidewater Ecosystem Assessment. Versar, Inc. Description: v.; 28 cm. Vols. for July 1984/Dec. 1999- issued in 2 vols. Current Frequency: Annual Cancelled/Invalid LCCN: 00274138 Notes: Description based on: July 1984/Dec. 1997. Latest issue consulted: July 1984/Dec. 2000. Subjects: Water quality--Chesapeake Bay (Md. and Va.)--Statistics --Periodicals. Water quality biological assessment--Chesapeake Bay (Md. and Va.)--Periodicals. Pollution--Environmental aspects--Chesapeake Bay (Md. and Va.) LC Classification: TD225.C43 C4675

Chesapeake Bay, issues, resources, status, and management: proceedings of a seminar held September 23, 1985, Washington, D.C. / edited by Samuel E. McCoy. Published/Created: [Washington, D.C.]:

U.S. Dept. of Commerce, National Oceanic and Atmospheric Administration, NOAA Estuarine Programs Office, [1988] Related Names: McCoy, Samuel E. United States. NOAA Estuarine Programs Office. Description: vii, 185 p.: ill.; 28 cm. Notes: "July 1988"--Cover. Includes bibliographical references (p. 157-157). Subjects: Water--Pollution--Chesapeake Bay (Md. and Va.)--Congresses. Estuarine ecology--Chesapeake Bay (Md. and Va.) --Congresses. Chesapeake Bay (Md. and Va.)--Congresses. Series: NOAA estuary-of-the-month seminar series; no. 5 LC Classification: TD225.C43 C4464 1985 Dewey Class No.: 333.91/64 20 Geographic Area Code: n-us-md n-us-va

Chesapeake futures: choices for the 21st century / edited by Donald F. Boesch and Jack Greer. Published/Created: Edgewater, MD: Chesapeake Research Consortium, 2003. Related Names: Boesch, Donald F. Greer, Jack. Chesapeake Bay Program (U.S.). Scientific and Technical Advisory Committee. Description: ix, 160 p.: ill. (some col.); 28 cm. Notes: "An independent report by the Scientific and Technical Advisory Committee." Includes bibliographical references. Subjects: Estuarine pollution--Chesapeake Bay (Md. and Va.) Estuarine ecology--Chesapeake Bay (Md. and Va.) Water quality--Chesapeake Bay (Md. and Va.) Chesapeake Bay (Md. and Va.) Series: STAC publication; no. 03-001 LC Classification: QH545.W3 C455 2003 Dewey Class No.: 333.91/640916347 22

Chesapeake waters: four centuries of controversy, concern, and legislation / by Steven G. Davison ... [et al.]. Edition Information: 2nd ed. Published/Created: Centreville, Md.: Tidewater Publishers, 1997. Related Names: Davison, Steven G. (Steven Gebauer) Description: xiii, 272 p.: ill., map; 24 cm. ISBN: 0870335014 (hardcover) Notes: Previous ed. cataloged under: Capper, John, 1937- . Includes bibliographical references (p. 256-263) and index. Subjects: Water--Pollution--Health aspects--Chesapeake Bay Region (Md. and Va.)--History. Marine pollution--Health aspects--Chesapeake Bay (Md. and Va.)--History. Water--Pollution--Chesapeake Bay Region (Md. and Va.) --History. Marine pollution--Chesapeake Bay (Md. and Va.)--History. Water--Pollution--Chesapeake Bay Region (Md. and Va.) --Public opinion--History. Marine pollution--Chesapeake Bay (Md. and Va.)--Public opinion--History. Environmental policy--Chesapeake Bay Region (Md. and Va.) --History. Chesapeake Bay Region (Md. and Va.)--History. LC Classification: RA592.C45 C44 1997 Dewey Class No.: 363.739/4/0916347 21 Geographic Area Code: n-us-md n-us-va

Chlorinated hydrocarbons as a factor in the reproduction and survival of lake trout (Salvelinus namaycush) in Lake Michigan / Great Lakes Fishery Laboratory. Published/Created: Washington, D.C.: U.S. Dept. of the Interior, Fish and Wildlife Service, 1981. Related Names: Great Lakes Fishery Laboratory. Description: 42 p.: ill.; 26 cm. Notes: Includes index. Includes bibliographies. Subjects: Lake trout--Michigan, Lake--Effect of water pollution on. Chlorohydrocarbons--Physiological effect. Fishes--Michigan, Lake--Effect of water pollution on. Series: Technical papers of the U.S. Fish and Wildlife Service; 105 LC Classification: SH11 .A313 no. 105 QL638.S2 Dewey Class No.: 639 s 597/.55 19 Government Document No.: I 49.68:105 Geographic Area Code: nl----- n-us-mi

Clean air and water news. Serial Key Title: Clean air and water news Abbreviated Title: Clean air water news Published/Created: Chicago: Commerce Clearing House, c1969- Related Names: Commerce Clearing House. Description:

v.; 23 cm. Vol. 1, no. 1 (Jan. 2, 1969)-Current Frequency: Weekly Merger of: Clean air news 0578-4778 (DLC)sf 84001187 (OCoLC)3005743 Water control news 0511-3571 (DLC) 75003880 (OCoLC)1769492 ISSN: 0009-8612 Cancelled/Invalid LCCN: sn 85019616 Notes: Title from caption. SERBIB/SERLOC merged record Formed by the union of: Clean air news, and: Water control news. Subjects: Pollution--Periodicals. Air--Pollution--Law and legislation--United States --Periodicals. Noise control--Law and legislation--United States --Periodicals. Water--Pollution--Law and legislation--United States --Periodicals. Water resources development--Law and legislation--United States--Periodicals. Air Pollution--Periodicals. Water Pollution--Periodicals. LC Classification: TD180 .C54 NLM Class No.: W1 CL122N Dewey Class No.: 628/.05

Clean water act corporate counsel retreat and information exchange: course materials / American Bar Association, Section of Natural Resources, Energy and Environmental Law. Published/Created: Chicago, Ill.: American Bar Association, 1993- Related Names: American Bar Association. Section of Natural Resources, Energy, and Environmental Law. Description: v.; 28 cm. June 24, 1993-Current Frequency: Annual Continued by: Annual clean water act corporate counsel retreat and information exchange (DLC) 96640319 Notes: SERBIB/SERLOC merged record Subjects: Water--Pollution--Law and legislation--United States. LC Classification: KF3790.Z9 C585 Dewey Class No.: 344.73/046343 21

Clean Water Act permit guidance manual / Russell S. Frye ... [et al.]. Published/Created: New York, N.Y.: Executive Enterprises Publications Co, c1984. Related Names: Frye, Russell S.

Description: xvii, 644 p.; 28 cm. ISBN: 0880571365 (pbk.) Notes: Includes index. Subjects: Water--Pollution--Law and legislation--United States. Environmental law--United States. LC Classification: KF3790 .C55 1984 Dewey Class No.: 344.73/046343 347.30446343 19 Geographic Area Code: n-us---

Clean Water Act Section 303(d) list: Illinois' submittal for 1998. Published/Created: Springfield, Ill.: Illinois Environmental Protection Agency, Bureau of Water, Division of Water Pollution Control, Planning Section, [1998] Related Names: Illinois. Bureau of Water. Illinois. Division of Water Pollution Control. Planning Section. Description: 1 v. (various pagings): maps (some col.); 28 cm. Summary: The purpose of this report is to fulfill the requirements set forth in Section 303(d) of the Federal Clean Water Act (CWA) and the Water Quality Planning and Management regulation at 40 CFR Part 130. This report is submitted to the U.S. Environmental Protection Agency (USEPA) for review and approval of Illinois' list of water quality limited waters. It provides the state's supporting documentation required by 40 CFR Part 130.7 (b)(6) and rationale in fulfilling Section 303(d) requirements. Illinois has elected to provide its submittal of Section 303(d) requirements as a separate document as opposed to inclusion in the state's biennial 305(b) report. Notes: "April 1, 1998"--Cover. "IEPA/BOW/97-023."--Cover. Subjects: United States. Federal Water Pollution Control Act. Water quality management--Illinois. Water--Pollution--Law and legislation--Illinois. LC Classification: TD224.I3 .C58 1998 Dewey Class No.: 363.739/456/09773 21

Clean Water Act thirty-year retrospective: history and documents related to the federal statute. Published/Created: [Washington, D.C.]: Association of State and Interstate Water Pollution Control Administrators, c2004. Description: xxii, 785 p.: ill.; 28 cm. ISBN: 0615125220 Notes: Includes bibliographical references (p. 767-772) and index. Subjects: United States. Federal Water Pollution Control Act. Water--Pollution--Law and legislation--United States. Water quality management--United States--History. LC Classification: KF3790 .C552 2004

Clean Water Act update / Russell S. Frye ... [et al.]. Published/Created: New York, N.Y.: Executive Enterprises Publications, c1987. Related Names: Frye, Russell S. United States. Water Quality Act of 1987. 1987. Related Titles: Clean Water Act permit guidance manual. Description: iii, 198 p.; 28 cm. ISBN: 0880578319 (pbk.) Notes: Supplement and update of: Clean Water Act permit guidance manual. Subjects: Water--Pollution--Law and legislation--United States. LC Classification: KF3790 .C55 1984 Suppl. Dewey Class No.: 344.73/046343 347.30446343 19 Geographic Area Code: n-us---

Clean Water Act: corporate counsel retreat and information exchange: course materials, June 24, 1993, Chicago, Illinois. Published/Created: [Chicago]: American Bar Association, Section of Natural Resources, Energy, and Environmental Law, c1993. Related Names: American Bar Association. Section of Natural Resources, Energy, and Environmental Law. Description: 1 v. (various pagings); 28 cm. Subjects: Water--Pollution--Law and legislation--United States. LC Classification: KF3790.A2 C57 1993 Dewey Class No.: 344.73/046343 347.30446343 20 Geographic Area Code: n-us---

Clean Water Act: law and regulation: ALI-ABA course of study materials. Portion of Title: ALI-ABA course of study materials Serial Key Title: Clean Water Act Published/Created: Philadelphia, PA:

American Law Institute-American Bar Association Committee on Continuing Professional Education, c2002- Related Names: American Bar Association. Section of Environment, Energy, and Resources. Environmental Law Institute. American Law Institute-American Bar Association Committee on Continuing Professional Education. Description: v.: ill.; 28 cm. Oct. 23/25, 2002- Current Frequency: Annual ISSN: 1932-2178 Cancelled/Invalid LCCN: 2003265977 Notes: Vols. for 2002- cosponsored by the ABA Section of Environment, Energy, and Resources and the Environmental Law Institute. Latest issue consulted: Nov. 5/7, 2003. Subjects: United States. Federal Water Pollution Control Act. Water--Pollution--Law and legislation--United States. LC Classification: KF3790.Z9 C587 Dewey Class No.: 344 14

Clean water deskbook. Edition Information: 2nd ed., 1991 ed. Published/Created: Washington, D.C.: Environmental Law Reporter, Environmental Law Institute, 1991. Related Names: Environmental Law Institute. Related Titles: Environmental law reporter. Clean water desk book. Description: vi, 562 p.: ill.; 28 cm. ISBN: 0911937382 Notes: Includes bibliographical references. Subjects: Water--Pollution--Law and legislation--United States. LC Classification: KF3790 .C553 1991 Dewey Class No.: 344.73/046343 347.30446343 20

Clean water deskbook. Published/Created: Washington, D.C.: Environmental Law Reporter, Environmental Law Institute, 1988. Related Names: Environmental Law Institute. Related Titles: Environmental law reporter. Description: vi, 532 p.: ill., map; 28 cm. ISBN: 0911937242: Subjects: Water--Pollution--Law and legislation--United States. LC Classification: KF3790 .C553 1988 Dewey Class No.:

344.73/046343 347.30446343 19 Geographic Area Code: n-us---

Clean water handbook / by Patton, Boggs & Blow, Roy F. Weston, Inc.; editors, J. Gordon Arbuckle, Russell V. Randle; contributors, John C. Martin ... [et al.]. Published/Created: Rockville, MD, U.S.A.: Government Institutes, c1990. Related Names: Arbuckle, J. Gordon, 1942- Randle, Russell V., 1955- Martin, John C., 1953- Patton, Boggs & Blow. Roy F. Weston, inc. Description: viii, 446 p.; 28 cm. ISBN: 0865872104 (pbk.) Notes: "May 1990"--T.p. verso. Includes bibliographical references. Subjects: Water--Pollution--Law and legislation-- United States. LC Classification: KF3790 .C556 1990 Dewey Class No.: 344.73/046343 347.30446343 20 Geographic Area Code: n-us---

Clean Water/Clean Air Bond Act, annual report / New York State Department of Environmental Conservation. Published/Created: [Albany, N.Y.?]: The Dept., 1998- Related Names: New York (State). Dept. of Environmental Conservation. Description: v.: ill.; 28 cm. Vols. 1998-99- issued in 2 vols. [v.1] annual report, v.2. financial summary. June 1, 1998- Current Frequency: Annual Additional Formats: Electronic version (Scanned image TIFF format) available via New York State Library WWW site. Subjects: Air quality management--New York (State)--Finance --Periodicals. Water quality management--New York (State)-- Finance --Periodicals. Air--Pollution--Law and legislation--New York (State) -- Periodicals. Water--Pollution--Law and legislation--New York (State) -- Periodicals. Municipal bonds--New York (State)--Periodicals. LC Classification: HC107.N73 A43 Government Document No.: ENV 214-3 CLEWC 99-86 nydocs

Clean waters / New York State Water Pollution Control Board. Published/Created:

[Albany, N.Y.: N.Y. State Dept. of Health, 1952- Related Names: New York (State). Water Pollution Control Board. Description: v.; 25 cm. Vol. 1, no. 1 (May 1952)- Current Frequency: Quarterly Notes: Title from caption. SERBIB/SERLOC merged record Subjects: Water--Pollution--New York (State)--Periodicals. LC Classification: WMLC L 83/1119 Government Document No.: WAT 156-3 CLEWA 98-48 nydocs

Cleaner Production Challenge: a voluntary resource conservation effort, final report / Washington State Dept. of Ecology. Published/Created: [Olympia, Wash.]: Washington State Dept. of Ecology, Hazardous Waste & Toxics Reduction Program, [2004] Related Names: Washington (State). Dept. of Ecology. Washington (State). Hazardous Waste and Toxics Reduction Program. Description: 22 p.: ill., form; 28 cm. Summary: The Cleaner Production Challenge is a non-enforcement project designed to help industries reduce the amount of water used, wastewater produced and hazardous sludge generated. It will offer better ways to conserve energy and process chemicals. Notes: "November 2004." Additional Formats: Also available on the Internet. Subjects: Factory and trade waste--Washington (State)--Management. Waste minimization--Washington (State) Water use--Washington (State)--Management. Water--Pollution--Washington (State)--Prevention. Series: Publication; no. 04-04-025 Publication (Washington (State). Dept. of Ecology); 04-04-025. LC Classification: TD897.75.W2 C56 2004 Dewey Class No.: 363.72/8 22 Government Document No.: WA 574.5 Ec7cle p 2004 wadocs

Clean-up of former Soviet military installations: identification and selection of environmental technologies for use in Central and Eastern Europe / edited by Roy C. Herndon ... [et al.]. Published/Created:

Berlin; New York: Springer, c1995. Related Names: Herndon, Roy C. NATO Advanced Research Workshop "Identification and Selection of Technologies for Use at Former Soviet Military Installations in Central and Eastern Europe (1994: Visegrad, Hungary) Description: viii, 250 p.: ill.; 24 cm. ISBN: 3540590781 Notes: "Proceedings of the NATO Advanced Research Wrorkshop "Identification and Selection of Technologies for Use at Former Soviet Military Installations in Central and Eastern Europe", held at Visegrad, Hungary, June 21-23, 1994"-- T.p. verso. Includes bibliographical references and index. Subjects: Hazardous waste site remediation--Europe, Central --Congresses. Hazardous waste site remediation--Europe, Eastern --Congresses. Soil pollution--Europe, Central--Congresses. Soil pollution--Europe, Eastern--Congresses. Water--Pollution--Europe, Central--Congresses. Water--Pollution--Europe, Eastern--Congresses. Soviet Union--Armed Forces--Facilities--Congresses. Series: NATO ASI series. Partnership sub-series 2, Environment; vol. 1 LC Classification: TD1045.C36 C58 1995 Dewey Class No.: 363.73/84 20 Geographic Area Code: ec----- ee----- e-ur---

Clearwater [sound recording] / production, Pete Seeger [and] John Milliken. Published/Created: Poughkeepsie, N.Y.: Hudson River Sloop Restoration, [1974] Related Names: Milliken, John, producer. Seeger, Pete, 1919- Collier, Jimmy. Killen, Louis. McGinness, Tim. Willcox, Peter. Bok, Gordon. McLean, Don, 1945- Wyatt, Lorre. Wallace, Andy. Sullivan, Keith, singer. Kirkpatrick, Frederick Douglass. Hudson River Sloop Restoration, Inc. Related Titles: Sloop song. Description: 1 sound disc: analog, 33 1/3 rpm, stereo.; 12 in. Publisher No.: PS 1001 Hudson River Sloop Restoration Contents: Old Father

Hudson -- Love our river again -- Haul on the bowline -- Strike the bell -- Mrs. MacDonald's lament -- Tapestry -- Sweet Rosyanne -- Sailing up, sailing down -- Reuben Ranzo -- See Rock City -- Once a boat has broken from the shore -- You can't eat the oysters -- Good mornin' Brother Hudson -- Shenandoah -- Seaman's hymn. Notes: Title on pamphlet: Sloop song. Lyrics contributed by Gordon Bok, Jimmy Collier, Bud Foote, Brother Fred Kirkpatrick, A.L. Lloyd, Alan Lomax, Don McLean, Pete Seeger, Keith Sullivan, and Lorre Wyatt. Studio and production assistance provided by Sound House Records, Newburgh, N.Y. Pamphlet published by Hudson River Sloop Restoration, Inc. (88 Market Street, Poughkeepsie, N.Y. 12601) inserted in container. Performer Notes: Lead vocals by Pete Seeger, Jimmy Collier, Louis Killen, Tim McGinness and Peter Willcox, Gordon Bok, Don McLean, Lorre Wyatt, Andy Wallace, Keith Sullivan, and Brother Fred Kirkpatrick; accompanying themselves or with various instrumentalists. Subjects: Sea songs. Water--Pollution--Hudson River (N.Y. and N.J.)--Songs and music. LC Classification: Hudson River Sloop Restoration PS 1001

Climate and the environment. Edition Information: North American ed. Published/Created: Milwaukee, WI: World Almanac Library, 2002. Description: 64 p.: col. ill., col. maps; 29 cm. ISBN: 0836850068 (lib. bdg.) Summary: Describes Earth's environment and various climates, as well as the damage being done by air, water, and soil pollution to the Earth and its inhabitants. Notes: Includes bibliographical references (p. 63) and index. Subjects: Climatology--Juvenile literature. Environmental sciences--Juvenile literature. Climatology. Environmental sciences. Pollution. Series: 21st century science LC Classification:

QC981.3 .C58 2002 Dewey Class No.: 551.6 21

Coan River Watershed total maximum daily load (TMDL) report for six shellfish areas listed due to bacteria contamination / Virginia Department of Environmental Quality. Published/Created: [Richmond, Va.]: Dept. of Environmental Quality, [2003] Related Names: Virginia. Dept. of Environmental Quality. Description: 1 v. (various pagings): ill. (some col.), maps (some col.); 28 cm. Notes: Title from cover. "July, 2003"--cover. Includes bibliographical references. c. 1-2, June 2005, dep., KFI Additional Formats: Report also available in PDF format at Virginia DEQ web site. Subjects: Water quality biological assessment--Virginia--Coan River Watershed (Northumberland County) Water--Pollution--Total maximum daily load--Virginia--Coan River Watershed (Northumberland County) Escherichia coli--Virginia--Coan River Watershed (Northumberland County) Shellfish populations--Virginia--Coan River Watershed (Northumberland County) LC Classification: TD224.V8 C63 2003

Coastal and estuarine risk assessment / edited by Michael C. Newman, Morris H. Roberts, Jr., Robert C. Hale. Published/Created: Boca Raton: Lewis Publishers, c2002. Related Names: Newman, Michael C. Roberts, Morris H. Hale, Robert C. Description: 347 p.: ill., maps; 24 cm. ISBN: 1566705568 (alk. paper) Contents: Machine generated contents note: Chapter 1 -- Overview of Ecological Risk Assessment in Coastal -- and Estuarine Environments1 -- Morris H. Roberts, Jr, Michael C. Newman, and Robert C. Hale -- Chapter 2 -- European Approaches to Coastal and Estuarine Risk Assessment15 -- Mark Crane, Neal Sorokin, James R. Wheeler, Albania Grosso, -- Paul Whitehouse, and David

Morritt -- Chapter 3 -- Emerging
Contaminants of Concern in Coastal -- and
Estuarine Environments 41 -- Robert C.
Hale and Mark J. La Guardia -- Chapter 4 -
- Enhancing Belief during Causality.
Assessments: Cognitive Idols -- or Bayes's
Theorem? 73 -- Michael C. Newman and
David A. Evans -- Chapter 5 --
Bioavailability, Biotransformation, and
Fate of Organic Contaminants -- in
Estuarine Animals 97 -- Richard E Lee --
Chapter 6 -- The Bioaccumulation of
Mercury, Methylmercury, and Other Toxic
-- Elements into Pelagic and Benthic
Organisms127 -- Robert PR Mason --
Chapter 7 -- Dietary Metals Exposure and
Toxicity to Aquatic Organisms: --
Implications for Ecological Risk
Assessment151 -- Christian E. Schlekat,
Byeong-Gweon Lee, and Samuel N.
Luoma -- Chapter 8 -- Endocrine
Disruption in Fishes and Invertebrates:
Issues for Saltwater -- Ecological Risk
Assessment 189 -- Kenneth M.Y. Leung,
James R. Wheeler, David Morritt and Mark
Crane -- Chapter 9 -- The Use of Toxicity
Reference Values (TRVs) to Assess the
Risks That -- Persistent Organochlorines
Pose to Marine Mammals 217 -- Paul D.
Jones, Kurunthachalam Kannan, Alan L.
Blankenship, -- and John P Giesy --
Chapter 10 -- Effects of Chronic Stress on
Wildlife Pop pulations: A Modeling --
Approach and Case Study247 -- Diane E.
Nacci, Timothy R. Gleason, Ruth Gutjahr-
Gobell, Marina Huber, -- and Wayne R.
Munns, Jr -- Chapter 11 -- Structuring
Population-Based Ecological Risk
Assessments -- in a Dynam ic Landscape
273 -- Christopher E. Mackay, Jenee A.
Colton, and Gary Bigham -- Chapter 12 --
Incremental Chemical Risks and Damages
in Urban Estuaries: -- Spatial and
Historical Ecosystem Analysis 297 -- Dave
Ludwig and Timothy J. Iannuzzi -- Chapter
13 -- Ecological Risk Assessment in
Coastal and Estuarine Environments327 --
Michael C. Newman, Robert C. Hale, and

Morris H. Roberts, Jr. Notes: Includes
bibliographical references. Subjects:
Marine pollution. Estuarine pollution.
Ecological risk assessment. Coastal
animals--Effect of water pollution on.
Estuarine animals--Effect of water
pollution on. Series: Environmental and
ecological risk assessment series LC
Classification: QH545.W3 C59 2002
Dewey Class No.: 577.7/27 21

Coastal discharges: engineering aspects and
    experience: proceedings of the conference /
    organized by the Institution of Civil
    Engineers and held in London on 7-9
    October 1980. Published/Created: London:
    T. Telford, 1981. Related Names:
    Institution of Civil Engineers (Great
    Britain) Description: 216 p.: ill.; 30 cm.
    ISBN: 072770124X Notes: Includes
    bibliographical references. Subjects:
    Sewage disposal in the ocean--Europe--
    Congresses. Factory and trade waste--
    Europe--Congresses. Water--Pollution--
    Europe--Congresses. Coastal engineering--
    Europe--Congresses. LC Classification:
    TD763 .C63 1981 Dewey Class No.:
    363.7/28 19 Geographic Area Code: e------

Colorado forest stewardship guidelines to
    protect water quality: best management
    practices (BMPs) for Colorado.
    Published/Created: [Fort Collins, Colo.:
    Colorado State Forest Service; Durango,
    CO: Colorado Timber Industry
    Association, 1998] Related Names: Logan,
    Robert S. (Robert Steven), 1945- Clinch,
    Bud. Colorado State Forest Service.
    Colorado Timber Industry Association.
    Description: 33 p.: ill.; 28 cm. Notes:
    Cover title. Adapted from original text
    authored by: Bob Logan ... Bud Clinch."--
    P. [2] of cover. "February 1998."--P. [2] of
    cover. Subjects: Forestry engineering--
    Colorado--Handbooks, manuals, etc.
    Water--Pollution--Colorado--Handbooks,
    manuals, etc. Forests and forestry--
    Colorado--Handbooks, manuals, etc. LC

Classification: SD388 .C654 1998
Government Document No.:
UCSU20/3.2/ST4/1998 codocs

Combating a deadly menace: early experiences
with a community-based arsenic mitigation
project in Bangladesh, June 1999-June
2000 / [research team, A.M.R. Chowdhury
... [et al.]]. Published/Created: Dhaka:
BRAC, Research and Evaluation Division,
2000. Related Names: Chowdhury, A. M.
Raza. BRAC. Research and Evaluation
Division. Description: vii, 116 p.: ill., maps
(some col.); 25 cm. Summary: Report of
project "Action Research into Community-
Based Arsenic Mitigation" implemented in
Narayanganj and Jessore districts. Notes:
Includes bibliographical references (p. 88-
89). Subjects: Groundwater pollution--
Bangladesh. Arsenic--Toxicology--
Bangladesh. Arsenic--Environmental
aspects--Bangladesh. Water--Purification--
Arsenic removal--Bangladesh. Series:
Research monograph series; no. 16
Research monograph series (BRAC.
Research and Evaluation Division) LC
Classification: TD427.A77 C66 2000
Overseas Acquisitions No.: B-E-2003-
314011; 26

Combined sewer overflow policy: federal Clean
Water Act & Pa. clean streams law
regulation comes to local government.
Published/Created: [Harrisburg, Pa.] (104
S. St., Harrisburg 17108-1027):
Pennsylvania Bar Institute, c1995. Related
Names: Pennsylvania Bar Institute.
Description: ix, 114 p.: ill.; 28 cm. Notes:
Includes bibliographical references (p. 59-
60). Subjects: Sewage disposal--Law and
legislation--Pennsylvania. Water--
Pollution--Law and legislation--
Pennsylvania. Combined sewer overflows-
-United States. Series: PBI; no. 1995-999
PBI (Series); no. 1995-999. LC
Classification: KFP359.S4 C66 1995
Dewey Class No.: 344.748/04622 21
Geographic Area Code: n-us-pa n-us---

Combined sewer overflow pollution abatement
/ prepared by Task Force on CSO Pollution
Abatement, under the direction of the
Facilities Development Subcommittee,
Technical Practice Committee.
Published/Created: Alexandria, VA: Water
Pollution Control Federation, 1989.
Related Names: Water Pollution Control
Federation. Task Force on CSO Pollution
Abatement. Water Pollution Control
Federation. Facilities Development
Subcommittee. Description: x, 272 p.: ill.;
26 cm. ISBN: 0943244323: Notes:
Includes bibliographical references (p.
262-266). Subjects: Combined sewer
overflows--Environmental aspects. Water--
Pollution. Sewerage--Management. Series:
Manual of practice; no. FD-17 Manual of
practice. FD; no. 17. LC Classification:
TD428.S47 C66 1989 Dewey Class No.:
628/.21 20

Commercial fishery investigations of the
Kentucky River / principal investigator,
John C. Williams; sponsored by United
States Department of Commerce, National
Oceanic and Atmospheric Administration,
National Marine Fisheries Service, and
Kentucky Department of Fish and Wildlife
Resources. Published/Created: Richmond:
Eastern Kentucky University, 1975.
Related Names: Williams, John C., 1925-
United States. National Marine Fisheries
Service. Kentucky. Dept. of Fish and
Wildlife Resources. Eastern Kentucky
University. Dept. of Biological Sciences.
Description: 3 v.; 29 cm. Contents: pt. 1.
Jobs 1 and 2: Fish population studies and
mussel bed surveys.--pt. 2. Job 3: Water
and silt analysis.--pt. 3. Jobs 4 and 5: Fish
and mussel tissue analysis. Notes:
"[Project] subcontracted to Department of
Biological Sciences, Eastern Kentucky
University, Richmond, Kentucky."
Includes bibliographies. Subjects:
Fisheries--Kentucky--Kentucky River.
Freshwater fishes--Kentucky--Kentucky
River. Freshwater mussels--Kentucky--

Kentucky River. Water--Pollution--Kentucky--Kentucky River. Fish as food--Analysis. LC Classification: SH222.K4 C65 Dewey Class No.: 639/.21/097693 Geographic Area Code: n-us-ky

Commonwealth of Virginia Shenandoah and Potomac River Basins tributary nutrient reduction strategy / Virginia Secretary of Natural Resources ... [et al.]. Portion of Title: Shenandoah and Potomac River Basins tributary nutrient reduction strategy Published/Created: [Virginia]: Commonwealth of Virginia, 1996. Related Names: Virginia. Secretary of Natural Resources. Description: viii, 90 p.: map; 28 cm. Notes: "December 1996." Subjects: Nutrient pollution of water--Shenandoah River Watershed (Va. and W. Va.) Water quality management--Shenandoah River Watershed (Va. and W. Va.) Nutrient pollution of water--Potomac River Watershed. Water quality management--Potomac River Watershed. LC Classification: TD427.N87 C66 1996 Dewey Class No.: 363.739/46/097559 21 Geographic Area Code: n-us-va n-us-wv

Community-level aquatic system studies: interpretation criteria / edited by Jeffrey M. Giddings ... et al. Published/Created: Pensacola, Fla.: Society of Environmental Toxicology and Chemistry, 2002. Related Names: Giddings, J. M. (Jeffrey M.) CLASSIC Workshop (1999: Schmallenberg, Germany) Description: xvi, 44 p.: ill.; 26 cm. ISBN: 188061149X (alk. paper) Notes: Cover title. Proceedings from the CLASSIC Workshop held at the Fraunhofer Institute, Schmallenberg, Germany, May 30-June 2, 1999. Includes bibliographical references (p. 41). Subjects: Ecological risk assessment--Congresses. Aquatic organisms--Effect of water pollution on --Congresses. Pesticides--Risk assessment--Congresses. LC Classification: QH541.15.R57 C65 2002 Dewey Class No.: 577.6/27 21

Comparative uptake and biodegradability of DDT and methoxychlor by aquatic organisms [by] Keturah A. Reinbold [and others] Published/Created: Urbana, Ill., Dept. of Registration and Education, Natural History Survey Division, 1971. Related Names: Reinbold, Keturah A. Description: 405-417 p. illus. 26 cm. Notes: Bibliography: p. 415. Subjects: Pesticides--Environmental aspects. DDT (Insecticide) Methoxychlor--Environmental aspects. Aquatic animals--Effect of water pollution on. Series: Illinois. Natural History Survey. Bulletin; v. 30, article 6 Bulletin (Illinois. Natural History Survey Division); v. 30, article 6. LC Classification: QH1 .I25 vol. 30, art. 6 QH545.P4 Dewey Class No.: 574/.08 s 632/.951

Compilation of E.P.A.Ê¾s sampling and analysis methods / edited by Lawrence H. Keith; compiled by William Mueller & David L. Smith. Published/Created: Boca Raton, Fla.: CRC Press, c1991. Related Names: Keith, Lawrence H., 1938- Mueller, William, 1929- Smith, David, 1949- Related Titles: Compilation of EPAÊ¾s sampling and analysis methods. E.P.A.Ê¾s sampling and analysis methods. EPAÊ¾s sampling and analysis methods. Description: 803 p.; 24 cm. ISBN: 0873714334 Notes: Also issued as a computer file. Subjects: Water--Pollution--Measurement--Handbooks, manuals, etc. Pollutants--Analysis--Handbooks, manuals, etc. Water--Sampling--Handbooks, manuals, etc. LC Classification: TD423 .C65 1991 Dewey Class No.: 628.1/61 20

Compliance and enforcement policy for the habitat protection and pollution prevention provisions of the Fisheries Act. Running Title: Habitat protection and pollution prevention provisions of the Fisheries Act Published/Created: [Ottawa]: Environment Canada, c2002. Related Names: Canada.

Environment Canada. Description: 43 p.; 28 cm. ISBN: 0662319087 Notes: Issued also in French under title: Politique de conformiteì et d'application des dispositions de la Loi sur les peì,ches pour la protection de l'habitat du poisson et la preì vention de la pollution. Distributed by the Government of Canada Depository Services Program. "November 2001." Subjects: Canada. Fisheries Act. Canada. Loi sur les peì,ches. Fish habitat improvement--Government policy--Canada Water--Pollution--Government policy--Canada. Fishery law and legislation--Canada. Fishes--Habitat--Environmental aspects--Canada. Fishery conservation--Canada. Peì,ches--Droit--Canada. Poissons--Habitat--Aspect de l'environnement--Canada. Peì,ches--Conservation--Canada. Eau--Pollution--Politique gouvernementale--Canada. LC Classification: SH157.8 .C648 2002 Canadian Class No.: COP.CA.2.2002-607 Dewey Class No.: 343.71/07692 21 Government Document No.: En40-659/2002E National Bibliographic Agency No.: 20029800846

Complying with the Edwards Aquifer rules: administrative guidance / prepared by Field Operations Division. Published/Created: Austin, TX: Texas Natural Resource Conservation Commission, [1999] Related Names: Texas Natural Resource Conservation Commission. Description: 1 v. (various pagings): maps; 28 cm. Notes: "RG-349 June 1999." Subjects: Water--Pollution--Law and legislation--Texas. Environmental law--Texas. Pollution prevention--Texas. Edwards Aquifer (Tex.) LC Classification: KFT1648.E39 C66 1999 Dewey Class No.: 344.764/046343 21 Government Document No.: N330.7 R339 NO.349 txdocs

Comprehensive coastal water quality monitoring program: report to the legislature / State Water Resources Control

Board, California Environmental Protection Agency. Published/Created: Sacramento: State Water Resources Control Board, [2001] Related Names: California Environmental Protection Agency. State Water Resources Control Board. Southern California Coastal Water Research Project. Related Titles: Pollutant mass emissions to the coastal ocean of California. Description: 1 v. (various pagings): ill., map; 28 cm. Notes: Cover title. "January 2001." Includes: Pollutant mass emissions to the coastal ocean of California. Westminster, Calif: Southern California Coastal Water Research Project ... [et al., 2000] (ii, 21 p.). Includes bibliographical references. Subjects: Water quality management--California--Pacific Coast. Coastal zone management--California. Water quality--California--Pacific Coast--Measurement. Environmental monitoring--California--Pacific Coast. Marine pollution--California--Measurement. LC Classification: TD224.C3 C68 2001

Comprehensive pollution survey & studies of Ganga River Basin in West Bengal, June 1980-December 1981 / by Center for Study of Man & Environment, Calcutta for West Bengal Prevention & Control of Water Pollution Board, Calcutta. Published/Created: New Delhi: Central Board for the Prevention and Control of Water Pollution, [1982] Related Names: Center for Study of Man & Environment (Calcutta, India) West Bengal Prevention & Control of Water Pollution Board. India. Central Board for the Prevention and Control of Water Pollution. Description: 52, lix p.: maps (some folded); 30 cm. Notes: Includes index. Bibliography: p. [lviii] Subjects: Water--Pollution--Environmental aspects--Ganges River Watershed (India and Bangladesh) Land use--Environmental aspects--Ganges River Watershed (India and Bangladesh) Human settlements--India--West Bengal. Series:

Assessment and development study of river basin series; ADSORBS 4 (1980-81 Assessment and development study of river basin series; 1980-81/4. LC Classification: TD303.G36 C66 1982 Dewey Class No.: 363.7/394/095414 19 Overseas Acquisitions No.: I E 46628 Geographic Area Code: a-ii--- a-bg---

Concentration of metal pollutants in bacterial biofilms and inhibition of oyster settlement and metamorphosis / Marianne Walch ... [et al.] Published/Created: [Baltimore: Maryland Dept. of Natural Resources, 1997?] Related Names: Walch, Marianne. Maryland. Tidewater Administration. Description: 80 p.; 28 cm. Notes: Final project report. "Contract no. CB93-004-004." Prepared for Tidewater Administration, Chesapeake Bay Research and Monitoring Division, Maryland Department of Natural Resources. "CBWP-MANTA-TR-97-3"-- Cover. "Literature cited": p. 73-80. Subjects: Metals--Environmental aspects--Chesapeake Bay (Md. and Va.) Oysters--Effect of water pollution on--Chesapeake Bay (Md. and Va.) LC Classification: QH545.M45 C65 1997

Concentrations and transport of atrazine in the Delaware River-Perry Lake system, northeast Kansas, July 1992 through September 1995 / by Larry M. Pope ... [et al.]. Published/Created: Washington, DC: U.S. G.P.O.; Denver, CO: For sale by the U.S. Geological Survey, Branch of Information Services, [1997] Related Names: Pope, Larry M. Kansas. State Conservation Commission. Kansas. Dept. of Agriculture. Kansas State University. Description: v, 43 p.: col. ill., col. maps; 28 cm. ISBN: 0607870451 (pbk.) Notes: "Prepared in cooperation with the Kansas State Conservation Commission, Kansas State University, and the Kansas Department of Agriculture." Includes bibliographical references (p. 42-43).

Subjects: Atrazine--Environmental aspects--Kansas--Delaware River Watershed. Atrazine--Environmental aspects--Kansas--Perry Lake Watershed (Jefferson County) Water--Pollution--Kansas--Delaware River Watershed. Water--Pollution--Kansas--Perry Lake Watershed. Series: U.S. Geological Survey water-supply paper; 2489 LC Classification: TD427.A86 C65 1997 Dewey Class No.: 363.738/4 21 Geographic Area Code: n-us-ks

Concentrations of selected herbicides, herbicide metabolites, and nutrients in outflow from selected midwestern reservoirs, April 1992 through September 1993 / by Elisabeth A. Scribner ... [et al.]; prepared as part of the Toxic Substances Hydrology Program. Published/Created: Lawrence, Kan.: U.S. Geological Survey; Denver, CO: Branch of Information Services [distributor], 1996. Related Names: Scribner, Elisabeth A. U.S. Geological Survey Toxic Substances Hydrology Program. Geological Survey (U.S.) Description: iv, 128 p.: ill., maps; 28 cm. Notes: Chiefly tables. Includes bibliographical references (p. 14, 18). Subjects: Herbicides--Environmental aspects--Middle West. Nutrient pollution of water--Middle West. Series: U.S. Geological Survey open-file report; 96-393 LC Classification: TD427.H36 C655 1996 Dewey Class No.: 363.738/4 21

Concentrations of selected herbicides, two triazine metabolites, and nutrients in storm runoff from nine stream basins in the midwestern United States, 1990-92 / by Elisabeth A. Scribner ... [et al.]. Published/Created: Lawrence, Kan.: U.S. Dept. of the Interior, U.S. Geological Survey; Denver, Colo.: U.S. Geological Survey, Earth Science Information Center, Open-File Reports Section [distributor], 1994. Related Names: Scribner, Elisabeth A. Description: iv, 144 p.: ill., map; 28 cm. Notes: Chiefly tables. Includes bibliographical references (p. 20-22).

Subjects: Herbicides--Environmental aspects--Middle West. Nutrient pollution of water--Middle West. Runoff--Environmental aspects--Middle West. Watersheds--Middle West. Series: U.S. Geological Survey open-file report; 94-396 LC Classification: TD427.H46 C66 1994

Concentrations of selected organochlorine compounds in fish tissue in the Mississippi Embayment Study Unit: Arkansas, Kentucky, Louisiana, Mississippi, Missouri, and Tennessee, 1995-99 / by Suzanne R. Femmer ... [et al.]. Published/Created: Reston, Va.: U.S. Dept. of the Interior, U.S. Geological Survey; Denver, CO: For sale by U.S. Geological Survey, Information Services, 2004. Related Names: Femmer, Suzanne R. National Water-Quality Assessment Program (U.S.) Geological Survey (U.S.) Description: viii, 36 p.: ill., col. maps; 28 cm. Notes: "National Water-Quality Assessment Program." Includes bibliographical references (p. 16-18). Subjects: Organochlorine compounds--Environmental aspects --Mississippi Embayment. Water quality biological assessment--Mississippi Embayment. Fishes--Effect of water pollution on--Mississippi Embayment. Series: Scientific investigations report; 2004-0059 LC Classification: TD427.O7 C66 2004 Dewey Class No.: 363.738/4982/0976 22 Geographic Area Code: n-usm--

Condition of freshwaters in Washington State for the year ... Variant Title: Condition of fresh waters in Washington State for the year <2002-> Published/Created: Olympia, WA: Washington State Dept. of Ecology Related Names: Environmental Assessment Program (Wash.). Freshwater Monitoring Unit. Washington (State). Dept. of Ecology. Description: v.: col. ill., col. maps; 28 cm. Cancelled/Invalid LCCN: 2002410212 Notes: Description based on: 2000; title from cover. Latest

issue consulted: 2003. Includes technical appendix. Issued by the Freshwater Monitoring Unit. Additional Formats: Also issued online. Subjects: Water quality--Washington (State)--Periodicals. Water--Pollution--Washington (State)--Periodicals. Water quality biological assessment--Washington (State) --Periodicals. Environmental monitoring--Washington (State)--Periodicals. Series: Publication Publication (Washington (State). Dept. of Ecology) LC Classification: TD224.W2 C66 Government Document No.: WA 574.5 Ec7con f2 wadocs

Conferences in connection with the International Water Conservancy Exhibition at Joìˆnkoìˆping, Sweden, 1-5 September, 1975. Published/Created: [s.l.: s.n., 1975?] Related Names: International Water Conservancy Exhibition, Joìˆnkoìˆping, Sweden, 1975. Description: 796 p. in various pagings: ill.; 21 cm. Notes: Cover title. English, French, or German. LC copy imperfect: pp. 95-97 bound incorrectly. Includes bibliographical references. Subjects: Water--Pollution--Congresses. Sewage--Purification--Congresses. LC Classification: TD419 .C65 Dewey Class No.: 628/.3 Language Code: engfreger National Bibliography No.: S***

Conferences in connection with the International Water Conservancy Exhibition at Joìˆnkoìˆping, Sweden August 28-September 3, 1972. Published/Created: [Joìˆnkoìˆping, 1973?] Related Names: International Water Conservancy Exhibition, Joìˆnkoìˆping, Sweden, 1972. Description: 1 v. (various pagings) illus. 21 cm. Notes: At head of title: Vaìˆrlden Vattnet och Vi. World water and we. Welt Wasser und Wir. Cover title. Includes bibliographical references. Subjects: Water--Pollution--Congresses. Sewage--Purification--Congresses. LC Classification: TD420

.C59 Dewey Class No.: 628.3 National
Bibliography No.: S***

Congressional Research Service reports
   [electronic resource]: redistributed as a
   service of the NLE. Variant Title: Title on
   HTML header: Congressional Research
   Service reports at the National Library for
   the Environment (NLE) Published/Created:
   Washington, DC: National Council for
   Science and the Environment, National
   Library for the Environment, Related
   Names: National Library for the
   Environment (U.S.) National Council for
   Science and the Environment (U.S.)
   Library of Congress. Congressional
   Research Service. Description: Probably
   begun after 1999. Computer File
   Information: Mode of access: World Wide
   Web. Summary: Contains reports and
   briefs on the following topics: agriculture,
   air, biodiversity, climate change,
   economics, energy, forests, marine,
   mining, natural resources, pesticides,
   pollution, population, public lands,
   stratospheric ozone, transportation, waste
   management, water, and wetlands. A
   search engine allows the user to sort by
   date, title, or code, and to search abstracts
   and reports by author's name, CRS code
   number, or topic. Includes links to new and
   updated reports, the NLE and the NCSE
   website, their directory, contact
   information, jobs and internships'
   opportunities, and to updates and
   announcements from the NCSE. Also
   contains links to the Science and Policy
   Program, the University Affiliates, the
   Minority Program, the National
   Conference on Science, Policy and the
   Environment, the Council of
   Environmental Deans and Directors, and
   the National Council on Science for
   Sustainable Forestry. Notes: Title from
   home page (viewed on Nov. 5, 2004).
   Point of entry to environmental
   information and data for the use of all
   participants in the environmental

enterprise. The documents were produced
by the Congressional Research Service,
branch of the Library of Congress, to
provide research reports to members of the
House and Senate. Subjects:
Environmental policy--United States.
Environmental protection--United States.
Environmental protection. Environmental
quality--United States. Environmental
quality. Environmental sciences. Natural
resources--United States. United States--
Environmental conditions. LC
Classification: GE180 Electronic File
Information:
http://www.ncseonline.org/NLE/CRS/

Connecticut River, 1980-1981 wastewater
   discharge data / prepared by Water Quality
   and Research Section, Massachusetts
   Division of Water Pollution Control.
   Published/Created: Westborough, Mass.:
   The Division, [1983] Related Names:
   Massachusetts. Water Quality and
   Research Section. Related Titles:
   Connecticut River wastewater discharge
   data, 1980-1981. Description: 48 p.: ill.; 28
   cm. Notes: Cover title: Connecticut River
   wastewater discharge data, 1980-1981.
   "April 1983." Subjects: Water--Pollution--
   Connecticut River Watershed. Sewage--
   Analysis. Sewage disposal plants--
   Massachusetts. LC Classification:
   TD225.C74 C627 1983 Dewey Class No.:
   363.7/3942/097442 19 Geographic Area
   Code: n-us-ma

Conservation and management of aquatic
   resources / editor Ashutosh Gautam.
   Published/Created: Delhi: Daya Pub.
   House, 1998. Related Names: Gautam,
   Ashutosh, 1965- Description: vii, 238 p.:
   ill., maps; 22 cm. ISBN: 8170351839
   Summary: With reference to India. Notes:
   Includes bibliographical references and
   index. Subjects: Aquatic ecology--India.
   Water--Pollution--Environmental aspects--
   India Aquatic resources conservation--
   India. Water quality management--India.

LC Classification: QH183 .C66 1998
Dewey Class No.: 333.91/1 21 Overseas
Acquisitions No.: I-E-98-907059; 49-13
Reproduction/Stock No.: Library of
Congress -- New Delhi Field Office
Rs480.00 Geographic Area Code: a-ii---

Contaminant boundary at the Faultless
underground nuclear test / prepared by
Greg Pohll ... [et al.]. Published/Created:
Las Vegas, Nev.: Desert Research Institute,
[2003] Related Names: Pohll, Greg.
University and Community College
System of Nevada. Desert Research
Institute. Description: viii, 52 p.: ill. (some
col.); 28 cm. Notes: Cover title. "April
2003." "Submitted to Nevada Site Office,
National Nuclear Security Administration,
U.S. Department of Energy, Las Vegas,
Nevada". "DOE/NV/13609-24." Includes
bibliographical references (p. 45-47).
Subjects: Underground nuclear explosions-
-Environmental aspects --Nevada Test Site.
Groundwater--Pollution--Nevada Test Site.
Radioactive pollution of water--Nevada
Test Site. Series: Publication; no. 45196
Desert Research Institute publication;
45196. LC Classification: TD427.R3 C657
2003

Contaminant effects on fisheries / edited by
Victor W. Cairns, Peter V. Hodson, Jerome
O. Nriagu. Published/Created: New York:
Wiley, c1984. Related Names: Cairns, V.
W. Hodson, Peter V. Nriagu, Jerome O.
Great Lakes Fishery Commission.
Description: xv, 333 p.: ill.; 25 cm. ISBN:
0471880140 Notes: Papers presented at a
workshop organized by the Great Lakes
Fishery Commission. "A Wiley-
Interscience publication." Includes
bibliographies and index. Subjects: Fishes-
-Effect of water pollution on. Water--
Pollution--Measurement. Series: Advances
in environmental science and technology;
v. 16 LC Classification: TD180 .A38 vol.
16 SH174 Dewey Class No.: 628 s
597/.05222 19

Contaminant fluxes through the coastal zone: a
symposium held in Nantes, 14-16 May
1984 / edited by Gunnar Kullenberg.
Published/Created: Copenhague K,
Danemark: Conseil international pour
l'exploration de la mer, [1986] Related
Names: Kullenberg, Gunnar. Symposium
on Contaminant Fluxes through the Coastal
Zone (1984: Institut scientifique et
technique des peìches maritimes)
Description: 485 p.: ill.; 27 cm. Notes:
English and French. "Symposium on
Contaminant Fluxes through the Coastal
Zone was held at the Institut scientifique et
technique des peìches maritimes in
Nantes"--P. 3. "Octobre 1986." Includes
bibliographies. Subjects: Water--Pollution-
-Congresses. Marine pollution--
Congresses. Pollutants--Congresses.
Coastal zone management--Congresses.
Series: Rapports et proceìes-verbaux des
reì unions, 0074-4336; v. 186 LC
Classification: GC1 .I66 vol. 186 TD419.5
Language Code: engfre

Contaminant levels in fish tissue from San
Francisco Bay: final report / San Francisco
Bay Regional Water Quality Control
Board, California Department of Fish and
Game, Marine Pollution Studies
Laboratory, State Water Resources Control
Board. Published/Created: [San Francisco]:
The Boards: The Dept., [1995] Related
Names: California Regional Water Quality
Control Board--San Francisco Bay Region.
California. Marine Pollution Studies
Laboratory. California. State Water
Resources Control Board. Description: ix,
135 p.: ill., maps; 28 cm. Notes: "June,
1995." Includes bibliographical references
(p. 47-49). Subjects: Fishes--Effect of
water pollution on--California--San
Francisco Bay. Fish as food--
Contamination--California--San Francisco
Bay. Fishes--California--San Francisco
Bay--Composition. Pollutants--California--
San Francisco Bay. LC Classification:
SH174 .C66 1995

Contaminant monitoring programmes using marine organisms: quality assurance and good laboratory practice / United Nations Environment Programme; prepared in co-operation with IOC, IAEA, FAO. Published/Created: Nairobi, Kenya: UNEP: Oceans and Coastal Areas Programme Activity Centre, United Nations Environment Programme [distributor]; Monaco: Marine Environmental Studies Laboratory, International Laboratory of Marine Radioactivity, IAEA [distributor], 1990. Related Names: United Nations Environment Programme. Intergovernmental Oceanographic Commission. International Atomic Energy Agency. Food and Agriculture Organization of the United Nations. Description: iii, 27 p.: ill.; 30 cm. Notes: "November 1990." Includes bibliographical references (p. 1-2). Subjects: Marine pollution--Measurement. Marine organisms--Effect of water pollution on. Series: Regional seas. Reference methods for marine pollution studies; no. 57 LC Classification: GC1085 .C664 1990

Contaminant problems and management of living Chesapeake Bay resources / edited by Shyamal K. Majumdar, Lenwood W. Hall, Jr., Herbert M. Austin. Published/Created: Easton, Pa.: Pennsylvania Academy of Science, c1987. Related Names: Majumdar, Shyamal K., 1938- Hall, Lenwood W. Austin, Herbert M. Description: xii, 573 p.: ill.; 24 cm. ISBN: 0960667075: Notes: Includes bibliographies and index. Subjects: Water--Pollution--Chesapeake Bay (Md. and Va.) Fishes--Effect of water pollution on--Chesapeake Bay (Md. and Va.) Fishery management--Chesapeake Bay (Md. and Va.) LC Classification: QH545.W3 C65 1987 Dewey Class No.: 363.7/3942/0916347 19 Geographic Area Code: n-us-md n-us-va

Contaminant trends in the Southern California Bight: inventory and assessment / Alan J. Mearns ... [et al.]. Published/Created: Seattle, Wash.: United States Dept. of Commerce, National Oceanic and Atmoshperic Administration, National Ocean Service, [1991] Related Names: Mearns, Alan J. Description: 1 v. (various pagings): ill., maps; 28 cm. Notes: "October 1991"--Cover. Includes bibliographical references. Subjects: Indicators (Biology)--Southern California Bight (Calif. and Mexico) Marine organisms--Effect of water pollution on--Southern California Bight (Calif. and Mexico) Mussels--Effect of water pollution on--Southern California Bight (Calif. and Mexico) Fishes--Effect of water pollution on--Southern California Bight (Calif. and Mexico) Marine sediments--Southern California Bight (Calif. and Mexico)--Analysis. Series: NOAA technical memorandum NOS ORCA; 62 LC Classification: QH91.57.B5 C66 1991 Dewey Class No.: 574.5/2636/0916432 20 Geographic Area Code: n-us-ca n-mx---

Contaminants and sediments / edited by Robert A. Baker. Published/Created: Ann Arbor, Mich.: Ann Arbor Science, c1980. Related Names: Baker, Robert Andrew, 1925- Description: 2 v.: ill.; 24 cm. ISBN: 025040270X (v. 1) 0250403072 (v. 2) Contents: v. 1. Fate and transport, case studies, modeling, toxicity.--v. 2. Analysis, chemistry, biology. Notes: Includes bibliographical references and indexes. Subjects: Dredging spoil--Environmental aspects--Congresses. Contaminated sediments--Congresses. Water--Pollution--Congresses. LC Classification: TD195.D72 C66 Dewey Class No.: 628.5/5 19

Contaminants in the Mississippi River, 1987-92 / edited by Robert H. Meade. Published/Created: [Washington, D.C.]: U.S. G.P.O.; Denver, CO: Free on application to U.S. Geology Service,

Information Services, Denver Federal Center, 1995. Related Names: Meade, Robert H., 1930- Description: xiii, 140 p.: col. ill., col. maps; 28 cm. Notes: Includes bibliographical references. Subjects: Water--Pollution--Mississippi River. Series: U.S. Geological Survey circular; 1133 LC Classification: TD223.4 .C657 1995 Dewey Class No.: 363.73/942/0977 20 Geographic Area Code: n-usm--

Contaminants in the sediments and biota from the Western Beaufort Sea / Nathalie J. Valette-Silver ... [et al.]. Published/Created: Silver Spring, Md.: U.S. Dept. of Commerce, National Oceanic and Atmospheric Administration, National Ocean Service, [1997] Related Names: Valette-Silver, Nathalie J. United States. National Ocean Service. National Status and Trends Program (U.S.) Description: 179 p.: ill.; 28 cm. Notes: "National Status and Trends Program for Marine Environmental Quality"--Cover. "December 1997." Includes bibliographical references (p. 52-74). Subjects: Contaminated sediments--Environmental aspects--Beaufort Sea. Marine organisms--Effect of water pollution on--Beaufort Sea. Water quality biological assessment--Beaufort Sea. Series: NOAA technical memorandum NOS ORCA; 108 LC Classification: QH545.C59 C66 1997 Dewey Class No.: 577.7/163/27 21 Geographic Area Code: r------

Contaminated soil sediment & water. Variant Title: Soil sediment & water Contaminated soil sediment and water Serial Key Title: Contaminated soil, sediment & water Abbreviated Title: Contam. soil sediment water Published/Created: Amherst, MA: Association for Environmental Health and Sciences, 2001- Related Names: Association for Environmental Health and Sciences. Description: v.: ill.; 28 cm. Feb. 2001- Current Frequency: 6 times a year (plus special issues) ISSN: 1533-4155

CODEN: CSSWAY Notes: Title from cover. Subjects: Soil pollution--Periodicals. Soil remediation--Periodicals. Contaminated sediments--Periodicals. Water--Pollution--Periodicals. Water--Purification--Periodicals. LC Classification: TD878 .C65522 Dewey Class No.: 363 13

Contamination of ground water by toxic organic chemicals. Published/Created: [Washington, D.C.]: Council on Environmental Quality: For sale by the Supt. of Docs., U.S. G.P.O., 1981. Related Names: Council on Environmental Quality (U.S.) Description: xxiv, 84 p.: ill., 1 map; 27 cm. Notes: "January 1981." S/N 041-011-00064-5 Item 856-E Includes bibliographical references. Subjects: Groundwater--Pollution--United States. Organic water pollutants--United States. Drinking water--Contamination--United States. LC Classification: TD223 .C689 Dewey Class No.: 363.7/394 19 Government Document No.: PrEx 14.2:C 76/2 Geographic Area Code: n-us---

Contamination of soils about Vanda Station, Antarctica / D.S. Sheppard ... [et al.] Published/Created: Lower Hutt, N.Z.: Institute of Geological & Nuclear Sciences, c1993. Related Names: Sheppard, D. S., 1951- Institute of Geological & Nuclear Sciences (N.Z.) Description: 140 p.: ill., maps; 30 cm. ISBN: 0478088108 0478088108 (pbk.): Notes: "May 1994." Includes bibliographical references. Subjects: Soil pollution--Antarctica--Vanda, Lake. Water--Pollution--Antarctica--Vanda, Lake. Series: Institute of Geological & Nuclear Sciences science report, 1171-9184; 94/20 Institute of Geological & Nuclear Sciences science report; 94/20. LC Classification: IN PROCESS (COPIED) (lcres)

Contributions to environmental geoscience: commemoration volume in honour of Prof. K.B. Powar / editors, A.M. Pathan, S.S.

Thigale. Edition Information: 1st ed. Published/Created: New Delhi: Aravali Books International, 2000. Related Names: Pathan, A. M. Thigale, S. S. (Satish Shripad), 1945- Powar, K. B. Description: 262 p.: ill., maps; 25 cm. ISBN: 8186880712 Summary: Comprises contributed articles in the Indian context. Notes: Series statement from jacket. Includes bibliographical references. Subjects: Environmental geology--India. Water--Pollution--India. Series: [Environmental studies series] Environmental studies series (New Delhi, India) LC Classification: QE38 .C68 2000 Dewey Class No.: 555.4 21 Overseas Acquisitions No.: I-E-99-953823; 49-91

Control of black flies in the Athabasca River: technical report / edited by W.O. Haufe and G.C.R. Croome. Published/Created: Edmonton, Alta.: Alberta Environment, Pollution Control Division, [1980] Related Names: Haufe, W. O. Croome, G. C. R. Description: xvii, 241 p.: ill.; 28 cm. Notes: "An interdisciplinary study for the chemical control of Simulium arcticum Malloch in relation to the bionomics of biting flies in the protection of human, animal, and industrial resources and its impact on the aquatic environment." "March 1980." Includes bibliographies. Subjects: Simulium arcticum--Control--Alberta--Athabasca River. Simulium arcticum--Control--Environmental aspects--Alberta --Athabasca River. Cattle--Parasites--Control--Alberta--Athabasca River. Cattle--Parasites--Control--Environmental aspects--Alberta --Athabasca River. Insecticides--Environmental aspects--Alberta--Athabasca River. Methoxychlor--Environmental aspects--Alberta--Athabasca River. Aquatic animals--Effect of insecticides on--Alberta --Athabasca River. Aquatic animals--Effect of water pollution on--Alberta --Athabasca River. Water--Pollution--Alberta--Athabasca

River. LC Classification: SF967.S57 C66 1980 Dewey Class No.: 636.089/6968 Geographic Area Code: n-cn-ab

Control of Pollution Act 1974, part II (pollution of water): code of good agricultural practice approved by the Secretary of State for Scotland for the purposes of section 31(2)(c) of the Act / Department of Agriculture and Fisheries for Scotland. Published/Created: Edinburgh: H.M.S.O., 1985. Related Names: Great Britain. Dept. of Agriculture and Fisheries for Scotland. Great Britain. Scottish Office. Description: 19 p.; 21 cm. ISBN: 0114924376 (pbk.): Notes: Bibliography: p. 17-18. Subjects: Water--Pollution--Law and legislation--Scotland. LC Classification: KDC684 .A33 1985 Dewey Class No.: 344.411/046343 344.110446343 19 Geographic Area Code: e-uk-st

Control of river pollution by industry / general report by Joseph Litwin, with national reports from Federal Republic of Germany, France, Netherlands, Poland, Sweden, United Kingdom, United States of America, and Yugoslavia. Published/Created: Brussels: International Institute of Administrative Sciences, 1965. Related Names: Litwin, Joìzef. International Institute of Administrative Sciences. Description: 227 p.; 23 cm. Notes: Issued also under title: La lutte contre la pollution des eaux par l'industrie. Includes bibliographical references. Subjects: Water--Pollution--Law and legislation. Series: Cases in comparative public administration LC Classification: K3590.4 .C65 Dewey Class No.: 344/.046343 342.446343 19

Controlling industrial laundry discharges in wastewater / prepared under the direction of the Pollution Prevention Committee. Published/Created: Alexandria, VA: Water Environment Federation, 1999. Related Names: WEF Pollution Prevention Committee. Description: xi, 119 p.: ill.; 26

cm. ISBN: 1572781572 Notes: Includes index. Subjects: Laundries--Waste disposal. Water--Pollution. Laundries--Waste disposal--United States--Case studies. Best management practices (Pollution prevention)--United States--Case studies. LC Classification: TD899.L37 C65 1999 Dewey Class No.: 628.1/683 21 Geographic Area Code: n-us---

Controlling pollution: principles and prospects: the government's response to the tenth report of the Royal Commission on Environmental Pollution. Published/Created: London: H.M.S.O., 1984. Related Names: Great Britain. Royal Commission on Environmental Pollution. Great Britain. Central Directorate on Environmental Pollution. Description: vii, 35 p.; 25 cm. ISBN: 0117517585 (pbk.): Notes: At head of title: Department of the Environment, Central Directorate of [i.e. on] Environmental Protection. Includes bibliographical references and index. Subjects: Pollution--Government policy--Great Britain. Environmental policy--Great Britain. Water quality management--Great Britain. Air quality management--Great Britain. Series: Pollution paper; no. 22 LC Classification: TD186.5.G7 C66 1984 Geographic Area Code: e-uk---

Controlling radionuclides and other contaminants in drinking water supplies: a workbook for small systems. Published/Created: Denver, CO: American Water Works Association, c1991. Related Names: American Water Works Association. Related Titles: Environmental pollution control alternatives. 1991. Description: vi, 111 p.: ill.; 28 cm. ISBN: 0898675839 Notes: The appendices (p. 32-110) consist of a document entitled "Environmental pollution control alternatives" which was published in 1990 by the USEPA's Center for Environmental Research Information. "1P-1M-20277-

9/91-MP"--P. 4 of cover. Includes bibliographical references (p. 111). Subjects: Radioactive pollution of water. Drinking water--Purification. Drinking water--Contamination. LC Classification: TD427.R3 C66 1991 Dewey Class No.: 363.6/1 20

Controlling toxic substances in agricultural drainage: emerging technologies and research needs: proceedings from the 1989 seminar, Sacramento, California, November 15-16, 1989 / co-sponsored by U.S. Committee on Irrigation and Drainage [and] Bureau of Reclamation; edited by T.E. Backstrom, Lydia J. Reid. Published/Created: Denver, CO: The Committee, c1990. Related Names: Backstrom, T. E. Reid, Lydia J. U.S. Committee on Irrigation and Drainage. United States. Bureau of Reclamation. Related Titles: USCID 1989 seminar. Description: v, 149 p.: ill.; 22 cm. ISBN: 0961825766 Notes: Spine title: USCID 1989 seminar. Includes bibliographical references. Subjects: Agricultural pollution--California--San Joaquin Valley --Congresses. Water--Pollution--Toxicology--California--San Joaquin Valley--Congresses. Irrigation water--Environmental aspects--California--San Joaquin Valley--Congresses. Drainage--Environmental aspects--California--San Joaquin Valley--Congresses. LC Classification: TD428.A37 C65 1990 Dewey Class No.: 628.1/684 20 Geographic Area Code: n-us-ca

Convention on the Protection and Use of Transboundary Watercourses and International Lakes: done at Helsinki, on 17 March 1992. Published/Created: Geneva: United Nations, 1992. Description: 18 p.; 30 cm. Notes: "E/ECE/1267." Subjects: Water resources development--Law and legislation--Europe. Water--Pollution--Law and legislation--Europe. Transboundary pollution--Law and legislation--Europe. International rivers--Europe. International lakes--Europe. LC Classification: KJC6094.A41992 A2 1992 Dewey Class No.: 346.404/691 344.064691 20 Geographic Area Code: e------

Cost-benefit analysis of treating saline groundwater / prepared by: AMEC Earth & Environmental Calgary, Alberta for Alberta Environment. Published/Created: [Edmonton]: Alberta Environment, 2007. Related Names: AMEC Earth & Environmental. Alberta. Alberta Environment. Description: v, 52 p.; 28 cm. ISBN: 9780778567233 (print) 0778567230 (print) 9780778567240 (online) 0778567249 (online) Computer File Information: System requirements for online access. Notes: "March 2007." Includes bibliographical references (p. 46-52). Additional Formats: Also available on the Internet. Subjects: Groundwater--Purification--Cost effectiveness. Groundwater--Alberta. Saline water conversion. Groundwater--Sodium content. Groundwater--Pollution. Geographic Area Code: n-cn-ab Electronic File Information: http://environment.gov.ab.ca/info/library/7816.pdf

Cotton production and water quality: economic and environmental effects of pollution prevention / Stephen R. Crutchfield ... [et al.]. Published/Created: Washington, DC: U.S. Dept. of Agriculture, Economic Research Service, [1992] Related Names: Crutchfield, Stephen R. United States. Dept. of Agriculture. Economic Research Service. Description: iii, 33 p.: ill., map; 28 cm. Notes: Cover title. "December 1992"--P. i. Includes bibliographical references (p. 23-24). Subjects: Agricultural chemicals--Environmental aspects--United States. Pollution prevention--Economic aspects--United States. Cotton growing--United States. Water quality--United States. Series: Agricultural economic report; no.

664 LC Classification: HD1751 .A91854 no. 664 TD427.A35 NAL Class No.: A281.9 Ag8A no.664 Dewey Class No.: 338.1 s 363.73/1 20

Cover crops for clean water: the proceedings of an international conference, West Tennessee Experiment Station, April 9-11, 1991, Jackson, Tennessee / W.L. Hargrove, editor. Published/Created: Ankeny, Iowa: Soil and Water Conservation Society, c1991. Related Names: Hargrove, W. L. (William Leonard) West Tennessee Experiment Station. Soil and Water Conservation Society (U.S.) Description: xi, 198 p.: ill.; 28 cm. ISBN: 0935734252: Notes: Includes bibliographical references. Subjects: Cover crops--Congresses. Water--Pollution--Congresses. Water quality management--Congresses. LC Classification: SB284 .C68 1991 Dewey Class No.: 631.4/52 20

CRC handbook of environmental control: cumulative series index for volumes I- / editors, Richard G. Bond, Conrad P. Straub; coordinating editor, Richard Prober. Published/Created: West Palm Beach, Fla.: CRC Press, c1978- Related Names: Bond, Richard G., 1916- Straub, Conrad P. Prober, Richard. Related Titles: Handbook of environmental control. Description: v.; 28 cm. ISBN: 084930279X Subjects: CRC handbook of environmental control--Indexes. Environmental engineering--Indexes. Air pollution--Prevention and control--Indexes. Environmental health--Indexes. Water pollution--Prevention and control--Indexes. LC Classification: TD145.C22 C17 NLM Class No.: WA670 H236 Index Dewey Class No.: 363.7/00212 19

Created and natural wetlands for controlling nonpoint source pollution / edited by Richard K. Olson. Published/Created: Boca Raton, FL: C.K. Smoley, c1993. Related Names: Olson, Richard K. United States. Environmental Protection Agency. Office of Research and Development. United States. Environmental Protection Agency. Office of Wetlands, Oceans, and Watersheds. Description: v, 216 p.: ill.; 25 cm. ISBN: 0873719433 (alk. paper) Notes: "U.S. EPA, Office of Research and Development, and Office of Wetlands, Oceans, and Watersheds." Includes bibliographical references. Subjects: Water quality management--United States. Nonpoint source pollution--United States. Wetland conservation--United States. Constructed wetlands--United States. LC Classification: TD223 .C73 1993 Dewey Class No.: 628.1/68 20 Geographic Area Code: n-us--- Electronic File Information: Publisher description http://www.loc.gov/catdir/enhancements/fy 0744/92031139-d .html

Critical assessment of radon progeny exposure while showering in radon-bearing water / prepared by Philip K. Hopke ... [et al.]; sponsored by AWWA Research Foundation. Published/Created: Denver, CO: AWWA Research Foundation and American Water Works Association, c1996. Related Names: Hopke, Philip K., 1944- AWWA Research Foundation. Description: xix, 73 p.: ill.; 28 cm. ISBN: 0898678307 Notes: "Sponsored by AWWA Research Foundation." Includes bibliographical references (p. 71-73). Subjects: Radon--Isotopes--Health aspects. Radiation dosimetry. Water--Pollution. Baths. LC Classification: RA1247.R33 C75 1996 Dewey Class No.: 363.17/99 20

Cryptosporidium spp. oocyst and Giardia spp. cyst occurrence, concentrations, and distribution in Wisconsin waters / by J.R. Archer ... [et al.]. Published/Created: Madison, WI: Wisconsin Dept. of Natural Resources, [1995] Related Names: Archer, J. R. Wisconsin. Dept. of Natural Resources. Description: xi, 81 p.: maps; 28 cm. Notes: "August 1995." "Publication No. WR420-95." Includes bibliographical

references (p. 37-43). Subjects: Cryptosporidium--Wisconsin. Giardia--Wisconsin. Water--Pollution--Wisconsin. LC Classification: TD427.C78 C79 1995 Government Document No.: ENV.2:C 79/1995 widocs

Current state of the lake report: a report of the findings of a regional information programme co-ordinated by the Journalists Environmental Association of Tanzania and Panos, London. Published/Created: [Nairobi: s.n., 1994] Related Names: Journalists Environmental Association of Tanzania. Panos, London. Description: 1 v. (various pagings): ill.; 30 cm. Notes: Cover title. "May, 1994." Subjects: Water--Pollution--Victoria, Lake. Victoria, Lake--Environmental conditions. LC Classification: TD319.E18 C87 1994

Cyanotoxins: occurrence, causes, consequences / Ingrid Chorus (Ed.). Published/Created: Berlin; New York: Springer, 2001. Related Names: Chorus, Ingrid. Description: xviii, 357 p.: ill.; 24 cm. ISBN: 3540649999 Notes: Includes bibliographical references. Subjects: Cyanobacterial toxins. Microcystins. Cyanobacteria. Freshwater microbiology. Bacterial pollution of water. LC Classification: QP632.C87 C95 2001 Dewey Class No.: 579.3/9165 21

Data profiles for chemicals for the evaluation of their hazards to the environment of the Mediterranean Sea / [by the Franklin Institute ... et al.]. Published/Created: Geneva: International Register of Potentially Toxic Chemicals, United Nations Environment Programme, 1978. Related Names: Franklin Institute (Munich, Germany) International Register of Potentially Toxic Chemicals (United Nations) Related Titles: Profils de donneì es pour l'eì valuation du danger des substances chimiques pour l'environnement de la Meì diterraneì e. Description: 2 v.: ill.; 28 cm. Notes: English and French. Title on added t.p.:

Profils de donneĺ es pour l'eĺ valuation du danger des substances chimiques pour l'environnement de la Meĺ diterraneĺ e. Includes bibliographies. Subjects: International Register of Potentially Toxic Chemicals (United Nations) Pollution--Environmental aspects--Mediterranean Region. Water--Pollution--Mediterranean Region. Hazardous wastes--Mediterranean Region. Series: IRPTC data profile series; no. 1 LC Classification: TD196.C45 D37 1978 Dewey Class No.: 363.1/79 19 Language Code: engfre Geographic Area Code: mm-----

Defining and assessing adverse environmental impact from power plant impingement and entrainment of aquatic organisms / editors, Douglas A. Dixon, John A. Veil, Joe Wisniewski. Published/Created: Lisse; Exton (PA): A.A. Balkema, c2003. Related Names: Dixon, Douglas A. Veil, John A. Wisniewski, Joe. American Fisheries Society. Meeting (2001: Phoenix, Ariz.) Description: viii, 291 p.: ill., maps; 25 cm. ISBN: 9058095177 Notes: Symposium held in conjunction with the Annual Meeting of the American Fisheries Society, Aug. 23rd 2001 in Phoenix, Arizona. Includes bibliographical references and index. Subjects: Electric power-plants--Waste disposal--Environmental aspects--Congresses. Electric power-plants--Water supply--Congresses. Water quality management--Congresses. Aquatic organisms--Effect of water pollution on --Congresses. LC Classification: TD195.E4 D44 2003 Dewey Class No.: 363.7394 21 National Bibliography No.: GBA3-24456

Degradability, ecotoxicity, and bio-accumulation: the determination of the possible effects of chemicals and waste on the aquatic environment. Published/Created: Hague, Netherlands: Staatsuitgeverij (Govt. Pub. Office), 1980. Description: 1 v. (various pagings): ill.; 31 cm. ISBN: 9012028019 (loose-leaf) Notes:

Includes bibliographies. Subjects: Water--Pollution--Environmental aspects. Biodegradation. LC Classification: QH545.W3 D43 Dewey Class No.: 628.1/6836/0287 19

Demonstration of a method to correlate measures of ambient toxicity and fish community diversity / by S. Ian Hartwell ... [et al.]. Published/Created: Annapolis, Md.: Maryland Dept. of Natural Resources, Tidewater Administration, Chesapeake Bay Research and Monitoring Division, [1995] Related Names: Hartwell, S. Ian. Maryland. Chesapeake Bay Research and Monitoring Division. Description: iii, 73 p.: ill., maps; 28 cm. Notes: "January 1995." "CBRM-TX-95-1"--Cover. Includes bibliographical references (p. 71-73). Subjects: Water--Pollution--Toxicology--Chesapeake Bay Watershed. Fishes--Effect of water pollution on--Chesapeake Bay Watershed. Analytical toxicology--Chesapeake Bay Watershed. Water--Pollution--Chesapeake Bay Watershed--Measurement. Chesapeake Bay Watershed. LC Classification: QH541.5.W3 D46 1995

Deposition of air pollutants to the Great Waters: second report to Congress. Published/Created: Research Triangle Park, NC: U.S. Environmental Protection Agency. Office of Air Quality Planning and Standards, [1997]. Related Names: United States. Environmental Protection Agency. Office of Air Quality Planning and Standards. Description: xiv, 218, 19, 7 p.: ill., maps; 28 cm. Notes: "June 1997." "EPA-453/R-97-011." Includes bibliographical references (p. 193-218). Subjects: Atmospheric deposition--Environmental aspects--United States. Water--Pollution--United States. LC Classification: TD427.A84 D47 1997 Dewey Class No.: 363.739/2 21

Detailed study of selenium and selected constituents in water, bottom sediment, soil, and biota associated with irrigation drainage in the San Juan River area, New Mexico, 1991-95 / by Carole L. Thomas ... [et al.]. Published/Created: Albuquerque, N.M.: U.S. Geological Survey, Water Resources Division; Denver, CO: Can be purchased from U.S. Geological Survey, Branch of Information Services, 1998. Related Names: Thomas, Carole L. Geological Survey (U.S.) Description: v, 84 p.: ill., col. maps; 28 cm. Notes: Cover has incorrect series title: U.S. Geological Survey open-file report. Includes bibliographical references (p. 82-84). Subjects: Selenium--Environmental aspects--San Juan River Region (Colo.-Utah) Drainage--Environmental aspects--San Juan River Region (Colo.-Utah) Irrigation water--Pollution--San Juan River Region (Colo.-Utah) Series: Water-resources investigations report; 98-4096 LC Classification: GB701 .W375 no. 98-4096a QH545.S45 Dewey Class No.: 553.7/0973 s 363.739/42/0979222259 21 Geographic Area Code: n-us-nm

Detailed study of selenium in glacial-lake deposits, wetlands, and biota associated with irrigation drainage in the southern Freezeout Lake area, west-central Montana, 1994-95 / by Eloise Kendy ... [et al.]. Published/Created: Helena, MT: U.S. Geological Survey; Denver, CO: U.S. Geological Survey, Branch of Information Services [distributor], 1999. Related Names: Kendy, Eloise. Geological Survey (U.S.) Description: v, 51 p.: ill., maps; 28 cm. Notes: "In cooperation with the U.S. Geological Survey, U.S. Fish and Wildlife Service, Bureau of Reclamation, Bureau of Indian Affairs." "May 1999"--T.p. verso. Includes bibliographical references (p. 49-51). Subjects: Selenium--Environmental aspects--Montana--Freezeout Lake Region. Water--Pollution--Montana--Freezeout Lake Region. Irrigation water--Environmental aspects--Montana--Freezeout Lake Region. Drainage--Environmental aspects--Montana--

Freezeout Lake Region. Series: Water-resources investigations report; 99-4019 LC Classification: GB701 .W375 no. 99-4019 TD427.S38

Detection methods for algae, protozoa, and helminths in fresh and drinking water / edited by Franca Palumbo, Giuliano Ziglio, Andreì van der Beken. Published/Created: Chichester, West Sussex; New York: J. Wiley, c2002. Related Names: Palumbo, Franca. Ziglio, G. Beken, Andreì van der. Description: xviii, 225 p.: ill.; 25 cm. ISBN: 0471899895 (alk. paper) Notes: Includes bibliographical references and index. Subjects: Waterborne infection. Water--Pollution--Health aspects. Algae--Health aspects. Protozoa--Health aspects. Helminths--Health aspects. Series: Water quality measurements series LC Classification: RA642.W3 D48 2002 Dewey Class No.: 628.1/61 21

Detection of caliciviruses in water samples / prepared by Xi Jiang ... [et al.]. Published/Created: Denver, CO: AWWA Research Foundation and American Water Works Association, c2001. Related Names: Jiang, Xi. AWWA Research Foundation. Description: xxiii, 122 p.: ill.; 28 cm. ISBN: 1583211144 (pbk.) Notes: "Subject area: monitoring and analysis"--Cover. Includes bibliographical references (p. 107-118). Subjects: Caliciviruses. Viral pollution of water. LC Classification: QR398.8 .D48 2001 Dewey Class No.: 628.1/61 21

Developing total maximum daily load projects in Texas: a guide for lead organizations. Published/Created: Austin, Tex.: Texas Natural Resource Conservation Commission, [1999] Related Names: Texas Natural Resource Conservation Commission. Description: 1 v. (various pagings): ill.; 28 cm. Notes: Cover title. "Cooperative project of the Texas Natural Resource Conservation Commission, the

Texas Institute for Applied Environmental Research, and the Texas A&M University System." "June 1999." Includes bibliographical references. Subjects: Water--Pollution--Total maximum daily load--Texas --Handbooks, manuals, etc. Watershed management--Government policy--Texas--Handbooks, manuals, etc. LC Classification: TD224.T4 D48 1999 Dewey Class No.: 363.739/4 21

Development and testing of a fish-based index of biological integrity to quantify the health of grassland streams in Alberta / Cameron Stevens ... [et al.]. Published/Created: [Edmonton]: Alberta Conservation Association, [2006] Related Names: Stevens, Cameron. Alberta Conservation Association. Description: ix, 56 p.: col. map; 28 cm. ISBN: 0778547868 (print) 9780778547860 (print) 0778547876 (online) 9780778547877 (online) Notes: Includes bibliographical references (p. 43-50). Subjects: Water quality bioassay--Alberta. Water quality biological assessment--Alberta. Environmental monitoring--Alberta. Water--Pollution--Measurement--Alberta. Fishes--Effect of water quality on--Alberta. Stream ecology--Alberta. Series: Publication Number; T/110 Conservation report series LC Classification: QH96.8.B5 D49 2006 Geographic Area Code: n-cn-ab

Development and testing of a groundwater management model for the Faultless Underground Nuclear Test, Central Nevada Test Area / prepared by Douglas P. Boyle ... [et al.] Published/Created: Las Vegas, Nev.: Desert Research Institute; Springfield, VA: Available for sale from NTIS, [2005] Related Names: Boyle, Douglas P. University and Community College System of Nevada. Division of Hydrologic Sciences. University and Community College System of Nevada. Desert Research Institute. Description: iv, 24 p.: ill. (some col.), col. maps; 28 cm.

Notes: "December 2005." "DOE/NV/13609-41". "Submitted to Nevada Site Office, National Nuclear Security Administration, U.S. Department of Energy, Las Vegas, Nevada." Includes bibliographical references (p. 24). Subjects: Groundwater--Nevada--Nevada Test Site. Groundwater flow--Nevada--Nevada Test Site. Radioactive pollution of water--Nevada Test Site. Series: Publication; no. 45212 LC Classification: GB1025.N4 D48 2005

Development of an expert system based on a tidal prism water quality model for small coastal basins in Virginia / by Albert Y. Kuo ... [et al.]. Published/Created: Gloucester Point, VA: School of Marine Science/Virginia Institute of Marine Science, The College of William and Mary, [1999] Related Names: Kuo, Albert. Description: iv, 61 leaves: ill. (some col.); 28 cm. Notes: "A report to the Virginia Coastal Resources Management Program, Virginia Dept. of Environmental Quality." "October 1999." Includes bibliographical references (leaves 59-61). Subjects: Water quality--Virginia--Mathematical models. Water--Pollution--Virginia--Mathematical models. Expert systems (Computer science)--Virginia. Series: Special report in applied marine science and ocean engineering; no. 357 LC Classification: TD224.V8 D486 1999 Dewey Class No.: 363.739/42/015118 21 Geographic Area Code: n-us-va

Development of land segmentation, stream-reach network, and watersheds in support of Hydrological Simulation Program--Fortran (HSPF) modeling, Chesapeake Bay watershed, and adjacent parts of Maryland, Delaware, and Virginia / by Sarah K. Martucci ... [et al.]; prepared in cooperation with the U.S. Environmental Protection Agency Chesapeake Bay Program Office ... [et al.]. Published/Created: Reston, Va.: U.S. Dept.

of the Interior, U.S. Geological Survey, 2006. Related Names: Martucci, Sarah K. Chesapeake Bay Program (U.S.) Geological Survey (U.S.) Description: iv, 15 p.: maps; 28 cm. + 1 CD-ROM (4 3/4 in.) Notes: Includes bibliographical references (p. 14-15). Subjects: Water--Pollution--Chesapeake Bay Watershed--Computer simulation. Water quality--Chesapeake Bay Watershed--Computer simulation. Hydrology--Chesapeake Bay Watershed--Computer simulation. Series: Scientific investigations report; 2005-5073 LC Classification: TD225.C43 .D48 2006

Development of methods for effects-driven cumulative effects assessment using fish populations: Moose River project / by Kelly R. Munkittrick ... [et al.]. Published/Created: Pensacola, Fla.: Society of Environmental Toxicology and Chemistry, c2000. Related Names: Munkittrick, Kelly R. Description: xviii, 236 p.: ill; 23 cm. ISBN: 1880611414 Notes: Includes bibliographical references (p. 215-221). Subjects: Fishes--Effect of water pollution on--Ontario--Moose River Watershed. Cumulative effects assessment (Environmental assessment) Series: SETAC technical publications series LC Classification: QL626.5.O6 D48 2000 Dewey Class No.: 597.176 21 Geographic Area Code: n-cn-on

Development of molecular methods to detect infectious viruses in water / prepared by Theresa L. Cromeans ... [et al.]; sponsored by Awwa Research Foundation. Published/Created: Denver, CO: Awwa Research Foundation, c2004. Related Names: Cromeans, Theresa L. AWWA Research Foundation. Description: xxv, 101 p.: ill.; 28 cm. Cancelled/Invalid LCCN: 2005274244 Notes: Includes bibliographical references (p. 93-98). Subjects: Viral pollution of water--Measurement. Sewage--Analysis. LC

Classification: TD427.V55 D48 2004 Dewey Class No.: 628.1/61 22

Development of practical methods to assess the presence of bacterial pathogens in water / by William A. Yanko ... [et al.]. Published/Created: Alexandria, VA: Water Environment Research Foundation, 2004. Related Names: Yanko, William A. Description: 1 v. (various pagings): ill.; 28 cm. ISBN: 1843396882 Notes: "97-HHE-1." Includes bibliographical references. Subjects: Bacterial pollution of water--Measurement. Water quality biological assessment. Water--Microbiology. LC Classification: TD427.B37 D48 2004 Dewey Class No.: 628.1/61 22

Diagnosis and prognosis--Barrier Island/Salt Marsh Estuaries, Southeast Atlantic Coast: issues, resources, status, and management: proceedings of a seminar held February 17, 1988, Washington, D.C. Published/Created: [Washington, D.C.]: U.S. Dept. of Commerce, National Oceanic and Atmospheric Administration, NOAA Estuarine Programs Office, [1989] Related Names: United States. NOAA Estuarine Programs Office. Related Titles: Barrier Island/Salt Marsh Estuaries, Southeast Atlantic Coast. Description: vii, 106 p.: ill.; 28 cm. Notes: Cover title: Barrier Island/Salt Marsh Estuaries, Southeast Atlantic Coast. "August 1989." Includes bibliographical references. Subjects: Water--Pollution--South Carolina. Water--Pollution--Georgia. Estuarine ecology--South Carolina. Estuarine ecology--Georgia. Series: NOAA estuary-of-the-month seminar series; no. 12 LC Classification: GC96 .D53 1989 Dewey Class No.: 333.91/64 20

Diazinon and chlorpyrifos loads in precipitation and urban and agricultural storm runoff during January and February 2001 in the San Joaquin River Basin, California / by Celia Zamora ... [et al.]; prepared in cooperation with California Department of

Pesticide Regulation. Published/Created: Sacramento, Calif.: U.S. Dept. of the Interior, U.S. Geological Survey; Denver, CO: Information Services [distributor], 2003. Related Names: Zamora, Celia. California Environmental Protection Agency. Dept. of Pesticide Regulation. Geological Survey (U.S.) Description: vi, 56 p.: col. ill., col. maps; 28 cm. Notes: Includes bibliographical references (p. 54-56). Additional Formats: Also available via Internet. Subjects: Water--Pollution--California--San Joaquin River Watershed. Diazinon--Environmental aspects--California--San Joaquin River Watershed. Chlorpyrifos--Environmental aspects--California--San Joaquin River Watershed. Series: Water-resources investigations report; 03-4091 LC Classification: GB701 .W375 no. 03-4091 Dewey Class No.: 553.7/0973 s 628.5/29 22

Diffuse pollution of water resources principles and case studies in the Southern African Region / editor, R. Hranova. Published/Created: London; New York: Taylor & Francis, c2006. Projected Publication Date: 0512 Related Names: Hranova, R. Description: p. cm. ISBN: 0415383919 (hardcover: alk. paper) Contents: Diffuse pollution: principles, definnitions and regulatory aspects / R. Hranova -- Monitoring, abatement, and management of diffuse pollution / R. Hranova -- Characteristics of an urban environment in the context of diffuse pollution control / R. Hranova -- Assesing and managing urban storm water quality / R. Hranova & M. Magombeyi -- Diffuse pollution in high-density (low-income) urban areas / R. Hranova -- Impacts on groundwater quality anhd water supply of the Epworth semi-formal settlement, Zimbabwe / D. Love ... [et al.] -- Impacts of a solid waste disposal site and a cemetery on the groundwater quality in Harare, Zimbabwe / W. Moyce ... [et al.] -- Sewage sludge disposal on land: impacts

on surface water quality / R. Hranova & M. Manjonjo -- Sewage sludge disposal on land: impacts on soils and groundater quality / R. Hranova & A. Amos -- Irrigation with ponds effluent: impacts on soils and groundwater / R. Hranova & W. Gwenzi -- Diffuse pollution of urban rivers: case studies in Malawi and Swaziland / R. Hranova, S. Nkambule & S. Mwandira -- Integrated management of diffuse pollution in the Southern African Region / R. Hranova. Notes: Includes bibliographical references. Subjects: Water--Pollution--Africa, Southern. Nonpoint source pollution--Africa, Southern. LC Classification: TD424.4.A356 D54 2006 Dewey Class No.: 628.1/68 22 Geographic Area Code: fs-----

Digest of environmental pollution and water statistics / Department of the Environment. Published/Created: London: H.M.S.O., 1980- Related Names: Great Britain. Dept. of the Environment. Description: v.: ill., maps; 30 cm. No. 3 (1980)- Merger of: Great Britain. Dept. of the Environment. Water Data Unit. Water data (DLC) 77644604 (OCoLC)3779072 Digest of environmental pollution statistics (DLC) 81645034 Continued by: Digest of environmental protection and water statistics (DLC) 88646983 (OCoLC)18203161 Notes: "A publication of the government statistical service." SERBIB/SERLOC merged record Merger of: Water data; and, Digest of environmental pollution statistics. Subjects: Pollution--Great Britain--Periodicals. Pollution--Great Britain--Statistics--Periodicals. Water-supply--Great Britain--Periodicals. Water-supply--Great Britain--Statistics--Periodicals. LC Classification: TD186.5.G7 D52 Dewey Class No.: 363.7/32/0941 19

Digest of water resources decisions, Pollution Control Hearings Board and appellate courts: cases through October 1, 2005 / editors, Maia Bellon ... [et al.]. Edition Information: 2005 ed. Published/Created: Olympia, WA: Distributed by Dept. of Ecology, Water Resources Program, [2006] Related Names: Bellon, Maia. Washington (State). Pollution Control Hearings Board. Washington (State). Dept. of Ecology. Washington (State). Dept. of Ecology. Water Resources Program. Description: vi, 239 p.; 28 cm. Notes: "February 2006." Includes index. Additional Formats: Also available on the Internet. Subjects: Washington (State). Pollution Control Hearings Board. Water resources development--Law and legislation --Washington (State)--Digests. Water rights--Washington (State)--Digests. Water-supply--Law and legislation--Washington (State) --Digests. Series: Publication; no. 06-11-002 Publication (Washington (State). Dept. of Ecology); 06-11-002. LC Classification: KFW446.A57 D54 2006 Government Document No.: WA 574.5 Ec7dig w1 2006 wadocs

Dioxin monitoring program, State of Maine. Published/Created: Augusta, Maine: Dept. of Environmental Protection, Related Names: Maine. Dept. of Environmental Protection. Description: v.; 28 cm. Notes: Description based on: 1992; title from cover. SERBIB/SERLOC merged record ACQN: aq 99007229 1988-90(mono);91-96; Subjects: Tetrachlorodibenzodioxin--Environmental aspects--Maine. Fishes--Effect of water pollution on--Maine. Water quality bioassay--Maine. Water--Pollution--Maine--Measurement. LC Classification: QH545.T44 D56

Disaster in Bangladesh: selected readings / editor, K. Nizamuddin. Published/Created: Dhaka: Disaster Research Training and Management Centre, Dept. of Geography and Environment, University of Dhaka, [2001] Related Names: Nizamuddin, K.

(Khondakar) University of Dhaka. Disaster Research Training and Management Centre. Description: 197 p.: maps: 22 cm. ISBN: 9843110528 Notes: Includes bibliographical references. Subjects: Natural disasters--Bangladesh. Water--Pollution--Bangladesh. Arsenic--Toxicology--Bangladesh. LC Classification: GB5011.91.B3 D57 2001 Overseas Acquisitions No.: B-E-2001-416400; 23

Discharge of industrial effluents to municipal sewerage systems; proceedings of symposium. Published/Created: [Maidstone] Institute of Water Pollution Control, 1971. Related Names: Institute of Water Pollution Control. Description: 124 p. illus. 25 cm. Notes: Arranged by The Institute of Water Pollution Control. Includes bibliographical references. Subjects: Factory and trade waste--Congresses. Water--Pollution--Congresses. LC Classification: TD896 .D57 Dewey Class No.: 363.6

Dissolved radon and uranium, and ground-water geochemistry in an area near Hylas, Virginia / by Mark R. Stanton ... [et al.]. Published/Created: Washington: U.S. G.P.O.; Denver, CO: For sale by USGS Information Services, 1996. Related Names: Stanton, Mark R. Description: iv, 23 p.: ill., map; 28 cm. Notes: Includes bibliographical references (p. 22-23). Subjects: Radioactive pollution of water--Virginia--Hylas Region. Radon--Environmental aspects. Uranium--Environmental aspects. Groundwater--Pollution--Virginia--Hylas Region. Mylonite--Virginia--Hylas Region. Series: U.S. Geological Survey bulletin; 2070 LC Classification: QE75 .B9 no. 2070 TD427.R3 Dewey Class No.: 557.3 s 363.73/8 20 Government Document No.: I 19013:B2070 Geographic Area Code: n-us-va

Distribution of nitrates and other water
pollutants under fields and corrals in the
middle South Platte Valley of Colorado [by
B. A. Stewart and others.
Published/Created: Washington]
Agricultural Research Service, U.S. Dept.
of Agriculture, 1967. Related Names:
Stewart, B. A. (Bobby Alton), 1932-
Description: v, 206 p. illus. 26 cm. Notes:
Cover title. Bibliography: p. 205-206.
Subjects: Nitrates. Water--Pollution--South
Platte River Valley (Colo. and Neb.)
Nitrates. Water, Underground--South
Platte Valley. Water--Pollution. Series:
[United States. Agricultural Research
Service] ARS 41-134 LC Classification:
S900.U5 no. 134 NAL Class No.: A56.9
R31 no. 134

Dose reconstruction project. Published/Created:
[Alameda, Calif.: ChemRisk, 1991]
Related Names: ChemRisk. Colorado.
Division of Disease Control and
Environmental Epidemiology. Colorado.
Dept. of Health. Description: 2 v. (various
pagings): ill.; 28 cm. Contents: v. 1.
Identification of chemicals and
radionuclides used at Rocky Flats: task 1
report (R1) -- v. 2. Selection of the
chemicals and radionuclides of concern:
task 2 report. Notes: This is a draft report
of the Toxicologic review and dose
reconstruction project, Project task 1. It
was released in January 1991 as part of
Briefing book 4. No final report published.
Phase 1 was conducted by ChemRisk for
the Colorado Dept. of Health, Division of
Disease Control and Environmental
Epidemiology. Includes bibliographical
references and statistics. Subjects: Rocky
Flats Plant (U.S.). Radioactive pollution of
the atmosphere. Radioactive pollution of
water--Colorado. Radiation dosimetry.
Series: Health studies on Rocky Flats,
Phase 1, Historical public exposures LC
Classification: RA569 .D674 1991
Government Document No.:
HE18/30.11/1/1991 codocs

Draft final report by the regional consultants on
tasks 11, 16, and 17: in support of the
national working groups of Kenya,
Tanzania, and Uganda for Regional Task
Force II (Water quality and land use,
including wetlands) / Andy Bullock ... [et
al.]; with additional contributions by
Nsubuga Senfuma, E. Nyaga.
Published/Created: Nairobi, Kenya: Lake
Victoria Environmental Management
Programme, [1995] Related Names:
Bullock, Andy. Senfuma, Nsubuga. Nyaga,
E. Lake Victoria Environmental
Management Programme. Regional Task
Force II. Description: 1 v. (various
pagings): ill., maps; 30 cm. Notes: "August
30th, 1995." Includes bibliographical
references. Subjects: Water quality
management--Victoria, Lake, Watershed.
Land use--Victoria, Lake, Watershed--
Management. Wetland management--
Victoria, Lake, Watershed. Nonpoint
source pollution--Victoria, Lake,
Watershed. LC Classification: TD319.E18
D73 1995

Drainage basin survey report, Housatonic River
drainage basin.: Recommended
classifications and assignment of standards
of quality and purity for designated waters
of New York State. / Prepared by Dept. of
Health for Water Resources Commission.
Published/Created: [Albany: The Dept.,
1963] Related Names: New York (State).
Dept. of Health. New York (State). Water
Resources Commission. Description: x, 71
p.: maps, diagrs., tables; 28 cm. Subjects:
Water-supply--New York (State)--
Housatonic River watershed. Stream
measurements--New York (State)--
Housatonic River watershed. Water--
Pollution--New York (State)--Housatonic
River watershed. Water--Analysis. LC
Classification: TD223.15 .N44

Drainage basin survey report, surface waters of
Nassau County: recommended
classifications and assignment of standards

of quality and purity for designated waters of New York State. Prepared for Water Resources Commission. Published/Created: [Albany: Dept. of Health, 1963] Related Names: New York (State). Dept. of Health. New York (State). Water Resources Commission. Description: ix, 113 p.: maps; 28 cm. Notes: Cover title. Subjects: Stream measurements--New York (State)--Nassau County. Water-supply--New York (State)--Nassau County. Water--Pollution--New York (State)--Nassau County. Water--Composition. LC Classification: GB1225.N7 A48

Drainage management in the San Joaquin Valley: a status report / San Joaquin Valley Drainage Implementation Program. Published/Created: Sacramento, CA: The Program: For additional copies, contact Bulletins & Reports, Dept. of Water Resources, [1998] Related Names: San Joaquin Valley Drainage Implementation Program. Description: xi, 65 p.: col. ill.; 28 cm. Notes: "February 1998." Includes bibliographical references (p. 63-65). Subjects: Drainage--California--San Joaquin Valley--Management. Irrigation--California--San Joaquin Valley--Management. Agricultural pollution--California--San Joaquin Valley. Water reuse--California--San Joaquin Valley. LC Classification: TC977.C2 D73 1998 Dewey Class No.: 631.6/2/097948 21

Drinking water and health / Safe Drinking Water Committee, Advisory Center on Toxicology, Assembly of Life Sciences, National Research Council. Published/Created: Washington, D.C.: National Academy of Sciences, 1977-<1989 > Related Names: Assembly of Life Sciences (U.S.). Safe Drinking Water Committee. National Research Council (U.S.). Commission on Life Sciences. Safe Drinking Water Committee. Related Titles: Pharmacokinetics in risk management.

Description: v. <1-6, 8-9 >; 23 cm. ISBN: 0309026199 (v. 1): Notes: Vol. 8 has title: Pharmacokinetics in risk management; 9: Selected issues in risk assessment. Vols. 1-2 by the Safe Drinking Water Committee, Assembly of Life Sciences. Vols. 5-6, 9 by the Safe Drinking Water Committee under the Assembly's later name, Commission on Life Sciences. Vols. 5-6, 9 published by: National Academy Press. Includes bibliographies and indexes. Subjects: Drinking water--Contamination. Drinking water--Health aspects. Water--Pollution--Toxicology. Water--Purification. Health risk assessment. Toxicology--Dose-response relationship. LC Classification: RA591 .D75 1977 Dewey Class No.: 363.1/929 19

Drinking water and human health. Published/Created: Chicago, Ill.: American Medical Association: Copies from Order Dept., American Medical Association, c1984. Related Names: American Medical Association. Description: viii, 202 p.: ill.; 28 cm. ISBN: 0899701809 (pbk.) Notes: Includes bibliographies. Subjects: Drinking water--Health aspects--United States--Congresses. Drinking water--Contamination--United States--Congresses. Water--Pollution--Toxicology--United States--Congresses. Waterborne infection--United States--Congresses. LC Classification: RA592.A1 D75 1984 Dewey Class No.: 363.1/92 19 Geographic Area Code: n-us---

Drinking water contamination by arsenic in rural areas of Bangladesh [microform]: possible solution and awareness building: a workshop report / edited by Hasan Sarwar, A.K.M. Ashrafuzzaman. Published/Created: Comilla: Bangladesh Academy for Rural Development, 1999. Related Names: Ashrafuzzaman, A. K. M. Sarwar, Hasan. Baì,,mì£laì,,desì a Pallìì,, Unnayì‡ana Ekaì,,dì£emìì,,. Description: 83 p. 23 cm. Cancelled ISBN: 9845591033

Notes: Includes bibliographical references. Additional Formats: Microfiche. New Delhi: Library of Congress Office; Washington, D.C.: Library of Congress Photoduplication Service, 2000. 2 microfiches. Master microform held by: DLC. Subjects: Water--Pollution--Bangladesh. LC Classification: Microfiche 2000/62019 (T) Overseas Acquisitions No.: B-E-99-945369; 024

Drinking water supply and agricultural pollution: preventive action by the water supply sector in the European Union and the United States / edited by Geerten J.I. Schrama. Published/Created: Boston, MA: Kluwer Academic Publishers, c1998. Related Names: Schrama, Geerten J. I. Description: xi, 375 p.: ill.; 25 cm. ISBN: 0792351045 (alk. paper) Notes: Includes bibliographical references. Subjects: Water quality management--Government policy--European Union Countries. Drinking water--Contamination--Government policy--European Union Countries. Agricultural pollution--Government policy--European Union Countries. Water quality management--Government policy--United States. Drinking water--Contamination--Government policy--United States. Agricultural pollution--Government policy--United States. Series: Environment & policy; v. 11 LC Classification: TD255 .D75 1998 Dewey Class No.: 363.739/456/094 21 Geographic Area Code: e------ n-us---

Drinking water: EPA should strengthen ongoing efforts to ensure that consumers are protected from lead contamination: report to congressional requesters. Variant Title: EPA should strengthen ongoing efforts to ensure that consumers are protected from lead contamination Running Title: Lead in drinking water Published/Created: Washington, D.C.: The Office, [2005] Related Names: United States. Government Accountability Office.

Description: 76 p.: ill.; 28 cm. Notes: Cover title. "January 2006." Includes bibliographical references. Additional Formats: Also available via Internet from GAO web site. Address as of 2/1/06: http://www.gao.gov/new.items/d06148.pdf . Subjects: Drinking water--Contamination--United States. Groundwater--Pollution--United States. Waste disposal in the ground--United States. Series: Report GAO-06-148 LC Classification: RA592.A1 D755 2005

Dynamics of environmental bioprocesses: modelling and simulation / Jonathan B. Snape ... [et al.]. Published/Created: Weinheim; New York: VCH, c1995. Related Names: Snape, Jonathan B. Description: xxiii, 492 p.: ill.; 25 cm. + 1 computer disk (3 1/2 in.) ISBN: 3527287051 (acid-free paper) Computer File Information: System requirements for accompanying computer disk: DOS-PC. Notes: Includes bibliographical references and index. Subjects: Bioremediation--Mathematical models. Water--Pollution--Mathematical models. Bioremediation--Computer simulation. Water--Pollution--Computer simulation. LC Classification: TD192.5 .D95 1995 Dewey Class No.: 628.5/01/5118 20 Electronic File Information: Publisher description http://www.loc.gov/catdir/description/wiley032/95007132. html Table of contents http://www.loc.gov/catdir/toc/wiley022/95007132.html

Dynamique de populations et qualiteì de l'eau: actes du symposium de l'Institut d'eì cologie du bassin de la Somme, Chantilly, 7-9 novembre 1979: ouvrage collectif / preì senteì par H. Hoestlandt, avec la participation de J. Daget ... [et al.]; preì face de Michel d'Ornano. Published/Created: [Paris]: Gauthier-Villars, c1981. Related Names: Hoestlandt, H. Institut d'eì cologie du bassin de la Somme. Description: xvi, 275 p.: ill.; 24

cm. ISBN: 2040120688 Notes: English and French, with summaries in both languages. Includes bibliographies. Subjects: Water--Pollution--Environmental aspects--Congresses. Population biology--Congresses. Series: Formation permanente en eì cologie et biologie LC Classification: QH545.W3 D96 Dewey Class No.: 574.5/263 19 Language Code: engfre National Bibliography No.: F***

Early warning and management of surface water taste-and-odor events / prepared by William D. Taylor ... [et al.]. Published/Created: Denver, CO: AWWA Research Foundation/American Water Works Association/IWA Pub., c2006. Related Names: Taylor, W. D. (William Dee) AWWA Research Foundation. American Water Works Association. IWA Publishing. Description: xxix, 237 p.: ill. (some col.); 28 cm. + 1 CD-ROM (4 3/4 in.) ISBN: 1583214259 Notes: Includes bibliographical references (p. 219-233) and index. Subjects: Drinking water--Sensory evaluation. Water quality--United States. Drinking water--United States--Quality control. Groundwater--Pollution--Health aspects--United States. Waterworks--United States. LC Classification: TD375 .E37 2006 Dewey Class No.: 628.1/61 22

Earth resources and environmental issues / [editors], A.K. Sinha, Pankaj Srivastava. Edition Information: 1st ed. Published/Created: Jaipur: ABD Publishers: Distribution, Oxford Book Co., 2000. Related Names: Sinha, Anshu K. Srivastava, Pankaj. Description: 215 p.: ill., maps; 25 cm. ISBN: 8185771073 Summary: Contributed articles presented at a national seminar held in 1995; with reference to India. Notes: Includes bibliographical references. Subjects: Mineral industries--Environmental aspects--India --Congresses. Economic development--Environmental aspects--India --Congresses. Mineral resources

conservation--India--Congresses. Water--Pollution--India--Congresses. Water-supply--India--Management--Congresses. LC Classification: TD195.M5 E25 2000 Overseas Acquisitions No.: I-E-99-952935; 549-13

East Fork Kaskaskia River (ILOK01): TMDL and implementation plan / prepared for Illinois Environmental Protection Agency by MWH. Cover Title: East Fork Kaskaskia River TMDL report Running Title: Development of TMDLs and implementation plans Published/Created: Springfield, IL: Illinois Environmental Protection Agency, Bureau of Water, 2003. Related Names: Illinois. Bureau of Water. United States. Environmental Protection Agency. Description: 1 v. (various pagings): ill. (some col.), maps (some col.); 28 cm. Notes: "August 2003." Reviewed by the United States Environmental Protection Agency. "IEPA/BOW/03-007"--Cover. Includes bibliographical references. Subjects: Water--Pollution--Total maximum daily load--Illinois --Kaskaskia River, East Fork, Watershed. Nutrient pollution of water--Illinois--Kaskaskia River, East Fork, Watershed. Sediment control--Illinois--Kaskaskia River, East Fork, Watershed. Water quality management--Illinois--Kaskaskia River, East Fork, Watershed. Kaskaskia River, East Fork, Watershed (Ill.) LC Classification: TD224.I3 E2378 2003

Ecobiology of polluted waters / editor Arvind Kumar. Portion of Title: Polluted waters Published/Created: Delhi: Daya Pub. House, 2006. Related Names: Kumar, Arvind, 1953- Description: ix, 282 p.: maps; 25 cm. ISBN: 8170353866 Notes: Includes bibliographical references and index. Subjects: Aquatic ecology. Water--Pollution--Toxicology. Water--Pollution--Environmental aspects--India. LC Classification: QH541.5.W3 E285 2006

Overseas Acquisitions No.: I-E-2006-453067; 35-91

Ecological assessments of effluent impacts on communities of indigenous aquatic organisms: a symposium / sponsored by ASTM Committee D-19 on Water, American Society for Testing and Materials, Ft. Lauderdale, Fla., 29-30 Jan. 1979; J.M. Bates and C.I. Weber, editors. Published/Created: Philadelphia, Pa. (1916 Race St., Philadelphia 19103): ASTM, c1981. Related Names: Bates, J. M. Weber, Cornelius I. American Society for Testing and Materials. Committee D-19 on Water. Description: 333 p.: ill.; 24 cm. Notes: Includes bibliographical references and index. Subjects: Aquatic organisms--Effect of water pollution on --Congresses. Series: ASTM special technical publication; 730 LC Classification: QH545.W3 E28 Dewey Class No.: 574.5/263 19

Ecological impacts of the oil industry: proceedings of international meeting organized by the Institute of Petroleum and held in London in November 1987 / edited by Brian Dicks. Published/Created: Chichester [England]; New York: Wiley, c1989. Related Names: Dicks, Brian (Brian M.) Institute of Petroleum (Great Britain) Description: ix, 316 p.: ill.; 24 cm. ISBN: 0471921939: Notes: "Published on behalf of the Institute of Petroleum, London." Includes bibliographies. Subjects: Oil spills--Environmental aspects--Congresses. Oil pollution of water--Congresses. LC Classification: QH545.O5 E255 1989 Dewey Class No.: 574.5/263 19

Ecology and pollution in human settlements: Nala Lai. Published/Created: Islamabad: LEAD, Pakistan: UNDP, [1999] Related Names: Leadership for Environment and Development, Pakistan. United Nations Development Programme. Description: ii, 24 p.: col. ill.; 27 cm. Notes: Includes bibliographical references (p. 18).

Subjects: Water--Pollution--Environmental aspects--Pakistan--Nala Lai Watershed. Floods--Pakistan--Nala Lai Watershed. Human settlements--Pakistan--Nala Lai Watershed. LC Classification: TD304.5 .E35 1999 Overseas Acquisitions No.: P-E-00283169; 12

Ecology and pollution of Indian lakes and reservoirs / edited by P.C. Mishra & R.K. Trivedy. Published/Created: New Delhi: Ashish Pub. House, 1993. Related Names: Mishra, P. C. (Pramod Chandra), 1953- Trivedy, R. K. Description: viii, 347 p.: ill.; 23 cm. ISBN: 8170245397: Notes: Includes bibliographical references. Subjects: Lake ecology--India. Water--Pollution--India. LC Classification: QH541.5.L3 E36 1993 Dewey Class No.: 574.5/26322/0954 20 Overseas Acquisitions No.: I-E-70858

Ecology and pollution of Indian rivers / editor, R.K. Trivedy. Edition Information: 1st ed. Published/Created: New Delhi: Ashish Pub. House, 1988. Related Names: Trivedy, R. K. Description: xv, 447 p.: ill., maps; 22 cm. ISBN: 8170242150: Notes: Contributed articles. Includes indexes. Bibliography: p. [407]-423. Subjects: Water--Pollution--India. Stream ecology--India. Water quality--India. LC Classification: TD303.A1 E26 1988 Dewey Class No.: 363.7/3942/0954 20 Overseas Acquisitions No.: I E 59093 Geographic Area Code: a-ii---

Ecology of the Lakes of western New York / edited by Jay A. Bloomfield. Published/Created: New York: Academic Press, 1978. Related Names: Bloomfield, Jay A. Description: xiv, 473 p.: ill.; 24 cm. ISBN: 0121073025 Contents: Mayer, J. R. and others. Chautauqua Lake.--Bannister, T. T. and Bubeck, R. C. Limnology of Irondequoit Bay, Monroe County, New York.--Murphy, C. B., Jr. Onondaga Lake.--Mills, E. G. and others. Oneida Lake. Notes: Includes bibliographies and index.

Subjects: Lakes--New York (State), Western. Eutrophication--New York (State), Western. Water--Pollution--New York (State), Western. Series: Lakes of New York State; v. 2 LC Classification: GB1625.N7 L34 vol. 2 Dewey Class No.: 574.5/2632/09747 s 574.4/2632/09747 Geographic Area Code: n-us-ny

EcoNews. A radioactive accident on Navajo Lands. Part I / director, Darryl J.C. Leyden; director, executive producer, Nancy Pearlman. Portion of Title: Radioactive accident on Navajo Lands. Part I Published/Created: [United States: PBS, 1987]; United States: Educational Communications, Inc., 1988. United States. Related Names: Leyden, Darryl J.C., direction. Pearlman, Nancy, 1948, direction, production, editing, host. Educational Communications Collection (Library of Congress) Description: viewing copy. 1 videocassette of 1 (ca. 30 min.): sd. col.; 3/4 in. Contents: The river that harms / produced, directed & written by Colleen Keane; camera, David Tolley, Martin C. Baumann; camera & editing, Scott Drake; music composed and performed by Jonathan Merrill; photography, J.H. McGibbeny, Leslie Nelson; narrated by Joseph Campanella Notes: Copyright: unknown. Copyright notice on videocassette: Educational Communications, Inc.; 1988. Show #515. On cassette with: A radioactive accident on Navajo Lands, Part II. The video "The river that harms" courtesy of University of Southern California School of Journalism was shown. Sources used: paperwork in M/B/RS Acquisition file: Educational Communications, Inc. Performer Notes: Host: Nancy Pearlman. Credits: Camera, Bill West, Mohammed Al-Fadel, Carlos Masana; production assistants, Anna Harlowe, Bob Phillips, Lynn Cason, Dot Cannon, Julian Hanberg, Debbie Pepper, Mary Jane Parks, Cor Thorbridge, David Jayne; technical director, Kipenda Amisha;

videotape editor, Nancy Pearlman. Acquisition Source: viewing copy; Received: 11/15/2000 gift; Educational Communications Collection. Subjects: Nuclear accidents. Radioactive wastes. Radioactive pollution of water. Radioactive waste disposal in rivers, lakes, etc. Form/Genre: Nature--Series. migfg LC Classification: VBQ 6749 (viewing copy)

EcoNews. A radioactive accident on Navajo Lands. Part II / director, Darryl J.C. Leyden; director, executive producer, Nancy Pearlman. Portion of Title: Radioactive accident on Navajo Lands. Part II Published/Created: [United States: PBS, 1987]; United States: Educational Communications, Inc., 1988. United States. Related Names: Leyden, Darryl J.C., direction. Pearlman, Nancy, 1948, direction, production, editing, host. Educational Communications Collection (Library of Congress) Description: viewing copy. 1 videocassette of 1 (ca. 30 min.): sd. col.; 3/4 in. Contents: The river that harms / produced, directed & written by Colleen Keane; camera, David Tolley, Martin C. Baumann; camera & editing, Scott Drake; music composed and performed by Jonathan Merrill; photography, J.H. McGibbeny, Leslie Nelson; narrated by Joseph Campanella Notes: Copyright: unknown. Copyright notice on videocassette: Educational Communications, Inc.; 1988. Show #516. On cassette with: A radioactive accident on Navajo Lands, Part I. The video "The river that harms" courtesy of University of Southern California School of Journalism was shown. Sources used: paperwork in M/B/RS Acquisition file: Educational Communications, Inc. Performer Notes: Host: Nancy Pearlman. Credits: Camera, Bill West, Mohammed Al-Fadel, Carlos Masana; production assistants, Anna Harlowe, Bob Phillips, Lynn Cason, Dot Cannon, Julian Hanberg, Debbie Pepper, Mary Jane Parks, Cor Thorbridge, David

Jayne; technical director, Kipenda Amisha; videotape editor, Nancy Pearlman. Acquisition Source: viewing copy; Received: 11/15/2000 gift; Educational Communications Collection. Subjects: Nuclear accidents. Radioactive wastes. Radioactive pollution of water. Radioactive waste disposal in rivers, lakes, etc. Form/Genre: Nature--Series. migfg LC Classification: VBQ 6749 (viewing copy)

EcoNews. Garbage archeaology [and] Hall of fame [and] Yakety yak, take it back music video [and] Ocean trash [and] Milk carton recycling / executive producer, Nancy Pearlman. Variant Title: Title on inventory University of Arizona's garbage project with student reporters Ketara Gadahn and Adam Gadahn [and] Music video "Yakety yak, take it back [and] Update on ocean trash [and] Milk carton recycling Portion of Title: Garbage archeaology Hall of fame Yakety yak, take it back music video Ocean trash Milk carton recycling Published/Created: United States: Educational Communications, Inc., 1994. United States. Related Names: Pearlman, Nancy, 1948, production, editing, host. Educational Communications Collection (Library of Congress) Description: viewing copy. 1 videocassette of 1 (ca. 30 min.): sd. col.; 3/4 in. Notes: Copyright: unknown. Copyright notice on videocassette: Educational Communications, Inc.; 1994. Show #1202. Sources used: paperwork in M/B/RS Acquisition file: Educational Communications, Inc. Portions of the program was taped on location at the University of Arizona. Performer Notes: Host: Nancy Pearlman. Acquisition Source: viewing copy; Received: 11/15/2000 gift; Educational Communications Collection. Subjects: Ragpickers. Water--Pollution. Recycling (Waste, etc.) Form/Genre: Nature--Series. migfg LC Classification: VBQ 6914 (viewing copy)

EcoNews. Izaak Walton League's "Save our streams" program / director, executive producer, Nancy Pearlman. Variant Title: Alternative title on videocassette: Monitoring streams Portion of Title: Izaak Walton League's "Save our streams" program Published/Created: United States: Educational Communications, Inc., 1991. United States. Related Names: Pearlman, Nancy, 1948, direction, production, editing, host. Educational Communications Collection (Library of Congress) Related Titles: Save our streams (Motion picture) Description: viewing copy. 1 videocassette of 1 (ca. 30 min.): sd. col.; 3/4 in. Contents: Save our streams / the Izaak Walton League of America presents; Kinetoscope productions. Notes: Copyright: unknown. Copyright notice on videocassette: Educational Communications, Inc.; 1991. Show #816. The video "Save our streams" was shown. Sources used: paperwork in M/B/RS Acquisition file: Educational Communications, Inc.; copyright data base. Performer Notes: Host: Nancy Pearlman. Credits: Production assistants, Anna Harlowe, Lynn Cason, Bob Phillips, Julian Hanberg, Pauline Mullins, Debbie Kramer, Dave Dismore, Allan Ellstrom; videotape editor, Nancy Pearlman. Acquisition Source: viewing copy; Received: 11/15/2000 gift; Educational Communications Collection. Subjects: Water--Pollution. Stream conservation. Form/Genre: Nature--Series. migfg LC Classification: VBQ 6822 (viewing copy)

EcoNews. Ocean trash [and] Global warming / director, John Stealey, Nancy Pearlman; executive producer, Nancy Pearlman. Portion of Title: Ocean trash Global warming Published/Created: United States: Educational Communications, Inc., 1990. United States. Related Names: Stealey, John R., direction. Pearlman, Nancy, 1948, direction, production, host. Educational Communications Collection (Library of

Congress) Related Titles: Trashing the oceans (Motion picture) Marine refuse project (Motion picture) Greenhouse crisis: the American response (Motion picture) Description: viewing copy. 1 videocassette of 1 (ca. 30 min.): sd. col.; 3/4 in. Contents: Trashing the oceans / Saltwater Productions; National Oceanic Atmospheric Administration; executive producer for NOAA, Robert B. Amdur, Alan R. Bunn -- Marine refuse project / produced by National Oceanic Atmospheric Administration; written & directed by Lee Reilly; shot & edited by Westcom Productions, Inc. (Eugene, Oregon) -- Greenhouse crisis: the American response / Union of Concerned Scientist presents; produced by Mediawrights (Boston, MA). Copyright notice: Union of Concerned Scientist; 1989. Notes: Copyright: unknown. Copyright notice on videocassette: Educational Communications, Inc.; 1990. Show #712. The videos "Trashing the oceans," "Marine refuse project," and "Greenhouse crisis: the American response" was shown. Sources used: paperwork in M/B/RS Acquisition file: Educational Communications, Inc. Performer Notes: Host: Nancy Pearlman. Credits: Production assistants, Anna Harlowe, Lynn Cason, Elaine Stansfield, Bob Phillips, Julian Hanberg, Rick Hupp, Debbie Pepper; camera operators, Ken Kitchen, Paul Leach. Acquisition Source: viewing copy; Received: 11/15/2000 gift; Educational Communications Collection. Subjects: Water pollution. Ships--Waste disposal. Global warming. Greenhouse effect, Atmospheric. Form/Genre: Nature--Series. migfg LC Classification: VBQ 6796 (viewing copy)

EcoNews. Pollution from non-point sources. Part I / director, Joe Curran; executive producer, Nancy Pearlman. Portion of Title: Pollution from non-point sources. Part I Published/Created: United States

Educational Communications, Inc., 1989. United States. Related Names: Curran, Joseph, direction. Pearlman, Nancy, 1948, production, editing, host. Educational Communications Collection (Library of Congress) Related Titles: Riders on the storm: the challenge of non-point source pollution (Motion picture) Description: viewing copy. 1 videocassette of 1 (ca. 30 min.): sd. col.; 3/4 in. Contents: Riders on the storm: the challenge of non-point source pollution / produced by the Nebraska ETV Network in cooperation with the Soil Conservation Society of America; producer/director/writer, Bill Ganzel; videographers, John Beck, Ralph Hammack, Terry Hatch; production assistant, Brad Pace; narrated by James B. Levy. Copyright notice: KUON-TV; 1986. Notes: Copyright: unknown. Copyright notice on videocassette: Educational Communications, Inc.; 1989. Show #606. The video "Riders on the storm: the challenge of non-point source pollution" was shown. Sources used: paperwork in M/B/RS Acquisition file: Educational Communications, Inc. Performer Notes: Host: Nancy Pearlman. Credits: Technical director, Phil Martin; cameras, Chris Bancroft, David Mitchell; production assistants, Jodi Birch, Anna Harlowe, Bob Phillips, Julian Hanberg, Debbie Pepper, Barbara Winebright, Dave Dismore, Bruce Hemphill, Gail Koffman, Craig Lewis, Larry La Sota, Linda Nation, Dean Ossiander, Rick Hupp, Lynn Cason, John Owen; videotape editor, Nancy Pearlman. Acquisition Source: viewing copy; Received: 11/15/2000 gift; Educational Communications Collection. Subjects: Water pollution. Nonpoint source pollution. Form/Genre: Nature--Series. migfg LC Classification: VBQ 6768 (viewing copy)

EcoNews. Pollution from non-point sources. Part II / director, Joe Curran; executive producer, Nancy Pearlman. Portion of

Title: Pollution from non-point sources. Part II Published/Created: United States Educational Communications, Inc., 1989. United States. Related Names: Curran, Joseph, direction. Pearlman, Nancy, 1948, production, editing, host. Educational Communications Collection (Library of Congress) Related Titles: Riders on the storm: the challenge of non-point source pollution (Motion picture) Description: viewing copy. 1 videocassette of 1 (ca. 30 min.): sd. col.; 3/4 in. Contents: Riders on the storm: the challenge of non-point source pollution / produced by the Nebraska ETV Network in cooperation with the Soil Conservation Society of America; producer/director/writer, Bill Ganzel; videographers, John Beck, Ralph Hammack, Terry Hatch; production assistant, Brad Pace; narrated by James B. Levy. Copyright notice: KUON-TV; 1986. Notes: Copyright: unknown. Copyright notice on videocassette: Educational Communications, Inc.; 1989. Show #607. The video "Riders on the storm: the challenge of non-point source pollution" was shown. Sources used: paperwork in M/B/RS Acquisition file: Educational Communications, Inc. Performer Notes: Host: Nancy Pearlman. Credits: Technical director, Phil Martin; cameras, Chris Bancroft, David Mitchell; production assistants, Jodi Birch, Anna Harlowe, Bob Phillips, Julian Hanberg, Debbie Pepper, Barbara Winebright, Dave Dismore, Bruce Hemphill, Gail Koffman, Craig Lewis, Larry La Sota, Linda Nation, Dean Ossiander, Rick Hupp, Lynn Cason, John Owen; videotape editor, Nancy Pearlman. Acquisition Source: viewing copy; Received: 11/15/2000 gift; Educational Communications Collection. Subjects: Water pollution. Nonpoint source pollution. Form/Genre: Nature--Series. migfg LC Classification: VBQ 6769 (viewing copy)

Economic and technical review. Published/Created: [Ottawa] Water Pollution Control Directorate. Related Names: Canada. Environmental Protection Service. Description: 28 cm. Notes: Numbers <EPS 3-WP-73-2-> issued by the Environmental Protection Service. Subjects: Water--Pollution--Canada. LC Classification: TD226 .E27 Dewey Class No.: 363.6/1

Economic costs and benefits of an antipollution project in Italy. Summary report of a preliminary evaluation. [A cura dell'] ENI, Ente nazionale idrocarburi. Special issue for the United Nations Conference on the Human Environment, Stockholm, June 5-16, 1972. Published/Created: Roma, So. Gra. Ro., [1972]. Related Names: Ente nazionale idrocarburi. United Nations Conference on the Human Environment (1972: Stockholm, Sweden) Description: 244 p. 22 cm. Notes: Includes bibliographical references. Subjects: Air--Pollution--Economic aspects--Italy--Cost effectiveness. Water--Pollution--Economic aspects--Italy--Cost effectiveness. LC Classification: HC310.A4 E25 Dewey Class No.: 363.6 National Bibliography No.: It73-June Geographic Area Code: e-it---

Economic impact of existing ammonia nitrogen water quality standard, IPCB chapter 3 rule 203(f) / Charles B. Muchmore ... [et al.]. Published/Created: Chicago, IL (309 W. Washington St., Chicago 60606): State of Illinois, Institute of Natural Resources, 1981. Related Names: Muchmore, Charles B. Description: xvi, 325 p.: ill.; 28 cm. Notes: "July, 1981." Bibliography: p. 197-202. Subjects: Water--Pollution--Law and legislation--Economic aspects --Illinois. Ammonia--Environmental aspects--Illinois. Series: Document / State of Illinois, Institute of Natural Resources; no. 81/23 Document (Illinois Institute of Natural Resources); no. 81/23. LC

Classification: HC107.I33 W323 Dewey Class No.: 338.4/33637394 19 Geographic Area Code: n-us-il

Economic impact of the selenium effluent standard in Illinois (R76-21) / by Ronald C. Flemal ... [et al.]. Published/Created: Chicago: State of Illinois, Institute for Environmental Quality, 1977. Related Names: Flemal, Ronald C. Illinois. Institute for Environmental Quality. Description: viii, 88 p.: ill.; 28 cm. Notes: "Project no. 80.092." Bibliography: p. 82-88. Subjects: Water--Pollution--Economic aspects--Illinois. Effluent quality--Standards--Illinois. Selenium--Environmental aspects--Illinois. Series: IIEQ documents; no. 77/34 LC Classification: TD224.I3 E26 Dewey Class No.: 363.6/1 Geographic Area Code: n-us-il

Economic impact study of proposed IPCB amendments to water pollution regulations, R80-6 / by Robert G. Ducharme, Inc. Published/Created: Chicago, Ill. (309 West Washington St., Chicago 60606): State of Illinois, Dept. of Energy and Natural Resources, [1982] Related Names: Robert G. Ducharme, Inc. Illinois. Pollution Control Board. Description: 54 p. in various pagings; 28 cm. Notes: "February, 1982." "Project no. 80.235." Subjects: Water--Pollution--Law and legislation--Economic aspects --Illinois. Series: Document / Illinois Department of Energy and Natural Resources; no. 82/06 Document (Illinois Dept. of Energy and Natural Resources); no. 82/06. LC Classification: HC107.I33 W324 1982 Dewey Class No.: 338.4/33637394/09773 19 Geographic Area Code: n-us-il

Economic impact study of regulations for activities within setback zones and regulated recharge areas (R89-5): final report / prepared by Environmental Science & Engineering, Inc.; principal investigators, Dan Gallagher, Robert Scott,

Jack Garbade. Published/Created: Springfield, Ill.: Illinois Dept. of Energy and Natural Resources, Office of Research and Planning, [1991] Related Names: Environmental Science and Engineering, Inc. Illinois. Dept. of Energy and Natural Resources. Office of Research and Planning. Description: 1 v. (various pagings); 28 cm. Notes: "Prepared for Illinois Department of Energy and Natural Resources, Office of Research and Planning." "ILENR/RE-EA-91-07." Includes bibliographical refernces (p. 10-1-10-4). Subjects: Drinking water--Law and legislation--Illinois. Water--Pollution--Law and legislation--Illinois. Environmental law--Compliance costs--Illinois. LC Classification: KFI1649 .A84 1991 Dewey Class No.: 363.73/946/09773 20 Geographic Area Code: n-us-il

Economic instruments for the virtual elimination of persistent toxic substances in the Great Lakes Basin. Published/Created: Windsor: International Joint Commission, [1994] Related Names: International Joint Commission. Description: x, 131 p.: ill.; 28 cm. ISBN: 1895085829 Notes: "December 1994." Includes bibliographical references (p. 41-48). Subjects: Water--Pollution--Economic aspects--Canada. Water--Pollution--Economic aspects--United States. Water--Pollution--Economic aspects--Great Lakes Region (North America) Environmental impact charges--Canada. Environmental impact charges--United States. Environmental impact charges--Great Lakes Region (North America) LC Classification: HC120.W32 E27 1994 Geographic Area Code: n-cn--- n-us--- nl-----

Economic instruments for water pollution. Published/Created: [London]: Dept. of the Environment, Transport and the Regions, c1997. Related Names: Great Britain. Dept. of the Environment, Transport and the Regions. Description: 76 p.; 30 cm. Notes: Includes bibliographical references (p. 71). In English. Subjects: Water--Pollution--Economic aspects--Great Britain. Environmental impact charges--Great Britain. LC Classification: HC260.W32 E27 1997

Economics of air and water pollution. Edited by William R. Walker. Published/Created: Blacksburg, Water Resources Research Center, Virginia Polytechnic Institute, 1969. Related Names: Walker, William R., ed. Virginia Polytechnic Institute. Water Resources Research Center. Description: 250 p. illus., maps. 23 cm. Notes: Papers from a seminar sponsored by the Water Resources Research Center, Virginia Polytechnic Institute. Includes bibliographical references. Subjects: Air--Pollution--Economic aspects--United States. Water--Pollution--Economic aspects--United States. Series: Virginia Polytechnic Institute, Blacksburg. Water Resources Research Center. Bulletin 26 LC Classification: TD201 .V57 no. 26 Dewey Class No.: 551.4/8/08 s Geographic Area Code: n-us---

Ecosystem degradation in India / [edited by] B.N. Sinha. Published/Created: New Delhi: Ashish Pub. House, 1990. Related Names: Sinha, Bichitrananda, 1931- India. University Grants Commission. National Workshop on Eco-system Degradation (1987: Bhubaneswar, India) Related Titles: Eco-system degradation in India. Description: xii, 477 p.: ill., maps; 23 cm. ISBN: 8170243580: Notes: Spine title: Eco-system degradation in India. Papers presented at the National Workshop on Eco-system Degradation, held at Bhubaneswar by University Grants Commission, 1987. Includes index. Includes bibliographical references. Subjects: Pollution--Economic aspects--India--Congresses. Environmental policy--India--Congresses. Water quality

management--India--Congresses. Conservation of natural resources--India--Congresses. LC Classification: HC440.Z9 P553 1990 Overseas Acquisitions No.: I E 63572 Geographic Area Code: a-ii---

Ecosystem of the Gulf of Riga between 1920 and 1990 / edited by E. Ojaveer. Published/Created: Tallinn: Estonian Academy Publishers, 1995. Related Names: Ojaveer, E. Eesti Teaduste Akadeemia. Description: 277 p.: ill., maps; 26 cm. ISBN: 9985500652 Notes: At head of title: Estonian Academy of Sciences. Includes bibliographical references. Subjects: Brackish water ecology--Riga, Gulf of (Latvia and Estonia) Aquatic organisms--Riga, Gulf of (Latvia and Estonia) Water--Pollution--Riga, Gulf of (Latvia and Estonia) Fishes--Riga, Gulf of (Latvia and Estonia) Series: Academia;5 Academia (Tallinn, Estonia); 5. LC Classification: QH178.R57 E26 1995 Dewey Class No.: 578.77/334 21 Geographic Area Code: e-ur-lv e-ur-er

Ecotoxicology and the aquatic environment: proceedings of a pre-conference symposium held in Toronto, Canada in conjunction with the 10th IAWPR Conference held in Toronto, Canada / editor P.M. Stokes. Published/Created: Oxford; New York: Pergamon Press, c1981. Related Names: Stokes, P. M. (Pamela M.) International Association on Water Pollution Research. Conference (10th: 1980: Toronto, Ont.) Description: v, 97 p.: ill.; 25 cm. ISBN: 0080290922 (pbk.) Notes: Includes bibliographies and index. Subjects: Water--Pollution--Environmental aspects--Congresses. Aquatic ecology--Congresses. Series: Water science and technology LC Classification: QH545.W3 E36 1981 Dewey Class No.: 628.1/61 19

Educational systems for operators of water pollution control facilities. Edited by John H. Austin [and] John Kesler.

Published/Created: [n.p., 1969 or 70]
Related Names: Austin, John H., 1926- ed.
Kesler, John, ed. United States. Federal
Water Pollution Control Administration.
Clemson University. Description: ix, 411
p. illus., map. 28 cm. Notes: Proceedings
of a conference held Nov. 3-5, 1969, in
Atlanta, Ga. "[Sponsored by] Federal
Water Pollution Control Administration, in
cooperation with Clemson University."
Includes bibliographical references.
Subjects: Water--Pollution--Study and
teaching--United States --Congresses. LC
Classification: TD424.3 .E35 Dewey Class
No.: 628.1/68/071 Geographic Area Code:
n-us---

Effect of mass-bathing on water quality of
Brahma Sarovar (Kurukshetra) during solar
eclipse, September 1987 [microform].
Published/Created: New Delhi: Central
Board for the Prevention and Control of
Water Pollution, 1988. Related Names:
India. Central Board for the Prevention and
Control of Water Pollution. Related Titles:
Brahma Sarovar (Kurukshetra) during solar
eclipse, September 1987. Description: 38
leaves; 29 cm. Summary: Study conducted
during the peak bathing time, between 5.30
A.M. to 9.00 A.M. on solar eclipse day,
September 23, 1987. Notes: Master
microform held by: DLC. Bibliography:
leaves 34-38. Additional Formats:
Microfiche. New Delhi: Library of
Congress Office; Washington, D.C.:
Library of Congress Photoduplication
Service, 1988. 1 microfiche; 11 x 15 cm.
Subjects: Water--Pollution--India--
Kurukshetra. Water quality--India--
Kurukshetra. Bathing customs--
Environmental aspects--India--
Kurukshetra. Eclipes, Solar--1987.
Kumbha Melaì,, (Hindu festival) Series:
Assessment and development study of
river basin series; ADSORBS/18/1988
Assessment and development study of
river basin series; 1988/18. LC
Classification: TD304 Microfiche

88/60180 Overseas Acquisitions No.: I E
58680 Geographic Area Code: a-ii---

Effectiveness of urban wastewater treatment
policies in selected countries: an EEA pilot
study / European Environment Agency.
Published/Created: Copenhagen: European
Environment Agency, 2005. Related
Names: European Environment Agency.
Description: 51 p.: ill.; 28 cm. ISBN:
9291677639 Notes: "TH-AL-05-002-EN-
C"--P. [4] of cover. Includes
bibliographical references (p. 48-50).
Additional Formats: Also available in an
online version from the European
Environment Agency web site. Subjects:
Sewage--Purification--European Union
countries. Water--Pollution--European
Union countries. Water-supply--European
Union countries. Water--Government
policy--European Union countries. Series:
EEA report, 1725-9177; no. 2/2005 EEA
report; no. 2005/2. LC Classification:
TD745 .E43 2005 Dewey Class No.:
363.72/84 22

Effects of best-management practices in Otter
Creek in the Sheboygan River priority
watershed, Wisconsin, 1990-2002 / by
Steven R. Corsi ... [et al.]; in cooperation
with the Wisconsin Department of Natural
Resources. Published/Created: Reston, Va.:
U.S. Geological Survey, 2005. Related
Names: Corsi, Steven R. Geological
Survey (U.S.) Wisconsin. Dept. of Natural
Resources. Description: vi, 26 p.: ill.,
maps; 28 cm. Notes: Includes
bibliographical references (p. 25-26).
Subjects: Water quality management--
Wisconsin--Otter Creek (Sheboygan
County) Best management practices
(Pollution prevention)--Wisconsin --Otter
Creek (Sheboygan County) Series:
Scientific investigations report; 2005-5009
LC Classification: TD365 .E33 2005

Effects of best-management practices in the
Black Earth Creek Priority Watershed,
Wisconsin, 1984-98 / by D.J. Graczyk ...

[et al.]; in cooperation with the Wisconsin Department of Natural Resources. Published/Created: Reston, Va.: U.S. Dept. of the Interior, U.S. Geological Survey; Denver, CO: U.S. Geological Survey, Information Services [distributor], 2003. Related Names: Graczyk, David J. Wisconsin. Dept. of Natural Resources. Geological Survey (U.S.) Description: vi, 24 p.: ill., maps; 28 cm. Notes: Includes bibliographical references (p. 23-24). Subjects: Water quality management--Wisconsin--Black Earth Creek Watershed. Nonpoint source pollution--Wisconsin--Black Earth Creek Watershed. Best management practices (Pollution prevention)--Wisconsin --Black Earth Creek Watershed. Series: Water-resources investigations report; 03-4163 LC Classification: GB701 .W375 no. 03-4163

Effects of biologically treated bleached kraft mill effluent on cold water stream productivity in experimental stream channels: fifth progress report. Published/Created: New York, N.Y. (260 Madison Ave., New York 10016): National Council of the Paper Industry for Air and Stream Improvement, [1989] Related Names: National Council of the Paper Industry for Air and Stream Improvement (U.S.) Description: 3, 127, 10, 19 p.: ill.; 28 cm. Notes: Cover title. "May 1989." Includes bibliographical references (p. 122-127). Subjects: Salmonidae--Effect of water pollution on--Idaho--Lewiston. Aquatic organisms--Effect of water pollution on--Idaho --Lewiston. Wood-pulp industry--Waste disposal--Environmental aspects --Idaho--Lewiston. Freshwater productivity--Idaho--Lewiston. Sewage--Purification--Biological treatment--Environmental aspects--Idaho--Lewiston. Water quality bioassay--Idaho--Lewiston. Stream ecology--Idaho--Lewiston. Series: NCASI technical bulletin; no. 566 Technical bulletin (National Council of the Paper Industry for

Air and Stream Improvement (U.S.): 1981); no. 566. LC Classification: TD899.P3 N34 no. 566 QL638.S2 Dewey Class No.: 676/.2 s 597/.55 20 Geographic Area Code: n-us-id

Effects of biologically treated bleached kraft mill effluent on cold water stream productivity in experimental stream channels: fifth progress report. Published/Created: New York, N.Y. (260 Madison Ave., New York 10016): National Council of the Paper Industry for Air and Stream Improvement, [1989] Related Names: National Council of the Paper Industry for Air and Stream Improvement (U.S.) Description: 3, 127, 10, 19 p.: ill.; 28 cm. Notes: Cover title. "May 1989." Includes bibliographical references (p. 122-127). Subjects: Salmonidae--Effect of water pollution on--Idaho--Lewiston. Aquatic organisms--Effect of water pollution on--Idaho --Lewiston. Wood-pulp industry--Waste disposal--Environmental aspects --Idaho--Lewiston. Freshwater productivity--Idaho--Lewiston. Sewage--Purification--Biological treatment--Environmental aspects--Idaho--Lewiston. Water quality bioassay--Idaho--Lewiston. Stream ecology--Idaho--Lewiston. Series: NCASI technical bulletin; no. 566 Technical bulletin (National Council of the Paper Industry for Air and Stream Improvement (U.S.): 1981); no. 566. LC Classification: TD899.P3 N34 no. 566 QL638.S2 Dewey Class No.: 676/.2 s 597/.55 20 Geographic Area Code: n-us-id

Effects of fluvial tailings deposits on soils and surface- and ground-water quality, and implications for remediation--upper Arkansas River, Colorado, 1992-96 / by Katherine Walton-Day ... [et al.]; prepared in cooperation with the U.S. Environmental Protection Agency and the Bureau of Reclamation. Published/Created: Denver, CO: U.S. Dept. of the Interior,

U.S. Geological Survey, 2000. Related
Names: Walton-Day, Katherine. United
States. Environmental Protection Agency.
United States. Bureau of Reclamation.
Geological Survey (U.S.) Description: iv,
100 p.: ill., maps; 28 cm. Notes: Includes
bibliographical references (p. 48).
Subjects: Mineral industries--Waste
disposal--Environmental aspects --
Colorado--Leadville Region. Tailings
(Metallurgy)--Environmental aspects--
Colorado --Leadville Region. Water--
Pollution--Arkansas River. Water--
Pollution--Colorado--Leadville Region.
Soil--Pollution--Colorado--Leadville
Region. Series: Water-resources
investigations report; 99-4273 LC
Classification: GB701 .W375 no. 99-4273
TD899.M47 Dewey Class No.: 553.7/0973
s 363.739/42/097889 21 Geographic Area
Code: n-us-co

Effects of fluvial tailings deposits on soils and
surface- and ground-water quality, and
implications for remediation--upper
Arkansas River, Colorado, 1992-96 / by
Katherine Walton-Day ... [et al.]; prepared
in cooperation with the U.S.
Environmental Protection Agency and the
Bureau of Reclamation. Published/Created:
Denver, CO: U.S. Dept. of the Interior,
U.S. Geological Survey, 2000. Related
Names: Walton-Day, Katherine. United
States. Environmental Protection Agency.
United States. Bureau of Reclamation.
Geological Survey (U.S.) Description: iv,
100 p.: ill., maps; 28 cm. Notes: Includes
bibliographical references (p. 48).
Subjects: Mineral industries--Waste
disposal--Environmental aspects --
Colorado--Leadville Region. Tailings
(Metallurgy)--Environmental aspects--
Colorado --Leadville Region. Water--
Pollution--Arkansas River. Water--
Pollution--Colorado--Leadville Region.
Soil--Pollution--Colorado--Leadville
Region. Series: Water-resources
investigations report; 99-4273 LC

Classification: GB701 .W375 no. 99-4273
TD899.M47 Dewey Class No.: 553.7/0973
s 363.739/42/097889 21 Geographic Area
Code: n-us-co

Effects of land use on fresh waters: agriculture,
forestry, mineral exploitation, urbanisation
/ editor, J.F. de L.G. Solbeì .
Published/Created: Chichester, West
Sussex, England: Published for the Water
Research Centre by E. Horwood; New
York, N.Y., USA: Distributors, J. Wiley,
1986. Related Names: Solbeì , J. F. de L.
G. Description: 568 p.: ill.; 25 cm. ISBN:
0745800548: Notes: Includes
bibliographies and index. Subjects: Water--
Pollution. Land use--Environmental
aspects. Pollution--Environmental aspects.
Series: Ellis Horwood series in water and
wastewater technology Ellis Horwood
series, water and wastewater technology.
LC Classification: TD423 .E34 1986
Dewey Class No.: 363.7/394 19

Effects of land use on fresh waters: agriculture,
forestry, mineral exploitation, urbanisation
/ editor, J.F. de L.G. Solbeì .
Published/Created: Chichester, West
Sussex, England: Published for the Water
Research Centre by E. Horwood; New
York, N.Y., USA: Distributors, J. Wiley,
1986. Related Names: Solbeì , J. F. de L.
G. Description: 568 p.: ill.; 25 cm. ISBN:
0745800548: Notes: Includes
bibliographies and index. Subjects: Water--
Pollution. Land use--Environmental
aspects. Pollution--Environmental aspects.
Series: Ellis Horwood series in water and
wastewater technology Ellis Horwood
series, water and wastewater technology.
LC Classification: TD423 .E34 1986
Dewey Class No.: 363.7/394 19

Effects of petroleum on arctic and subarctic
marine environments and organisms /
edited by Donald C. Malins.
Published/Created: New York: Academic
Press, 1977. Related Names: Malins,
Donald C. Description: 2 v.: ill.; 24 cm.

ISBN: 0124669018 (v. 1) Contents: v. 1. Nature and fate of petroleum.--v. 2. Biological effects. Notes: Includes bibliographical references and indexes. Subjects: Oil pollution of the sea--Arctic regions. Oil spills--Environmental aspects. Aquatic animals--Effect of water pollution on. LC Classification: GC1085 .E34 Dewey Class No.: 363.6 Geographic Area Code: r------

Effects of petroleum on arctic and subarctic marine environments and organisms / edited by Donald C. Malins. Published/Created: New York: Academic Press, 1977. Related Names: Malins, Donald C. Description: 2 v.: ill.; 24 cm. ISBN: 0124669018 (v. 1) Contents: v. 1. Nature and fate of petroleum.--v. 2. Biological effects. Notes: Includes bibliographical references and indexes. Subjects: Oil pollution of the sea--Arctic regions. Oil spills--Environmental aspects. Aquatic animals--Effect of water pollution on. LC Classification: GC1085 .E34 Dewey Class No.: 363.6 Geographic Area Code: r------

Effects of reducing nutrient loads to surface waters within the Mississippi River Basin and the Gulf of Mexico: topic 4, report for the integrated assessment on hypoxia in the Gulf of Mexico / Patrick L. Brezonik ... [et al.]. Portion of Title: Report for the integrated assessment on hypoxia in the Gulf of Mexico Published/Created: Silver Spring, Md.: U.S. Dept. of Commerce, National Oceanic and Atmospheric Administration, Coastal Ocean Program, [1999] Related Names: Brezonik, Patrick L. Description: xvii, 130 p.: ill., maps; 28 cm. Notes: "May 1999." Includes bibliographical references (p. 118-130). Subjects: Nutrient pollution of water--Mississippi River Watershed. Water quality management--Mississippi River Watershed. Nutrient pollution of water--Mexico, Gulf of. Water--Dissolved

oxygen--Mexico, Gulf of. Hypoxia (Water) Series: NOAA Coastal Ocean Program decision analysis series; no. 18 LC Classification: TD427.N87 E44 1999 Dewey Class No.: 363.739/4/0916364 21 Geographic Area Code: n-usm-- nm-----

Effects of reducing nutrient loads to surface waters within the Mississippi River Basin and the Gulf of Mexico: topic 4, report for the integrated assessment on hypoxia in the Gulf of Mexico / Patrick L. Brezonik ... [et al.]. Portion of Title: Report for the integrated assessment on hypoxia in the Gulf of Mexico Published/Created: Silver Spring, Md.: U.S. Dept. of Commerce, National Oceanic and Atmospheric Administration, Coastal Ocean Program, [1999] Related Names: Brezonik, Patrick L. Description: xvii, 130 p.: ill., maps; 28 cm. Notes: "May 1999." Includes bibliographical references (p. 118-130). Subjects: Nutrient pollution of water--Mississippi River Watershed. Water quality management--Mississippi River Watershed. Nutrient pollution of water--Mexico, Gulf of. Water--Dissolved oxygen--Mexico, Gulf of. Hypoxia (Water) Series: NOAA Coastal Ocean Program decision analysis series; no. 18 LC Classification: TD427.N87 E44 1999 Dewey Class No.: 363.739/4/0916364 21 Geographic Area Code: n-usm-- nm-----

Effects of temperature on aquatic organism sensitivity to selected chemicals / John Cairns, Jr. ... [et al.]. Published/Created: Blacksburg: Virginia Water Resources Research Center, Virginia Polytechnic Institute and State University, 1978. Related Names: Cairns, John, 1923- Description: viii, 88 p.: graphs; 23 cm. Notes: Bibliography: p. 43-52. Subjects: Aquatic organisms--Effect of water pollution on. Aquatic organisms--Effect of temperature on. Series: Bulletin - Virginia Water Resources Research Center, Virginia Polytechnic Institute and State

University; 106 Bulletin (Virginia Water Resources Research Center); 106. LC Classification: TD201 .V57 no. 106 QH545.W3 Dewey Class No.: 333.91/009755 s 574.2/4 19

Effects of temperature on aquatic organism sensitivity to selected chemicals / John Cairns, Jr. ... [et al.]. Published/Created: Blacksburg: Virginia Water Resources Research Center, Virginia Polytechnic Institute and State University, 1978. Related Names: Cairns, John, 1923- Description: viii, 88 p.: graphs; 23 cm. Notes: Bibliography: p. 43-52. Subjects: Aquatic organisms--Effect of water pollution on. Aquatic organisms--Effect of temperature on. Series: Bulletin - Virginia Water Resources Research Center, Virginia Polytechnic Institute and State University; 106 Bulletin (Virginia Water Resources Research Center); 106. LC Classification: TD201 .V57 no. 106 QH545.W3 Dewey Class No.: 333.91/009755 s 574.2/4 19

Effects of temperature on the toxicity of oil refinery waste, sodium chlorate, and treated sewage to fathead minnows / by Curt C. Shifrer ... [et al.] Published/Created: Logan: Utah Water Research Laboratory, College of Engineering, Utah State University, 1974. Related Names: Shifrer, Curt C. Description: x, 79 p.: ill.; 28 cm. Notes: Bibliography: p. 43-44. Subjects: Fishes-- Effect of water pollution on. Toxicity, Effect of temperature on. Petroleum refineries--Environmental aspects. Sewage disposal plants--Environmental aspects. Fathead minnow. Series: PRWG; 105-4 LC Classification: TD224.U8 U85 no. 105-4 SH174 Dewey Class No.: 628.1/08 s 597/.53

Effects of temperature on the toxicity of oil refinery waste, sodium chlorate, and treated sewage to fathead minnows / by Curt C. Shifrer ... [et al.]

Published/Created: Logan: Utah Water Research Laboratory, College of Engineering, Utah State University, 1974. Related Names: Shifrer, Curt C. Description: x, 79 p.: ill.; 28 cm. Notes: Bibliography: p. 43-44. Subjects: Fishes-- Effect of water pollution on. Toxicity, Effect of temperature on. Petroleum refineries--Environmental aspects. Sewage disposal plants--Environmental aspects. Fathead minnow. Series: PRWG; 105-4 LC Classification: TD224.U8 U85 no. 105-4 SH174 Dewey Class No.: 628.1/08 s 597/.53

Effects of thermal pollution on productivity and stability of estuarine communities [by] J. A. Mihursky [and others. Published/Created: College Park, Water Resources Research Center, University of Maryland] 1971. Related Names: Mihursky, J. A. University of Maryland, College Park. Water Resources Research Center. Description: 65 p. illus. 28 cm. Notes: Report of research conducted by the Water Resources Research Center, University of Maryland. Subjects: Thermal pollution of rivers, lakes, etc.--Maryland -- PatuxentRiver Estuary. Aquatic ecology-- Maryland--Patuxent River Estuary. Aquatic animals--Effect of water pollution on. LC Classification: TD427.H4 E34 Dewey Class No.: 574.5/2632 Geographic Area Code: n-us-md

Effects of thermal pollution on productivity and stability of estuarine communities [by] J. A. Mihursky [and others. Published/Created: College Park, Water Resources Research Center, University of Maryland] 1971. Related Names: Mihursky, J. A. University of Maryland, College Park. Water Resources Research Center. Description: 65 p. illus. 28 cm. Notes: Report of research conducted by the Water Resources Research Center, University of Maryland. Subjects: Thermal pollution of rivers, lakes, etc.--Maryland --

PatuxentRiver Estuary. Aquatic ecology--
Maryland--Patuxent River Estuary.
Aquatic animals--Effect of water pollution
on. LC Classification: TD427.H4 E34
Dewey Class No.: 574.5/2632 Geographic
Area Code: n-us-md

Effects of water chemistry on the
bioavailability and toxicity of waterborne
cadmium, copper, nickel, lead, and zinc to
freshwater organisms / written by Joseph S
Meyer ... [et al.]. Published/Created:
Pensacola, FL: Society of Environmental
Toxicology and Chemistry/SETAC Press,
c2007. Related Names: Meyer, Joseph S.
Description: xxi, 328 p.: ill.; 23 cm. ISBN:
9781880611531 (alk. paper) 1880611538
(alk. paper) Notes: Includes bibliographical
references (p. 259-294) and index.
Subjects: Freshwater organisms--Effect of
metals on. Metals--Toxicology.
Bioavailability. Water--Pollution--
Toxicology. Water chemistry--
Environmental aspects. LC Classification:
QL120 .E34 2007 Dewey Class No.:
578.76 22

Effects of water chemistry on the
bioavailability and toxicity of waterborne
cadmium, copper, nickel, lead, and zinc to
freshwater organisms / written by Joseph S
Meyer ... [et al.]. Published/Created:
Pensacola, FL: Society of Environmental
Toxicology and Chemistry/SETAC Press,
c2007. Related Names: Meyer, Joseph S.
Description: xxi, 328 p.: ill.; 23 cm. ISBN:
9781880611531 (alk. paper) 1880611538
(alk. paper) Notes: Includes bibliographical
references (p. 259-294) and index.
Subjects: Freshwater organisms--Effect of
metals on. Metals--Toxicology.
Bioavailability. Water--Pollution--
Toxicology. Water chemistry--
Environmental aspects. LC Classification:
QL120 .E34 2007 Dewey Class No.:
578.76 22

Effluent + water treatment journal. Portion of
Title: Effluent and water treatment journal

Serial Key Title: Effluent & water
treatment journal Abbreviated Title: Effl.
water treat. j. Published/Created: [Harrow,
Middlesex: Thunderbird Enterprises, -
1986. Description: 26 v.: ill.; 28 cm. Began
with: Vol. 1, no. 1 (May 1961). Current
Frequency: Monthly, <Mar. 1962->
Former Frequency: Bimonthly, 1961-
ISSN: 0013-2217 Cancelled/Invalid
LCCN: sn 79007156 CODEN: EWTJAG
Notes: Description based on: Vol. 16, no. 5
(May 1976); title from cover.
SERBIB/SERLOC merged record Indexed
selectively by: Chemical abstracts 0009-
2258 Subjects: Water--Purification--
Periodicals. Sewage--Purification--
Periodicals. Factory and trade waste--
Periodicals. Sewage--Periodicals. Water
Pollution--prevention & control--
Periodicals. LC Classification: TD511 .E35
NLM Class No.: W1 EF383 NAL Class
No.: TD201.E3

Effluent + water treatment journal. Portion of
Title: Effluent and water treatment journal
Serial Key Title: Effluent & water
treatment journal Abbreviated Title: Effl.
water treat. j. Published/Created: [Harrow,
Middlesex: Thunderbird Enterprises, -
1986. Description: 26 v.: ill.; 28 cm. Began
with: Vol. 1, no. 1 (May 1961). Current
Frequency: Monthly, <Mar. 1962->
Former Frequency: Bimonthly, 1961-
ISSN: 0013-2217 Cancelled/Invalid
LCCN: sn 79007156 CODEN: EWTJAG
Notes: Description based on: Vol. 16, no. 5
(May 1976); title from cover.
SERBIB/SERLOC merged record Indexed
selectively by: Chemical abstracts 0009-
2258 Subjects: Water--Purification--
Periodicals. Sewage--Purification--
Periodicals. Factory and trade waste--
Periodicals. Sewage--Periodicals. Water
Pollution--prevention & control--
Periodicals. LC Classification: TD511 .E35
NLM Class No.: W1 EF383 NAL Class
No.: TD201.E3

Effluent toxicity status in water polluting industries. Part 1, Dye & dye intermediate, bulk drugs, and textile industries. Published/Created: Delhi: Central Pollution Control Board, Ministry of Environment & Forests, Govt. of India, 2002. Related Names: India. Central Pollution Control Board. Description: 49 p.: ill. (some col.), map; 29 cm. Summary: With reference to India. Subjects: Factory and trade waste--Toxicology--India. Water--Pollution--Toxicology--India. Water quality management--India. Factory and trade waste--Purification--India. Series: Programme objective series; PROBES91/2002-2003 Programme objective series; PROBES/2002-2003/91. LC Classification: TD897.8.I4 E33 2002 Overseas Acquisitions No.: I-E-2004-327435; 49-90

Effluent toxicity status in water polluting industries. Part 1, Dye & dye intermediate, bulk drugs, and textile industries. Published/Created: Delhi: Central Pollution Control Board, Ministry of Environment & Forests, Govt. of India, 2002. Related Names: India. Central Pollution Control Board. Description: 49 p.: ill. (some col.), map; 29 cm. Summary: With reference to India. Subjects: Factory and trade waste--Toxicology--India. Water--Pollution--Toxicology--India. Water quality management--India. Factory and trade waste--Purification--India. Series: Programme objective series; PROBES91/2002-2003 Programme objective series; PROBES/2002-2003/91. LC Classification: TD897.8.I4 E33 2002 Overseas Acquisitions No.: I-E-2004-327435; 49-90

Elmia international conferences on air and water conservation. Published/Created: Joìˆnkoìˆping, Sweden: Elmia, 1981. Related Names: Elmia AB. Description: 196 p.: ill.; 30 cm. Notes: Cover title. Includes bibliographical references.

Subjects: Air--Pollution--Congresses. Water--Pollution--Congresses. Environmental protection--Congresses. Water conservation--Congresses. LC Classification: TD881 .E43 1981 Dewey Class No.: 363.7/3 19

Elmia international conferences on air and water conservation. Published/Created: Joìˆnkoìˆping, Sweden: Elmia, 1981. Related Names: Elmia AB. Description: 196 p.: ill.; 30 cm. Notes: Cover title. Includes bibliographical references. Subjects: Air--Pollution--Congresses. Water--Pollution--Congresses. Environmental protection--Congresses. Water conservation--Congresses. LC Classification: TD881 .E43 1981 Dewey Class No.: 363.7/3 19

Emerging issues in water and infectious disease. Portion of Title: Water and infectious disease. Published/Created: Geneva: World Health Organization, c2003. Related Names: World Health Organization. Description: 22 p.: ill.; 30 cm. ISBN: 9241590823 Notes: Cover title. Includes bibliographical references (p. 20). Subjects: Waterborne infection. Water--Pollution--Health aspects. Pathogenic microorganisms. LC Classification: RA642.W3 E49 2003

Emerging issues in water and infectious disease. Portion of Title: Water and infectious disease. Published/Created: Geneva: World Health Organization, c2003. Related Names: World Health Organization. Description: 22 p.: ill.; 30 cm. ISBN: 9241590823 Notes: Cover title. Includes bibliographical references (p. 20). Subjects: Waterborne infection. Water--Pollution--Health aspects. Pathogenic microorganisms. LC Classification: RA642.W3 E49 2003

Engineering equipment and automation means for waste-water management in ECE countries / prepared under the auspices of the ECE Working Party on Engineering Industries and Automation. Published/Created: New York: United Nations, 1984. Related Names: ECE Working Party on Engineering Industries and Automation. United Nations. Economic Commission for Europe. Description: 2 v.: ill.; 30 cm. Contents: pt. 1. A report on prevailing practices and recent experience in production and use of engineering equipment and automation means for preventing water pollution -- pt. 2. National experiences of ECE member countries in production and use of engineering equipment and automation means for preventing water pollution. Notes: At head of title: Economic Commission for Europe. Bibliography: v. 1, p. 109-111. Subjects: Sewage disposal plants--Europe--Equipment and supplies. Sewage disposal plants--Europe--Automation. Water--Pollution--Europe. LC Classification: TD746 .E54 1984 Dewey Class No.: 628.3/094 19 Geographic Area Code: ew-----

Engineering equipment and automation means for waste-water management in ECE countries / prepared under the auspices of the ECE Working Party on Engineering Industries and Automation. Published/Created: New York: United Nations, 1984. Related Names: ECE Working Party on Engineering Industries and Automation. United Nations. Economic Commission for Europe. Description: 2 v.: ill.; 30 cm. Contents: pt. 1. A report on prevailing practices and recent experience in production and use of engineering equipment and automation means for preventing water pollution -- pt. 2. National experiences of ECE member countries in production and use of engineering equipment and automation means for preventing water pollution. Notes: At head of title: Economic Commission for Europe. Bibliography: v. 1, p. 109-111. Subjects: Sewage disposal

plants--Europe--Equipment and supplies. Sewage disposal plants--Europe--Automation. Water--Pollution--Europe. LC Classification: TD746 .E54 1984 Dewey Class No.: 628.3/094 19 Geographic Area Code: ew-----

Enteric virus detection in water by nucleic acid methods / prepared by Mark D. Sobsey ... [et al.]; sponsored by AWWA Research Foundation. Published/Created: Denver, CO: AWWA Research Foundation and American Water Works Association, c1996. Related Names: Sobsey, Mark D. AWWA Research Foundation. Description: xxi, 132 p.: ill.; 28 cm. ISBN: 0898678889 Notes: "1P-5C-90711-1/97-CM." Includes bibliographical references (p. 117-128) and index. Subjects: Viral pollution of water. Enteroviruses--Analysis. Nucleic acid probes. LC Classification: TD427.V55 E57 1996 Dewey Class No.: 628.1/61 21

Enteric virus detection in water by nucleic acid methods / prepared by Mark D. Sobsey ... [et al.]; sponsored by AWWA Research Foundation. Published/Created: Denver, CO: AWWA Research Foundation and American Water Works Association, c1996. Related Names: Sobsey, Mark D. AWWA Research Foundation. Description: xxi, 132 p.: ill.; 28 cm. ISBN: 0898678889 Notes: "1P-5C-90711-1/97-CM." Includes bibliographical references (p. 117-128) and index. Subjects: Viral pollution of water. Enteroviruses--Analysis. Nucleic acid probes. LC Classification: TD427.V55 E57 1996 Dewey Class No.: 628.1/61 21

Environment & change. Variant Title: Environment and change Serial Key Title: Environment & change Abbreviated Title: Environ. change Published/Created: [London, Maddox Editorial Ltd.] Description: 1 v. ill. 30 cm. v. 2, no. 1-6; Sept. 1973-Feb. 1974. Continues: Environment this month ISSN: 0301-3715

CODEN: EVCHAK Notes:
SERBIB/SERLOC merged record
Subjects: Human ecology--Periodicals. Air
Pollution--Periodicals. Ecology--
Periodicals. Environment--Periodicals.
Environmental Health--Periodicals. Water
Pollution--Periodicals. LC Classification:
GF1 .E56 NLM Class No.: W1 EN98NP
Dewey Class No.: 301.31/05

Environment & change. Variant Title:
Environment and change Serial Key Title:
Environment & change Abbreviated Title:
Environ. change Published/Created:
[London, Maddox Editorial Ltd.]
Description: 1 v. ill. 30 cm. v. 2, no. 1-6;
Sept. 1973-Feb. 1974. Continues:
Environment this month ISSN: 0301-3715
CODEN: EVCHAK Notes:
SERBIB/SERLOC merged record
Subjects: Human ecology--Periodicals. Air
Pollution--Periodicals. Ecology--
Periodicals. Environment--Periodicals.
Environmental Health--Periodicals. Water
Pollution--Periodicals. LC Classification:
GF1 .E56 NLM Class No.: W1 EN98NP
Dewey Class No.: 301.31/05

Environment 2000 position paper: report to the
Environment Liason Forum: pollution
issues of Lake Chivero and catchment,
June 1996 / co-ordinated by Heather
Bailey; assisted by Tendai Kajese,
Emmanual Koro. Published/Created:
[Harare: s.n., 1996] Related Names:
Bailey, Heather. Kajese, Tendai. Koro,
Emmanual. Environment Liason Forum
(Zimbabwe) Description: 30 leaves: ill.,
maps; 30 cm. Notes: Cover title. Includes
bibliographical references (leaf 29).
Subjects: Water--Pollution--Zimbabwe--
Chivero, Lake, Watershed. Environmental
degradation--Zimbabwe--Chivero, Lake,
Watershed. LC Classification: TD319.R45
E54 1996 Dewey Class No.:
363.739/4/096891 21

Environment 2000 position paper: report to the
Environment Liason Forum: pollution

issues of Lake Chivero and catchment,
June 1996 / co-ordinated by Heather
Bailey; assisted by Tendai Kajese,
Emmanual Koro. Published/Created:
[Harare: s.n., 1996] Related Names:
Bailey, Heather. Kajese, Tendai. Koro,
Emmanual. Environment Liason Forum
(Zimbabwe) Description: 30 leaves: ill.,
maps; 30 cm. Notes: Cover title. Includes
bibliographical references (leaf 29).
Subjects: Water--Pollution--Zimbabwe--
Chivero, Lake, Watershed. Environmental
degradation--Zimbabwe--Chivero, Lake,
Watershed. LC Classification: TD319.R45
E54 1996 Dewey Class No.:
363.739/4/096891 21

Environment protection and pollution control in
the Ganga / editor, P.K. Agrawal.
Published/Created: New Delhi: M D
Publications, 1994. Related Names:
Agrawal, P. K. (Pramod Kumar), 1950-
Description: 173 p.: ill.; 22 cm. ISBN:
8185880395: Summary: Contributed
articles. Notes: Includes bibliographical
references. Subjects: Water--Pollution--
Ganges River (India and Bangladesh)
Environmental protection--India. LC
Classification: TD304.G36 E54 1994
Dewey Class No.: 363.73/942/09541 20
Overseas Acquisitions No.: I-E-73484
Geographic Area Code: a-ii---

Environment protection and pollution control in
the Ganga / editor, P.K. Agrawal.
Published/Created: New Delhi: M D
Publications, 1994. Related Names:
Agrawal, P. K. (Pramod Kumar), 1950-
Description: 173 p.: ill.; 22 cm. ISBN:
8185880395: Summary: Contributed
articles. Notes: Includes bibliographical
references. Subjects: Water--Pollution--
Ganges River (India and Bangladesh)
Environmental protection--India. LC
Classification: TD304.G36 E54 1994
Dewey Class No.: 363.73/942/09541 20
Overseas Acquisitions No.: I-E-73484
Geographic Area Code: a-ii---

Environmental aspects of dredging / Edited by
R.N. Bray. Published/Created: London,
UK: Taylor & Francis, 2008. Projected
Publication Date: 0801 Related Names:
Bray, R. N. (Richard Nicholas)
Description: p. cm. ISBN: 9780415450805
(hardback: alk. paper) Notes: Includes
bibliographical references and index.
Subjects: Dredging--Environmental
aspects. Water--Pollution. LC
Classification: TD195.D72 E55 2008
Dewey Class No.: 627/.730286 22

Environmental aspects of dredging / Edited by
R.N. Bray. Published/Created: London,
UK: Taylor & Francis, 2008. Projected
Publication Date: 0801 Related Names:
Bray, R. N. (Richard Nicholas)
Description: p. cm. ISBN: 9780415450805
(hardback: alk. paper) Notes: Includes
bibliographical references and index.
Subjects: Dredging--Environmental
aspects. Water--Pollution. LC
Classification: TD195.D72 E55 2008
Dewey Class No.: 627/.730286 22

Environmental aspects of geology and
engineering in Oklahoma; a symposium of
the Oklahoma Academy of Science. Co-
sponsored by Oklahoma State University,
Stillwater, Oklahoma, December 4, 1970.
William D. Rose, editor.
Published/Created: Norman, Oklahoma
Geological Survey, University of
Oklahoma, 1971. Related Names: Rose,
William D. Oklahoma Academy of
Science. Oklahoma State University.
Description: v, 70 p. illus. 25 cm. Notes:
Includes bibliographical references.
Subjects: Civil engineering--
Environmental aspects--Oklahoma --
Congresses. Energy development--
Environmental aspects--Oklahoma --
Congresses. Water--Pollution--Oklahoma--
Congresses. Engineering geology--
Oklahoma--Congresses. Series: Annals of
the Oklahoma Academy of Science;
publication no. 2 LC Classification:

TD195.C54 E56 Dewey Class No.:
333.7/2/09766 19 Geographic Area Code:
n-us-ok

Environmental aspects of geology and
engineering in Oklahoma; a symposium of
the Oklahoma Academy of Science. Co-
sponsored by Oklahoma State University,
Stillwater, Oklahoma, December 4, 1970.
William D. Rose, editor.
Published/Created: Norman, Oklahoma
Geological Survey, University of
Oklahoma, 1971. Related Names: Rose,
William D. Oklahoma Academy of
Science. Oklahoma State University.
Description: v, 70 p. illus. 25 cm. Notes:
Includes bibliographical references.
Subjects: Civil engineering--
Environmental aspects--Oklahoma --
Congresses. Energy development--
Environmental aspects--Oklahoma --
Congresses. Water--Pollution--Oklahoma--
Congresses. Engineering geology--
Oklahoma--Congresses. Series: Annals of
the Oklahoma Academy of Science;
publication no. 2 LC Classification:
TD195.C54 E56 Dewey Class No.:
333.7/2/09766 19 Geographic Area Code:
n-us-ok

Environmental assessment for the Gallinas
Municipal Watershed wildland-urban
interface project: Pecos/Las Vegas Ranger
District, Santa Fe National Forest. Portion
of Title: Gallinas municipal watershed
wildland-urban interface project
Published/Created: Pecos, NM: United
States Dept. of Agriculture, [2006] Related
Names: United States. Forest Service.
Southwestern Region. Description: iii, 226
p.: ill., maps; 28 cm. Notes: Title from
cover. "05/25/2005." "Printed on recycled
paper -- March 2006."--inside cover.
Includes bibliographic references: p. 221-
226. 79 Additional Formats: Also available
in PDF format on the web at:
http://www.fs.fed.us/r3/sfe/projects/project
s/gallinasEA /gallinasEA.pdf Subjects:

Wildfires--Prevention and control Fuel reduction (Wildfire prevention)--Santa Fe National Forest (N.M.). Water--Pollution--Prevention. Water-supply--New Mexico--Las Vegas. Water-supply--New Mexico--Santa Fe National Forest. Water-supply--New Mexico--San Miguel County. LC Classification: SD421.32.N6 E59 2006 Dewey Class No.: 634.9/61809789 22 Government Document No.: A 13.91: G 13x (NM)

Environmental assessment for the Gallinas Municipal Watershed wildland-urban interface project: Pecos/Las Vegas Ranger District, Santa Fe National Forest. Portion of Title: Gallinas municipal watershed wildland-urban interface project Published/Created: Pecos, NM: United States Dept. of Agriculture, [2006] Related Names: United States. Forest Service. Southwestern Region. Description: iii, 226 p.: ill., maps; 28 cm. Notes: Title from cover: "05/25/2005." "Printed on recycled paper -- March 2006."--inside cover. Includes bibliographic references: p. 221-226. 79 Additional Formats: Also available in PDF format on the web at: http://www.fs.fed.us/r3/sfe/projects/projects/gallinasEA /gallinasEA.pdf Subjects: Wildfires--Prevention and control Fuel reduction (Wildfire prevention)--Santa Fe National Forest (N.M.). Water--Pollution--Prevention. Water-supply--New Mexico--Las Vegas. Water-supply--New Mexico--Santa Fe National Forest. Water-supply--New Mexico--San Miguel County. LC Classification: SD421.32.N6 E59 2006 Dewey Class No.: 634.9/61809789 22 Government Document No.: A 13.91: G 13x (NM)

Environmental coastal regions / editor, C.A. Brebbia. Published/Created: Boston: Computational Mechanics Publications: Southampton: WIT Press, c1998. Related Names: Brebbia, C. A. Wessex Institute of Technology. International Conference on Environmental Coastal Regions (2nd: 1998: Cancun, Mexico) Description: 436 p.: ill., maps; 24 cm. ISBN: 185312527X Notes: Includes bibliographical references and index. Subjects: Coasts--Environmental conditions--Congresses. Water--Pollution--Congresses. Environmental impact analysis--Congresses. Series: Environmental studies, 1462-6098; v. 1 Environmental studies (Southampton, England) LC Classification: GB450.2 .E58 1998 Dewey Class No.: 363.739/4 21

Environmental coastal regions / editor, C.A. Brebbia. Published/Created: Boston: Computational Mechanics Publications: Southampton: WIT Press, c1998. Related Names: Brebbia, C. A. Wessex Institute of Technology. International Conference on Environmental Coastal Regions (2nd: 1998: Cancun, Mexico) Description: 436 p.: ill., maps; 24 cm. ISBN: 185312527X Notes: Includes bibliographical references and index. Subjects: Coasts--Environmental conditions--Congresses. Water--Pollution--Congresses. Environmental impact analysis--Congresses. Series: Environmental studies, 1462-6098; v. 1 Environmental studies (Southampton, England) LC Classification: GB450.2 .E58 1998 Dewey Class No.: 363.739/4 21

Environmental coastal regions III / editors, G.R. Rodrigguez, C.A. Brebbia, E. Peì rez-Martell. Published/Created: Southampton: Boston: WIT Press, c2000. Related Names: Rodriì guez, G. R. Brebbia, C. A. Peì rez-Martell, E. International Conference on Environmental Problems in Coastal Regions (3rd: 2000) Description: xi, 442 p.: ill., maps; 24 cm. ISBN: 1853128279 Notes: Includes bibliographical references. Subjects: Coasts--Environmental conditions--Congresses. Water--Pollution--Congresses. Environmental impact analysis--

Congresses. Series: Environmental studies, 1462-6098; v. 5 Environmental studies (Southampton, England); v. 5. LC Classification: GB450.2 .E583 2000 Dewey Class No.: 363.739/4 21

Environmental coastal regions III / editors, G.R. Rodrigguez, C.A. Brebbia, E. Peì rez-Martell. Published/Created: Southampton: Boston: WIT Press, c2000. Related Names: Rodriì guez, G. R. Brebbia, C. A. Peì rez-Martell, E. International Conference on Environmental Problems in Coastal Regions (3rd: 2000) Description: xi, 442 p.: ill., maps; 24 cm. ISBN: 1853128279 Notes: Includes bibliographical references. Subjects: Coasts--Environmental conditions-- Congresses. Water--Pollution--Congresses. Environmental impact analysis-- Congresses. Series: Environmental studies, 1462-6098; v. 5 Environmental studies (Southampton, England); v. 5. LC Classification: GB450.2 .E583 2000 Dewey Class No.: 363.739/4 21

Environmental control seminar proceedings, Rotterdam, Warsaw, Bucharest, May 25- June 4, 1971. Published/Created: Washington, Bureau of International Commerce; for sale by the Supt. of Docs., U.S. Govt. Print. Off., 1971. Related Names: United States. Bureau of International Commerce. Description: viii, 297 p. illus. 27 cm. Notes: "A United States Department of Commerce publication." Includes bibliographical references. Subjects: Environmental engineering--Congresses. Water-- Pollution--Congresses. Air--Pollution-- Congresses. LC Classification: TD172.5 .E56 Dewey Class No.: 614.7 Government Document No.: C 42.2:En8

Environmental control seminar proceedings, Rotterdam, Warsaw, Bucharest, May 25- June 4, 1971. Published/Created: Washington, Bureau of International Commerce; for sale by the Supt. of Docs.,

U.S. Govt. Print. Off., 1971. Related Names: United States. Bureau of International Commerce. Description: viii, 297 p. illus. 27 cm. Notes: "A United States Department of Commerce publication." Includes bibliographical references. Subjects: Environmental engineering--Congresses. Water-- Pollution--Congresses. Air--Pollution-- Congresses. LC Classification: TD172.5 .E56 Dewey Class No.: 614.7 Government Document No.: C 42.2:En8

Environmental directions. 1978-10-01, no. 71 [sound recording]. Published/Created: Los Angeles: Educational Communications, 1978. Related Names: Pearlman, Nancy, 1948- spk Engelbrecht, Richard S. spk Canham, Robert. spk Educational Communications Collection (Library of Congress) Description: On side B of 1 sound cassette (ca. 30 min.): analog. Summary: Dr. Richard Engelbrecht, professor of environmental engineering at the University of Illinois, and Robert Canham, executive director of the Water Pollution Control Federation, discuss water quality and waste water/sewage facilities. Notes: With: Environmental directions. 1978-09-24 Recorded in Los Angeles, Calif., for radio broadcast on Oct. 1, 1978. Performer Notes: Host, Nancy Pearlman; guests, Richard Engelbrecht, Robert Canham. Credits: Produced by Educational Communications, Inc. Subjects: Water quality management. Sewage--Purification. Form/Genre: Informational programs-- Radio. radfg LC Classification: RYK 5286

Environmental directions. 1978-10-01, no. 71 [sound recording]. Published/Created: Los Angeles: Educational Communications, 1978. Related Names: Pearlman, Nancy, 1948- spk Engelbrecht, Richard S. spk Canham, Robert. spk Educational Communications Collection (Library of Congress) Description: On side B of 1 sound cassette (ca. 30 min.): analog.

Summary: Dr. Richard Engelbrecht, professor of environmental engineering at the University of Illinois, and Robert Canham, executive director of the Water Pollution Control Federation, discuss water quality and waste water/sewage facilities. Notes: With: Environmental directions. 1978-09-24 Recorded in Los Angeles, Calif., for radio broadcast on Oct. 1, 1978. Performer Notes: Host, Nancy Pearlman; guests, Richard Engelbrecht, Robert Canham. Credits: Produced by Educational Communications, Inc. Subjects: Water quality management. Sewage--Purification. Form/Genre: Informational programs--Radio. radfg LC Classification: RYK 5286

Environmental directions. 1982-11-28, no. 285 [sound recording]. Published/Created: Los Angeles: Educational Communications, 1982. Related Names: Pearlman, Nancy, 1948- spk Crow, Sonia F. spk Educational Communications Collection (Library of Congress) Description: On side B of 1 sound cassette (ca. 30 min.): analog. Summary: Sonia F. Crow, administrator of the Environmental Protection Agency, Region 9 (San Francisco), discusses Federal regulatory programs for air and water quality, toxic waste, and pollution. Notes: With: Environmental directions. 1982-11-21, no. 284 Recorded in Los Angeles, Calif., for radio broadcast on Nov. 28, 1982. Performer Notes: Host, Nancy Pearlman; guest, Sonia F. Crow. Credits: Produced by Educational Communications, Inc. Subjects: Environmental policy--United States. Environmental law--United States. Form/Genre: Informational programs--Radio. radfg LC Classification: RYK 5415

Environmental directions. 1982-11-28, no. 285 [sound recording]. Published/Created: Los Angeles: Educational Communications, 1982. Related Names: Pearlman, Nancy, 1948- spk Crow, Sonia F. spk Educational Communications Collection (Library of

Congress) Description: On side B of 1 sound cassette (ca. 30 min.): analog. Summary: Sonia F. Crow, administrator of the Environmental Protection Agency, Region 9 (San Francisco), discusses Federal regulatory programs for air and water quality, toxic waste, and pollution. Notes: With: Environmental directions. 1982-11-21, no. 284 Recorded in Los Angeles, Calif., for radio broadcast on Nov. 28, 1982. Performer Notes: Host, Nancy Pearlman; guest, Sonia F. Crow. Credits: Produced by Educational Communications, Inc. Subjects: Environmental policy--United States. Environmental law--United States. Form/Genre: Informational programs--Radio. radfg LC Classification: RYK 5415

Environmental directions. 1987-08-31, no. 528 [sound recording]. Published/Created: Los Angeles: Educational Communications, 1987. Related Names: Pearlman, Nancy, 1948- spk Carrick, Roger L. spk Educational Communications Collection (Library of Congress) Description: 1 sound tape reel (30 min., 18 sec.): analog, 7 1/2 ips, full track, mono.; 7 in. Summary: Roger Carrick discusses laws to protect citizens from toxics disposal of chemicals, safe drinking water, and Califorina's Toxic Enforcement Act of 1986. Notes: Recorded in Los Angeles, Calif., for radio broadcast on Aug. 31, 1987. Performer Notes: Host, Nancy Pearlman; guest, Roger Carrick. Credits: Produced by Educational Communications, Inc. Subjects: Hazardous substances--Law and legislation--California. Water--Pollution--Law and legislation--California. Form/Genre: Informational programs--Radio. radfg LC Classification: RXC 4807 (master)

Environmental directions. 1987-08-31, no. 528 [sound recording]. Published/Created: Los Angeles: Educational Communications, 1987. Related Names: Pearlman, Nancy, 1948- spk Carrick, Roger L. spk

Educational Communications Collection (Library of Congress) Description: 1 sound tape reel (30 min., 18 sec.): analog, 7 1/2 ips, full track, mono.; 7 in. Summary: Roger Carrick discusses laws to protect citizens from toxics disposal of chemicals, safe drinking water, and Califorina's Toxic Enforcement Act of 1986. Notes: Recorded in Los Angeles, Calif., for radio broadcast on Aug. 31, 1987. Performer Notes: Host, Nancy Pearlman; guest, Roger Carrick. Credits: Produced by Educational Communications, Inc. Subjects: Hazardous substances--Law and legislation--California. Water--Pollution--Law and legislation--California. Form/Genre: Informational programs--Radio. radfg LC Classification: RXC 4807 (master)

Environmental directions. 1998-04-26, no. 1061 sound recording]. Published/Created: Los Angeles: Educational Communications, 1998. Related Names: Pearlman, Nancy, 1948- spk Feuer, Gail Ruderman. spk Davis, Martha. spk Educational Communications Collection (Library of Congress) Description: 1 sound tape reel (29 min., 57 sec.): analog, 7 1/2 ips, full track, mono.; 7 in. Summary: Gail Ruderman Feuer, senior attorney, Natural Resources Defense Council, discusses national air quality standards. Martha Davis, former executive dierector of the Mono Lake Committee, discusses Mono Lake as a national scenic area and water conservation programs. Notes: Recorded in Los Angeles, Calif., for radio broadcast on Apr. 26, 1998. Performer Notes: Host, Nancy Pearlman; guests, Gail Ruderman Feuer, Martha Davis. Credits: Produced by Educational Communications, Inc. Subjects: Air quality--Environmental aspects--United States. Air--Pollution--United States. Water conservation. Mono Lake (Calif.) Form/Genre: Informational programs--Radio. radfg LC Classification: RXC 4324 (master)

Environmental directions. 1998-04-26, no. 1061 sound recording]. Published/Created: Los Angeles: Educational Communications, 1998. Related Names: Pearlman, Nancy, 1948- spk Feuer, Gail Ruderman. spk Davis, Martha. spk Educational Communications Collection (Library of Congress) Description: 1 sound tape reel (29 min., 57 sec.): analog, 7 1/2 ips, full track, mono.; 7 in. Summary: Gail Ruderman Feuer, senior attorney, Natural Resources Defense Council, discusses national air quality standards. Martha Davis, former executive dierector of the Mono Lake Committee, discusses Mono Lake as a national scenic area and water conservation programs. Notes: Recorded in Los Angeles, Calif., for radio broadcast on Apr. 26, 1998. Performer Notes: Host, Nancy Pearlman; guests, Gail Ruderman Feuer, Martha Davis. Credits: Produced by Educational Communications, Inc. Subjects: Air quality--Environmental aspects--United States. Air--Pollution--United States. Water conservation. Mono Lake (Calif.) Form/Genre: Informational programs--Radio. radfg LC Classification: RXC 4324 (master)

Environmental directions. 1999-06-06, no. 1119 sound recording]. Published/Created: Los Angeles: Educational Communications, 1999. Related Names: Pearlman, Nancy, 1948- spk Kadas, Mike. spk Cuse, Arthur. spk Educational Communications Collection (Library of Congress) Description: 1 sound tape reel (29 min., 39 sec.): analog, 7 1/2 ips, full track, mono.; 7 in. Summary: Mike Kadas, mayor, City of Missoula, Montana, discusses growth in Missoula, growth in Montana's rural areas, Blackfoot River gold mining, and septic pollution in Bitterroot River. Arthur Cuse, president of Bullshot Systems, Inc, discusses substituting adhesive lube for grease in fifth-wheel trucks to reduce water pollution. Notes: Recorded in Los Angeles,

Calif., for radio broadcast on June 06,
1999. Performer Notes: Host, Nancy
Pearlman; guests, Mike Kadas, Arthur
Cuse. Credits: Produced by Educational
Communications, Inc. Subjects:
Environmental protection--Montana.
Trucks--Wheels--Maintenance and repair.
Water pollution. Montana--Environmental
conditions. Form/Genre: Informational
programs--Radio. radfg LC Classification:
RXC 4329 (master)

Environmental directions. 1999-06-06, no.
1119 sound recording]. Published/Created:
Los Angeles: Educational
Communications, 1999. Related Names:
Pearlman, Nancy, 1948- spk Kadas, Mike.
spk Cuse, Arthur. spk Educational
Communications Collection (Library of
Congress) Description: 1 sound tape reel
(29 min., 39 sec.): analog, 7 1/2 ips, full
track, mono.; 7 in. Summary: Mike Kadas,
mayor, City of Missoula, Montana,
discusses growth in Missoula, growth in
Montana's rural areas, Blackfoot River
gold mining, and septic pollution in
Bitterroot River. Arthur Cuse, president of
Bullshot Systems, Inc, discusses
substituting adhesive lube for grease in
fifth-wheel trucks to reduce water
pollution. Notes: Recorded in Los Angeles,
Calif., for radio broadcast on June 06,
1999. Performer Notes: Host, Nancy
Pearlman; guests, Mike Kadas, Arthur
Cuse. Credits: Produced by Educational
Communications, Inc. Subjects:
Environmental protection--Montana.
Trucks--Wheels--Maintenance and repair.
Water pollution. Montana--Environmental
conditions. Form/Genre: Informational
programs--Radio. radfg LC Classification:
RXC 4329 (master)

Environmental effects of thermal discharges;
the elements in formulating a rational
public policy. Presented at the winter
annual meeting of the American Society of
Mechanical Engineers, New York,

December 1, 1970. Published/Created:
New York [1970] Related Names:
American Society of Mechanical
Engineers. Heat Transfer Division.
Description: 44 p. illus. 28 cm. Notes:
Symposium organized by the ASME
Technical Committee on Heat Transfer in
Biotechnology. Includes bibliographical
references. Subjects: Thermal pollution of
rivers, lakes, etc.--Congresses. Water--
Pollution--United States--Congresses. LC
Classification: TD427.H4 E58 Dewey
Class No.: 333.9/1 Geographic Area Code:
n-us---

Environmental effects of thermal discharges;
the elements in formulating a rational
public policy. Presented at the winter
annual meeting of the American Society of
Mechanical Engineers, New York,
December 1, 1970. Published/Created:
New York [1970] Related Names:
American Society of Mechanical
Engineers. Heat Transfer Division.
Description: 44 p. illus. 28 cm. Notes:
Symposium organized by the ASME
Technical Committee on Heat Transfer in
Biotechnology. Includes bibliographical
references. Subjects: Thermal pollution of
rivers, lakes, etc.--Congresses. Water--
Pollution--United States--Congresses. LC
Classification: TD427.H4 E58 Dewey
Class No.: 333.9/1 Geographic Area Code:
n-us---

Environmental impacts of poor quality ground
water for irrigation: final report /
conducted by Department of Agricultural
Engineering. Published/Created:
[Saskatchewan]: The Dept., [1989] Related
Names: University of Saskatchewan. Dept.
of Agricultural Engineering. Description:
x, 143 p.: ill., map; 28 cm. Notes: Includes
bibliographical references (p. 108-115).
Subjects: Irrigation water--Pollution--
Saskatchewan. Irrigation water--Quality--
Saskatchewan. Groundwater--Pollution--
Saskatchewan. Groundwater--Quality--

Saskatchewan. LC Classification: S618.47 .E58 1989 Geographic Area Code: n-cn-sn

Environmental impacts of poor quality ground water for irrigation: final report / conducted by Department of Agricultural Engineering. Published/Created: [Saskatchewan]: The Dept., [1989] Related Names: University of Saskatchewan. Dept. of Agricultural Engineering. Description: x, 143 p.: ill., map; 28 cm. Notes: Includes bibliographical references (p. 108-115). Subjects: Irrigation water--Pollution-- Saskatchewan. Irrigation water--Quality-- Saskatchewan. Groundwater--Pollution-- Saskatchewan. Groundwater--Quality-- Saskatchewan. LC Classification: S618.47 .E58 1989 Geographic Area Code: n-cn-sn

Environmental impacts of smelters / edited by Jerome O. Nriagu. Published/Created: New York: Wiley, c1984. Related Names: Nriagu, Jerome O. Description: xii, 608 p.: ill.; 24 cm. ISBN: 0471880434: Notes: "A Wiley-Interscience publication." Includes bibliographies and index. Subjects: Smelting furnaces--Environmental aspects. Water--Pollution--Environmental aspects. Water--Pollution--Toxicology. Series: Advances in environmental science and technology; v. 15 LC Classification: TD180 .A38 vol. 15 TD428.M47 Dewey Class No.: 628 s 574.5/222 19 Electronic File Information: Publisher description http://www.loc.gov/catdir/description/wile y034/83021761. html Table of Contents http://www.loc.gov/catdir/toc/onix05/8302 1761.html

Environmental impacts of smelters / edited by Jerome O. Nriagu. Published/Created: New York: Wiley, c1984. Related Names: Nriagu, Jerome O. Description: xii, 608 p.: ill.; 24 cm. ISBN: 0471880434: Notes: "A Wiley-Interscience publication." Includes bibliographies and index. Subjects: Smelting furnaces--Environmental aspects. Water--Pollution--Environmental aspects. Water--Pollution--Toxicology. Series:

Advances in environmental science and technology; v. 15 LC Classification: TD180 .A38 vol. 15 TD428.M47 Dewey Class No.: 628 s 574.5/222 19 Electronic File Information: Publisher description http://www.loc.gov/catdir/description/wile y034/83021761. html Table of Contents http://www.loc.gov/catdir/toc/onix05/8302 1761.html

Environmental implications of the spill from Water Reservoir No. 2 at Key Lake. Published/Created: [Saskatchewan?: s.n., 1984] Description: 35 leaves in various foliations: ill., maps; 28 cm. Notes: "March 1984." Subjects: Key Lake Mining Corporation. Uranium mines and mining-- Waste disposal--Environmental aspects-- Saskatchewan. Waste spills-- Environmental aspects--Saskatchewan. Water--Pollution--Environmental aspects-- Saskatchewan. LC Classification: TD195.U7 E58 1984 Dewey Class No.: 363.7/3942/0971241 19 Geographic Area Code: n-cn-sn

Environmental implications of the spill from Water Reservoir No. 2 at Key Lake. Published/Created: [Saskatchewan?: s.n., 1984] Description: 35 leaves in various foliations: ill., maps; 28 cm. Notes: "March 1984." Subjects: Key Lake Mining Corporation. Uranium mines and mining-- Waste disposal--Environmental aspects-- Saskatchewan. Waste spills-- Environmental aspects--Saskatchewan. Water--Pollution--Environmental aspects-- Saskatchewan. LC Classification: TD195.U7 E58 1984 Dewey Class No.: 363.7/3942/0971241 19 Geographic Area Code: n-cn-sn

Environmental issues: a TAPPI Press anthology of published papers, 1990 (other than the Environmental Conference) / Thomas W. Joyce, editor. Published/Created: Atlanta, Ga.: TAPPI, c1991. Related Names: Joyce, T. W. Description: xxxii, 538 p.: ill.; 28 cm. ISBN: 089852251X Notes: Includes bibliographical references. Subjects: Paper industry--Environmental aspects. Pollution--Environmental aspects. Wood-pulp industry--Environmental aspects. Water--Pollution--Environmental aspects. Air--Pollution--Environmental aspects. LC Classification: TD195.P37 E58 1991 Dewey Class No.: 676/.042 20

Environmental issues: a TAPPI Press anthology of published papers, 1990 (other than the Environmental Conference) / Thomas W. Joyce, editor. Published/Created: Atlanta, Ga.: TAPPI, c1991. Related Names: Joyce, T. W. Description: xxxii, 538 p.: ill.; 28 cm. ISBN: 089852251X Notes: Includes bibliographical references. Subjects: Paper industry--Environmental aspects. Pollution--Environmental aspects. Wood-pulp industry--Environmental aspects. Water--Pollution--Environmental aspects. Air--Pollution--Environmental aspects. LC Classification: TD195.P37 E58 1991 Dewey Class No.: 676/.042 20

Environmental law update. Published/Created: Mechanicsburg, PA: Pennsylvania Bar Institute, c2006. Related Names: Pennsylvania Bar Institute. Description: xvii, 321 p.: ill.; 28 cm. Notes: PBI (Series) no. 2006-4609. Subjects: Environmental law--Pennsylvania. Water--Pollution--Law and legislation--Chesapeake Bay Watershed. Water transfer--Law and legislation--United States. Series: PBI; no. 2006-4609 LC Classification: KFP354.Z9 E578 2006 Geographic Area Code: n-us---n-us-pa

Environmental law update. Published/Created: Mechanicsburg, PA: Pennsylvania Bar Institute, c2006. Related Names: Pennsylvania Bar Institute. Description: xvii, 321 p.: ill.; 28 cm. Notes: PBI (Series) no. 2006-4609. Subjects: Environmental law--Pennsylvania. Water--Pollution--Law and legislation--Chesapeake Bay Watershed. Water transfer--Law and legislation--United States. Series: PBI; no. 2006-4609 LC Classification: KFP354.Z9 E578 2006 Geographic Area Code: n-us---n-us-pa

Environmental management of agricultural watersheds: a selection of papers presented at a conference held in Smolenice, CSSR / G. Golubev, editor. Published/Created: Laxenburg, Austria: International Institute for Applied Systems Analysis; Springfield, VA, USA: National Technical Information Service [distributor], 1983. Related Names: Golubev, G. N. (Gennadiiì† Nikolaevich) International Institute for Applied Systems Analysis. CìŒeskoslovenskaì akademie veìŒd. Related Titles: Agricultural watersheds: a selection of papers presented at a conference held in Smolenice, CSSR. Description: viii, 279 p.: ill.; 24 cm. ISBN: 3704500593 (pbk.) Notes: The conference, held April 23-27, 1979, was organized jointly by the Czechoslovakian Academy of Sciences and the International Institute for Applied Systems Analysis. Includes bibliographies. Subjects: Watershed management--Congresses. Agricultural pollution--Congresses. Water--Pollution--Congresses. Environmental management--Congresses. Series: IIASA collaborative proceedings series; CP-83-S1 LC Classification: TC401 .E584 1983 Dewey Class No.: 628.1/684 19

Environmental management of agricultural watersheds: a selection of papers presented at a conference held in Smolenice, CSSR / G. Golubev, editor. Published/Created: Laxenburg, Austria: International Institute for Applied Systems Analysis; Springfield, VA, USA: National Technical Information Service [distributor], 1983. Related Names: Golubev, G. N. (Gennadiiì† Nikolaevich) International Institute for Applied Systems Analysis. CìŒeskoslovenskaì akademie veìŒd. Related Titles: Agricultural watersheds: a selection of papers presented at a conference held in Smolenice, CSSR. Description: viii, 279 p.: ill.; 24 cm. ISBN:

3704500593 (pbk.) Notes: The conference, held April 23-27, 1979, was organized jointly by the Czechoslovakian Academy of Sciences and the International Institute for Applied Systems Analysis. Includes bibliographies. Subjects: Watershed management--Congresses. Agricultural pollution--Congresses. Water--Pollution--Congresses. Environmental management--Congresses. Series: IIASA collaborative proceedings series; CP-83-S1 LC Classification: TC401 .E584 1983 Dewey Class No.: 628.1/684 19

Environmental management through biotechnology: microorganisms and enzymes / Leslie Burk, project analyst. Published/Created: Norwalk, CT: Business Communications Co., c2002. Related Names: Burk, Leslie. Business Communications Co. Description: 1 v. (various pagings): ill.; 28 cm. ISBN: 1569651930 Notes: "May 2002"--T.p. verso. Includes bibliographical references (p. 209-215). Subjects: Pollution control industry--United States. Biotechnology industries--United States. Microorganism industry--United States. Enzymes industry--United States. Water purification equipment industry--United States. Refuse disposal industry--United States. Market surveys--United States. Series: Business opportunity report; E-103 LC Classification: HD9718.U62 E59 2002 Geographic Area Code: n-us---

Environmental management through biotechnology: microorganisms and enzymes / Leslie Burk, project analyst. Published/Created: Norwalk, CT: Business Communications Co., c2002. Related Names: Burk, Leslie. Business Communications Co. Description: 1 v. (various pagings): ill.; 28 cm. ISBN: 1569651930 Notes: "May 2002"--T.p. verso. Includes bibliographical references (p. 209-215). Subjects: Pollution control industry--United States. Biotechnology

industries--United States. Microorganism industry--United States. Enzymes industry--United States. Water purification equipment industry--United States. Refuse disposal industry--United States. Market surveys--United States. Series: Business opportunity report; E-103 LC Classification: HD9718.U62 E59 2002 Geographic Area Code: n-us---

Environmental policies for agricultural pollution control / edited by J.S. Shortle and D.G. Abler. Published/Created: Wallingford, Oxon, UK; New York: CABI Pub., c2001. Related Names: Shortle, J. S. (James S.) Abler, David Gerrard, 1960- Description: x, 224 p.: ill.; 24 cm. ISBN: 0851993990 (alk. paper) Notes: Includes bibliographical references (p. 183-211) and index. Subjects: Agricultural pollution--Government policy. Water--Pollution--Government policy. Environmental policy--Economic aspects. LC Classification: TD428.A37 E58 2001 Dewey Class No.: 363.739/45 21

Environmental policies for agricultural pollution control / edited by J.S. Shortle and D.G. Abler. Published/Created: Wallingford, Oxon, UK; New York: CABI Pub., c2001. Related Names: Shortle, J. S. (James S.) Abler, David Gerrard, 1960- Description: x, 224 p.: ill.; 24 cm. ISBN: 0851993990 (alk. paper) Notes: Includes bibliographical references (p. 183-211) and index. Subjects: Agricultural pollution--Government policy. Water--Pollution--Government policy. Environmental policy--Economic aspects. LC Classification: TD428.A37 E58 2001 Dewey Class No.: 363.739/45 21

Environmental pollution, water / general editor, S.G. Misra; editors, D. Prasad, H.S. Gaur. Published/Created: New Delhi: Venus Pub. House, 1992. Related Names: Misra, S. G. Prasad, D., 1948- Gaur, H. S. (Hari Shankar), 1952- Description: xvi, 359 p.: ill.; 22 cm. ISBN: 8172380046: Summary:

Most in the Indian context. Notes: Includes index. Includes bibliographical references (p. 205-212). Subjects: Water--Pollution. Water quality management. Water--Pollution--India. Water quality management--India. Series: Environmental pollution and hazards series; 005 LC Classification: TD420 .E58 1992 Dewey Class No.: 363.73/94 20 Overseas Acquisitions No.: I-E-68717 Geographic Area Code: a-ii---

Environmental pollution, water / general editor, S.G. Misra; editors, D. Prasad, H.S. Gaur. Published/Created: New Delhi: Venus Pub. House, 1992. Related Names: Misra, S. G. Prasad, D., 1948- Gaur, H. S. (Hari Shankar), 1952- Description: xvi, 359 p.: ill.; 22 cm. ISBN: 8172380046: Summary: Most in the Indian context. Notes: Includes index. Includes bibliographical references (p. 205-212). Subjects: Water--Pollution. Water quality management. Water--Pollution--India. Water quality management--India. Series: Environmental pollution and hazards series; 005 LC Classification: TD420 .E58 1992 Dewey Class No.: 363.73/94 20 Overseas Acquisitions No.: I-E-68717 Geographic Area Code: a-ii---

Environmental problems along the border. Published/Created: San Diego, Calif.: Institute of Public and Urban Affairs, San Diego State University, c1979. Related Names: San Diego State University. Institute of Public and Urban Affairs. Description: 120 p.: maps; 28 cm. Notes: Includes bibliographies. Subjects: Pollution--Southwestern States. Water resources development--Southwestern States. Water resources development--Mexico. Pollution--Mexico. Mexican-American Border Region. Series: Occasional papers - Border-State University Consortium for Latin America; no. 7 Border-State University Consortium for Latin America. Occasional papers - Border-State University Consortium for Latin America; no. 7. LC Classification: TD181.S63 E58 Dewey Class No.: 333.7/0978 Geographic Area Code: n-mx-- - n-usp-- n-usu--

Environmental problems along the border. Published/Created: San Diego, Calif.: Institute of Public and Urban Affairs, San Diego State University, c1979. Related Names: San Diego State University. Institute of Public and Urban Affairs. Description: 120 p.: maps; 28 cm. Notes: Includes bibliographies. Subjects: Pollution--Southwestern States. Water resources development--Southwestern States. Water resources development--Mexico. Pollution--Mexico. Mexican-American Border Region. Series: Occasional papers - Border-State University Consortium for Latin America; no. 7 Border-State University Consortium for Latin America. Occasional papers - Border-State University Consortium for Latin America; no. 7. LC Classification: TD181.S63 E58 Dewey Class No.: 333.7/0978 Geographic Area Code: n-mx-- - n-usp-- n-usu--

Environmental protection: standards, compliance, and costs / editor, T.J. Lack. Published/Created: Chichester, West Sussex: Published for the Water Research Centre by E. Horwood; New York: Halsted Press [distributor], 1984. Related Names: Lack, T. J. (Timothy John), 1944- Water Research Centre (Great Britain) Description: 329 p.: ill.; 24 cm. ISBN: 0853127409: 0470200952 (Halsted Press) Notes: Papers from a conference sponsored by the Water Research Centre, held at Keel University, Staffordshire, in Oct. 1983. Includes bibliographies and index. Subjects: Water--Pollution--Great Britain. Environmental protection--Great Britain. Water--Pollution--Toxicology. LC Classification: TD257 .E58 1984 Dewey

Class No.: 363.7/394 19 Geographic Area Code: e-uk---

Environmental protection: standards, compliance, and costs / editor, T.J. Lack. Published/Created: Chichester, West Sussex: Published for the Water Research Centre by E. Horwood; New York: Halsted Press [distributor], 1984. Related Names: Lack, T. J. (Timothy John), 1944- Water Research Centre (Great Britain) Description: 329 p.: ill.; 24 cm. ISBN: 0853127409: 0470200952 (Halsted Press) Notes: Papers from a conference sponsored by the Water Research Centre, held at Keel University, Staffordshire, in Oct. 1983. Includes bibliographies and index. Subjects: Water--Pollution--Great Britain. Environmental protection--Great Britain. Water--Pollution--Toxicology. LC Classification: TD257 .E58 1984 Dewey Class No.: 363.7/394 19 Geographic Area Code: e-uk---

Environmental radiological surveillance report on Oregon surface waters, 1961-1993. Published/Created: Portland, Or.: Oregon Health Division, Radiation Protection Services, [1994] Related Names: Oregon. Radiation Protection Services. Description: 2 v.: col. ill.; 28 cm. Contents: v. 1. [Report]--v. 2. Data tables. Notes: "December 1994." Includes bibliographical references (v. 1, p. 23). Subjects: Radioactive pollution of water--Oregon-- Statistics. LC Classification: TD427.R3 E58 1994 Dewey Class No.: 363.17/992/09795 21 Government Document No.: HR/H34.7En8/2:961-93 ordocs

Environmental radiological surveillance report on Oregon surface waters, 1961-1993. Published/Created: Portland, Or.: Oregon Health Division, Radiation Protection Services, [1994] Related Names: Oregon. Radiation Protection Services. Description: 2 v.: col. ill.; 28 cm. Contents: v. 1. [Report]--v. 2. Data tables. Notes:

"December 1994." Includes bibliographical references (v. 1, p. 23). Subjects: Radioactive pollution of water--Oregon-- Statistics. LC Classification: TD427.R3 E58 1994 Dewey Class No.: 363.17/992/09795 21 Government Document No.: HR/H34.7En8/2:961-93 ordocs

Environmental radionuclide concentrations in the vicinity of the Calvert Cliffs Nuclear Power Plant. Published/Created: Annapolis, Md.: Maryland Power Plant Siting Program -[1994] Related Names: Maryland Power Plant Siting Program. Maryland Power Plant Research Program. Description: v.: ill.; 28 cm. -1991/1994. Merger of: Environmental radionuclide concentrations in the vicinity of the Peach Bottom Atomic Power Station (DLC) 2004210136 (OCoLC)55215782 Environmental radionuclide concentrations in the vicinity of the Calvert Cliffs Nuclear Power Plant and Peach Bottom Atomic Power Station (DLC) 2004210140 (OCoLC)55530232 Cancelled/Invalid LCCN: 83621806 87623168 89620351 00274132 Notes: Description based on: 1978/1980. Issued by: Maryland Power Plant Research Program, 1981/1984- 1991/1994. Merged with: Environmental radionuclide concentrations in the vicinity of the Calvert Cliffs Nuclear Power Plant; to form: Environmental radionuclide concentrations in the vicinity of the Calvert Cliffs Nuclear Power Plant and Peach Bottom Atomic Power Station. Subjects: Calvert Cliffs Nuclear Power Plant (Md.)-- Periodicals. Nuclear power plants-- Environmental aspects--Maryland -- Calvert Cliffs Region--Periodicals. Radioactive pollution of water--Maryland-- Calvert Cliffs Region--Periodicals. River sediments--Susquehanna River Watershed- -Periodicals. Fishes--Effect of radiation on- -Maryland--Calvert Cliffs Region-- Periodicals. LC Classification: TD428.A86 E58

Environmental radionuclide concentrations in the vicinity of the Calvert Cliffs Nuclear Power Plant and Peach Bottom Atomic Power Station. Published/Created: Annapolis, Md.: Maryland Power Plant Research Program, 1997- Related Names: Maryland Power Plant Research Program. Description: v.: ill.; 28 cm. Some issues are combined. 1995- Current Frequency: Irregular Merger of: Environmental radionuclide concentrations in the vicinity of the Calvert Cliffs Nuclear Power Plant (DLC) 2004210137 (OCoLC)55216673 Environmental radionuclide concentrations in the vicinity of the Peach Bottom Atomic Power Station (DLC) 2004210136 (OCoLC)55215782 Cancelled/Invalid LCCN: 00274135 00326829 2001330719 Notes: Latest issue consulted: 1998/1999. Merger of: Environmental radionuclide concentrations in the vicinity of the Calvert Cliffs Nuclear Power Plant; and: Environmental radionuclide concentrations in the vicinity of the Peach Bottom Atomic Power Station. Subjects: Calvert Cliffs Nuclear Power Plant (Md.)--Periodicals. Peach Bottom Atomic Power Station (Pa.)--Periodicals. Nuclear power plants--Environmental aspects--Maryland --Calvert Cliffs Region--Periodicals. Nuclear power plants--Environmental aspects--Susquehanna River Watershed--Periodicals. Radioactive pollution of water--Maryland--Calvert Cliffs Region--Periodicals. Radioactive pollution of water--Susquehanna River Watershed --Periodicals. River sediments--Maryland--Calvert Cliffs Region --Periodicals. River sediments--Susquehanna River Watershed--Periodicals. Fishes--Effect of radiation on--Maryland--Calvert Cliffs Region--Periodicals. Fishes--Effect of radiation on--Susquehanna River Watershed --Periodicals. LC Classification: TD428.A86 E582

Environmental radionuclide concentrations in the vicinity of the Calvert Cliffs Nuclear

Power Plant. Published/Created: Annapolis, Md.: Maryland Power Plant Siting Program -[1994] Related Names: Maryland Power Plant Siting Program. Maryland Power Plant Research Program. Description: v.: ill.; 28 cm. -1991/1994. Merger of: Environmental radionuclide concentrations in the vicinity of the Peach Bottom Atomic Power Station (DLC) 2004210136 (OCoLC)55215782 Environmental radionuclide concentrations in the vicinity of the Calvert Cliffs Nuclear Power Plant and Peach Bottom Atomic Power Station (DLC) 2004210140 (OCoLC)55530232 Cancelled/Invalid LCCN: 83621806 87623168 89620351 00274132 Notes: Description based on: 1978/1980. Issued by: Maryland Power Plant Research Program, 1981/1984-1991/1994. Merged with: Environmental radionuclide concentrations in the vicinity of the Calvert Cliffs Nuclear Power Plant; to form: Environmental radionuclide concentrations in the vicinity of the Calvert Cliffs Nuclear Power Plant and Peach Bottom Atomic Power Station. Subjects: Calvert Cliffs Nuclear Power Plant (Md.)--Periodicals. Nuclear power plants--Environmental aspects--Maryland --Calvert Cliffs Region--Periodicals. Radioactive pollution of water--Maryland--Calvert Cliffs Region--Periodicals. River sediments--Susquehanna River Watershed--Periodicals. Fishes--Effect of radiation on--Maryland--Calvert Cliffs Region--Periodicals. LC Classification: TD428.A86 E58

Environmental radionuclide concentrations in the vicinity of the Calvert Cliffs Nuclear Power Plant and Peach Bottom Atomic Power Station. Published/Created: Annapolis, Md.: Maryland Power Plant Research Program, 1997- Related Names: Maryland Power Plant Research Program. Description: v.: ill.; 28 cm. Some issues are combined. 1995- Current Frequency: Irregular Merger of: Environmental radionuclide concentrations in the vicinity of the Calvert Cliffs Nuclear Power Plant (DLC) 2004210137 (OCoLC)55216673 Environmental radionuclide concentrations in the vicinity of the Peach Bottom Atomic Power Station (DLC) 2004210136 (OCoLC)55215782 Cancelled/Invalid LCCN: 00274135 00326829 2001330719 Notes: Latest issue consulted: 1998/1999. Merger of: Environmental radionuclide concentrations in the vicinity of the Calvert Cliffs Nuclear Power Plant; and: Environmental radionuclide concentrations in the vicinity of the Peach Bottom Atomic Power Station. Subjects: Calvert Cliffs Nuclear Power Plant (Md.)--Periodicals. Peach Bottom Atomic Power Station (Pa.)--Periodicals. Nuclear power plants--Environmental aspects--Maryland --Calvert Cliffs Region--Periodicals. Nuclear power plants--Environmental aspects--Susquehanna River Watershed--Periodicals. Radioactive pollution of water--Maryland--Calvert Cliffs Region--Periodicals. Radioactive pollution of water--Susquehanna River Watershed --Periodicals. River sediments--Maryland--Calvert Cliffs Region --Periodicals. River sediments--Susquehanna River Watershed--Periodicals. Fishes--Effect of radiation on--Maryland--Calvert Cliffs Region--Periodicals. Fishes--Effect of radiation on--Susquehanna River Watershed --Periodicals. LC Classification: TD428.A86 E582

Environmental radionuclide concentrations in the vicinity of the Peach Bottom Atomic Power Plant. Published/Created: Annapolis, Md.: Maryland Power Plant Research Program Related Names: Maryland Power Plant Research Program. Maryland Power Plant Research Program. Description: v.: ill.; 28 cm. Ceased in 1994. Current Frequency: Irregular Merger of: Environmental radionuclide concentrations in the vicinity of the Calvert Cliffs Nuclear Power Plant (DLC)

2004210137 (OCoLC)55216673 Environmental radionuclide concentrations in the vicinity of the Calvert Cliffs Nuclear Power Plant and Peach Bottom Atomic Power Station (DLC) 2004210140 (OCoLC)55530232 Cancelled/Invalid LCCN: 88621425 Notes: Description based on: 1987/1990. Merged with: Environmental radionuclide concentrations in the vicinity of the Calvert Cliffs Nuclear Power Plant; to form: Environmental radionuclide concentrations in the vicinity of the Calvert Cliffs Nuclear Power Plant and Peach Bottom Atomic Power Station. Subjects: Peach Bottom Atomic Power Station (Pa.)--Periodicals. Nuclear power plants--Environmental aspects--Susquehanna River Watershed--Periodicals. Radioactive pollution of water--Susquehanna River Watershed --Periodicals. River sediments--Susquehanna River Watershed--Periodicals. Fishes--Effect of radiation on--Susquehanna River Watershed --Periodicals. LC Classification: TD428.A86 E586

Environmental requirements of blue-green algae; proceedings of a symposium jointly sponsored by University of Washington and Federal Water Pollution Control Administration, Pacific Northwest Water Laboratory, September 23-24, 1966. Published/Created: Corvallis, Or., Pacific Northwest Water Laboratory, 1967. Related Names: University of Washington. Pacific Northwest Water Laboratory. Description: v, 111 p. illus. 27 cm. Notes: Includes bibliographies. Subjects: Cyanobacteria--Congresses. Cyanobacteria--Ecology--Congresses. Water--Pollution--Congresses. LC Classification: QK569.C96 E56 Dewey Class No.: 589/.8

Environmental sciences / edited by Frances S. Sterrett. Published/Created: New York, N.Y.: New York Academy of Sciences, 1987. Related Names: Sterrett, Frances S.

New York Academy of Sciences. Section of Environmental Sciences. Description: 245 p.: ill.; 24 cm. ISBN: 0897663993 0897664000 (pbk.) Notes: "The papers in this volume were presented at meetings of the Environmental Sciences Section of the New York Academy of Sciences, New York, New York, during the years 1984 and 1985"--P. [v]. Includes bibliographies and index. Subjects: Air--Pollution--Congresses. Radioactive pollution--Congresses. Water--Pollution--Congresses. Series: Annals of the New York Academy of Sciences, 0077-8923; v. 502 LC Classification: Q11 .N5 vol. 502 Q11.N5 Dewey Class No.: 500 s 363.7/3 19

Environmental status of lakes in Rajasthan. Published/Created: Jaipur: Rajasthan Pollution Prevention and Control Board, 1986. Related Names: Rajasthan Pollution Prevention and Control Board. Description: 87 p., [19] leaves of plates (2 folded): ill., maps; 29 cm. Notes: Bibliography: p. 70-71. Subjects: Water--Pollution--Environmental aspects--India--Rajasthan. Lakes--Environmental aspects--India--Rajasthan. Series: Environmental status document series on surface waters; ESDS/SW/1/86-87 Environmental status document series on surface waters; ESDS/SW/86-87/1. LC Classification: TD304.R35 E58 1986 Dewey Class No.: 363.7/394/09544 19 Overseas Acquisitions No.: I E 57071 Geographic Area Code: a-ii---

Environmental study during Kumbh-2001 at Prayag (Allahabad). Published/Created: Delhi: Central Pollution Control Board, Ministry of Environment & Forests, Govt. of India, 2003. Related Names: India. Central Pollution Control Board. Description: 124 p.: col. ill.; 29 cm. Notes: Includes statistical tables. Includes bibliographical references (p. 106). Subjects: Water--Pollution--India--Allahabad. Kumbha Melaì,, (Hindu

festival)--Environmental aspects. Water quality--India--Allahabad. Sanitation--Health aspects--India--Allahabad. Allahabad (India)--Environmental conditions. Series: Assessment and development study of river basin series; 39/2003-2004 Assessment and development study of river basin series; 2003-2004/39. LC Classification: TD304.A43 E68 2003 Overseas Acquisitions No.: I-E-2004-329007; 49-91

Environmental toxicology and water quality. Serial Key Title: Environmental toxicology and water quality (Print) Abbreviated Title: Environ. toxicol. water qual. (Print) Published/Created: New York, NY: John Wiley, c1991-c1998. Description: 8 v.: ill.; 23 cm. Vol. 6, no. 1 (Feb. 1991)-v. 13, no. 4 (1998). Current Frequency: Quarterly Continues: Toxicity assessment 0884-8181 (DLC) 86649504 (OCoLC)12437662 Continued by: Environmental toxicology 1520-4081 (DLC) 99111896 (OCoLC)39481374 ISSN: 1053-4725 Cancelled/Invalid LCCN: sn 90001740 CODEN: ETWQEZ Notes: Title from cover. SERBIB/SERLOC merged record Indexed selectively by: Chemical abstracts 0009-2258 1991- Additional Formats: Issued also online. Environmental toxicology and water quality (Online) 1098-2256 (DLC)sn 98009126 (OCoLC)38745918 Subjects: Water quality bioassay--Periodicals. Water--Pollution--Toxicology--Periodicals. Microbiological assay--Periodicals. Toxicity testing--Periodicals. Environmental Monitoring--Periodicals. Environmental Pollutants--Periodicals. Environmental Pollution--Periodicals. Water Pollutants--Periodicals. Water Pollution--Periodicals. LC Classification: QH90.57.B5 T69 NLM Class No.: W1 EN986KP NAL Class No.: RA1221.T69 Dewey Class No.: 628.1/61 20 National Bibliographic Agency No.: 9425700 DNLM SR0070689 DNLM

Environmental toxicology. Serial Key Title: Environmental toxicology (Print) Abbreviated Title: Environ. toxicol. (Print) Published/Created: New York, N.Y.: John Wiley & Sons, c1999- Description: v.: ill.; 28 cm. Vol. 14, no. 1 (1999)- Current Frequency: Six times a year, 2001- Former Frequency: Five no. a year, 1999-2000 Continues: Environmental toxicology and water quality 1053-4725 (DLC) 91649238 (OCoLC)22537100 ISSN: 1520-4081 Cancelled/Invalid LCCN: sn 98007852 CODEN: ETOXFH Notes: Title from cover. Published: New York, N.Y.: John Wiley & Sons, 1999-; published: Hoboken, N.J.: Wiley Periodicals, Inc., <2005-> Latest issue consulted: Vol. 20, no. 6 (Dec. 2005). Has occasional special issue. SERBIB/SERLOC merged record Indexed by: Chemical abstracts 0009-2258 Additional Formats: Issued also online. Environmental toxicology (Online) 1522-7278 (DLC) 2001212267 (OCoLC)44043525 Subjects: Water quality bioassay--Periodicals. Water--Pollution--Toxicology--Periodicals. Microbiological assay--Periodicals. Toxicity testing--Periodicals. Environmental toxicology--Periodicals. Environmental Pollution--Periodicals. Environmental Monitoring--Periodicals. Environmental Pollutants--Periodicals. LC Classification: QH90.57.B5 T69 NLM Class No.: W1 EN986KP NAL Class No.: RA1221.T69 Dewey Class No.: 628.1/61 20 Postal Registration No.: 005664 USPS National Bibliographic Agency No.: 100885357 DNLM

Environmentally devastated areas in river basins in Eastern Europe / edited by Alfons Georges Buekens, Vasily Victorovich Dragalov. Published/Created: Berlin; New York: Springer, c1998. Related Names: Buekens, Alfons Georges, 1942- Dragalov, Vasily Victorovich, 1950- North Atlantic Treaty Organization. Scientific Affairs Division. NATO Advanced Research Workshop on Integrated Rehabilitation of Environmentally Devastated Areas in River Basins in Eastern Europe (1994: Moscow, Russia) Description: viii, 386 p.: ill.; 24 cm. ISBN: 3540647511 Notes: "Proceedings of the NATO Advanced Research Workshop on Integrated Rehabilitation of Environmentally Devastated Areas in River Basins in Eastern Europe, held at Moscow, Russia, September 1994"--T.p. verso. "Published in cooperation with NATO Scientific Affairs Division." Includes bibliographical references. Subjects: Water quality management--Europe, Eastern--Congresses. Water quality management--Former Soviet republics --Congresses. Water quality management--Case studies--Congresses. Water--Pollution--Europe, Eastern--Congresses. Water--Pollution--Former Soviet republics--Congresses. Watersheds--Europe, Eastern--Congresses. Watersheds--Former Soviet republics--Congresses. Series: NATO ASI series. Partnership sub-series 2, Environment; vol. 45 LC Classification: TD255 .E58 1998 Dewey Class No.: 363.739/4/0947 21 Geographic Area Code: ee----- e-ur---

EPA's sampling and analysis methods database [computer file] / edited by Lawrence H. Keith; initial database compiled by William Mueller and David Smith; second edition database compiled by Lawrence H. Keith. Variant Title: E.P.A.'s sampling and analysis methods database Portion of Title: Sampling and analysis methods database Edition Information: Version 2.0. Published/Created: Boca Raton, Fl.: Lewis Publishers, c1996. Related Names: Keith, Lawrence H., 1938- Mueller, William, 1929- Smith, David, 1949- Description: 3 computer disks; 3 1/2 in.: + 1 manual. ISBN: 1566701635 Computer File Information: Computer data and program. System requirements: IBM-compatible PC; 1.5MB RAM; DOS 2.0 or higher; hard drive. IBM PC DOS Summary: An electronic reference source, presented in three volumes on the EPA's sampling and analytical methods. Helps users quickly locate appropriate methods. Provides method/analyte summaries without loss of information. Notes: Title from disk label. chemists, engineers; federal, state, and local officials; ppgulators, lawyers, environmentalists, and commercial insurers. Subjects: Water--Pollution--Measurement--Databases. Pollutants--Analysis--Databases. Water--Sampling--Databases. LC Classification: TD423 LC Copy: TD423 Copy 2. Dewey Class No.: 628.5 12

Episodal pollution: a case study, Union Territory of Goa. Published/Created: New Delhi: Central Board for the Prevention and Control of Water Pollution, [1980?] Related Names: India. Central Board for the Prevention and Control of Water Pollution. Description: 9 p., [3] leaves of plates: ill., maps; 28 cm. Notes: Cover title. Subjects: Sugar factories--India--Goa, Daman and Diu--Waste disposal --Case studies. Water--Pollution--India--Goa, Daman and Diu. Series: Programme objective series; PROBES/5/1979-80 Programme objective series; PROBES/1979-1980/5. LC Classification: TD899.S8 E65 1980 Dewey Class No.: 363.7/394/0954799 19 Geographic Area Code: a-ii---

Episodic acidification of streams in the northeastern United States: chemical and biological results of the episodic response project: episodic response project final report / by P.J. Wigington, Jr. ... [et al.]. Published/Created: [Cincinnati, OH: U.S. Environmental Protection Agency, Center for Environmental Research Information, 1993] Related Names: Wigington, P. J. Description: liii, 337 p.: ill., maps; 28 cm. Notes: "EPA/600/R-93/190." "September 1993." Includes bibliographical references. Subjects: Acid pollution of rivers, lakes,

etc. Acid pollution of rivers, lakes, etc.--United States. Water acidification--Environmental aspects. Water acidification--Environmental aspects--United States. LC Classification: TD427.A27 E64 1993 Dewey Class No.: 363.73/94 20

Episodic nutrient loading impacts on eutrophication of the southern Pamlico Sound: the effects of the 1999 hurricanes / by Robert R. Christian ... [et al.]. Published/Created: Raleigh, NC: Water Resources Research Institute of the University of North Carolina, [2004] Related Names: Christian, Robert R. (Robert Raymond), 1947- Geological Survey (U.S.) Water Resources Research Institute of the University of North Carolina. Description: xvii, 57 p.: ill., maps; 28 cm. Notes: "The research on which this report is based was financed in part by the United States Department of the Interior, Geological Survey, through the N.C. Water Resources Research Institute." "July 2004." "UNC-WRRI-349." Includes bibliographical references (p. 51-57). WRRI 70183 Subjects: Eutrophication--North Carolina--Pamlico Sound. Nutrient pollution of water--North Carolina--Pamlico Sound. Hurricanes--Environmental aspects--North Carolina--Pamlico Sound. Series: Report / Water Resources Research Institute of the University of North Carolina; no. 349 Report (Water Resources Research Institute of the University of North Carolina); no. 349. LC Classification: HD1694.N8 E65 2004 Dewey Class No.: 363.739/4720916348 22 Government Document No.: G67 10:349 ncdocs

Erosion impact assessment of land management activities, Evans Creek Basin, Oregon: Oregon 208 nonpoint source assessment project / David A. Rickert ... [et al.]. Published/Created: Portland: State of Oregon, Dept. of Environmental Quality,

Water Quality Program, 1978. Related Names: Rickert, David A., 1940- Oregon. Dept. of Environmental Quality. Description: xii, 95 p.: ill.; 28 cm. Notes: Part of illustrative matter in pocket. Bibliography: p. 83-85. Subjects: Erosion--Oregon--Evans Creek Watershed (Jackson County) Land use--Oregon--Evans Creek Watershed (Jackson County) Water quality--Oregon--Evans Creek Watershed (Jackson County) Nonpoint source pollution--Oregon--Evans Creek Watershed. LC Classification: QE581 .E74 Dewey Class No.: 333.73/13 19 Geographic Area Code: n-us-or

Estimating the hazard of chemical substances to aquatic life / sponsored by ASTM Committee D-19 on Water, American Society for Testing and Materials; John Cairns, Jr., K. L. Dickson, A. W. Maki, editors. Published/Created: Philadelphia: ASTM, c1978. Related Names: Cairns, John, 1923- Dickson, Kenneth L. Maki, Alan W., 1947- American Society for Testing and Materials. Committee D-19 on Water. Description: vii, 278 p.: ill.; 23 cm. ISBN: 0465700016: Notes: Includes index. Bibliography: p. 272-273. Subjects: Water--Pollution--Environmental aspects. Aquatic animals--Effect of water pollution on. Environmental impact analysis. Series: ASTM special technical publication; 657 ASTM special technical publication; 657. LC Classification: QH545.W3 E88 Dewey Class No.: 574.5/263

Estimating the susceptibility of surface water in Texas to nonpoint-source contamination by use of logistic regression modeling / by William A. Battaglin ... [et al.]. Published/Created: [Denver? Colo.]: U.S. Dept. of the Interior, U.S. Geological Survey, 2003. Related Names: Battaglin, William A. Geological Survey (U.S.) Description: iii, 24 p.: col. maps; 28 cm. Notes: Includes bibliographical references (p. 24). Additional Formats: Also available

via Internet. Subjects: Nonpoint source pollution--Texas. Water--Pollution--Texas--Mathematical models. Logistic regression analysis--Texas. Series: Water-resources investigations report; 03-4205 LC Classification: GB701 .W375 no. 03-4205 Dewey Class No.: 553.7/0973 s 628.1/68 22

Estrogens and xenoestrogens in the aquatic environment: an integrated approach for field monitoring and effect assessment / edited by Dick Vethaak, Marca Schrap, Pim de Voogt. Published/Created: Pensacola, Fla.: Society of Environmental Toxicology and Chemistry, c2006. Related Names: Vethaak, Andreÿ Dirk, 1954- Schrap, Saskia Marca, 1960- Voogt, Pim de. Description: xxix, 481 p.: ill.; 23 cm. ISBN: 9781880611852 (alk. paper) 1880611856 (alk. paper) Notes: Errata slip tipped in. Includes bibliographical references and index. Subjects: Estrogen--Environmental aspects. Environmental toxicology. Aquatic organisms--Effect of water pollution on. Endocrine toxicology. Environmental monitoring. LC Classification: TD427.E82 E82 2006 Dewey Class No.: 363.738 22

Estuarine water quality management: monitoring, modelling, and research / W. Michaelis, ed. Published/Created: Berlin; New York: Springer-Verlag, c1990. Related Names: Michaelis, W. (Walfried), 1931- GKSS-Forschungszentrum Geesthacht. International Symposium "Estuarine Water Quality Management--Monitoring, Modelling, and Research (1989: Reinbek, Germany) Description: xv, 478 p.: ill.; 25 cm- ISBN: 0387521410 (alk. paper) Notes: Summary of work reported at the International Symposium "Estuarine Water Quality Management--Monitoring, Modelling, and Research" held at Reinbek, 19-23 June 1989. Symposium sponsored by the GKSS Research Centre Geesthacht, and others.

Includes bibliographical references and indexes. Subjects: Water quality management--Congresses. Estuarine pollution--Congresses. Estuaries--Congresses. Series: Coastal and estuarine studies; 36 LC Classification: TD365 .E77 1990 Dewey Class No.: 363.73/94 20

European model code of safe practice for the prevention of ground and surface water pollution by oil from storage tanks and during the transport of oil. Published/Created: London: Applied Science Publishers, 1974. Description: viii, 24 p.; 26 cm. ISBN: 0853344574: Notes: "Prepared by a working group, formed following a meeting of representatives of European technical organisations, held at the Institute of Petroleum, London." Subjects: Petroleum--Storage. Petroleum--Transportation. Oil pollution of water. LC Classification: TP692.5 .E95 Dewey Class No.: 665/.54 National Bibliography No.: GB74-28926

European water pollution control / Thomas Kevin Swift, project analyst. Published/Created: Norwalk, CT: Business Communications Co., c1993. Related Names: Swift, T. Kevin. Business Communications Co. Description: x, 116 leaves; 28 cm. ISBN: 0893369608 Subjects: Water pollution control industry--Europe. Water pollution control equipment industry--Europe. Water purification equipment industry--Europe. Market surveys--Europe. Series: Business opportunity report; E-065 LC Classification: HD9718.5.W363 E854 1993 Geographic Area Code: e------

European water pollution control: official publication of the European Water Pollution Control Association (EWPCA). Published/Created: Amsterdam, The Netherlands: Elsevier Science Publishers, c1991- Related Names: European Water Pollution Control Association. Description: v.: ill.; 30 cm. Vol. 1, no. 1 (Jan. 1991)-

Current Frequency: Bimonthly Continued by: European water management (OCoLC)39447027 ISSN: 0925-5060 Notes: Title from cover. SERBIB/SERLOC merged record Subjects: Water quality management--Europe--Periodicals. Water--Pollution--Europe--Periodicals. LC Classification: TD255 .E96

Europe's water: an indicator-based assessment / S. Nixon ... [et al.]. Published/Created: Copenhagen: European Environment Agency; Lanham, MD: Bernan Associates [distributor], c2003. Related Names: Nixon, S. European Environment Agency. Description: 97 p.: col. ill., col. maps; 30 cm. ISBN: 9291675814 (alk. paper) Notes: "Topic report, No. 1/2003." "TH-53-03-443-EN-C"--P. [4] of cover. Includes bibliographical references. Subjects: Water--Pollution--European Union countries. Water quality management--European Union countries. Water levels--European Union countries. LC Classification: TD255 .E965 2003 Dewey Class No.: 363.739/4/094 22

Europe's water: an indicator-based assessment: summary. Published/Created: Copenhagen: European Environmental Agency, c2003. Related Names: European Environment Agency. Description: 23 p.: col. ill.; 30 cm. ISBN: 9291675768 Notes: "TH-53-03-346-EN-C"--P. [4] of cover. Subjects: Water--Pollution--European Union countries. Water quality management--European Union countries. Water levels--European Union countries. LC Classification: TD424.4.E85 E87 2003

Evaluating standards for protecting aquatic life in Washington's surface water quality standards: temperature criteria: draft discussion paper and literature summary / prepared by Water Quality Program, Washington State Department of Ecology, Watershed Management Section. Variant Title: Title from cover letter: Proposed temperature criteria, decision process for Ecology's proposed rule Edition Information: Rev., Dec. 2002. Published/Created: Olympia, Wash.: The Section, [2002] Related Names: Washington (State). Dept. of Ecology. Watershed Management Section. Description: 10, iii, 189 p.: ill., maps; 28 cm. Notes: Includes bibliographical references (p. 138-173). Additional Formats: Also available via the Internet. Subjects: Freshwater fishes--Effect of temperature on--Washington (State) Fishes--Effect of temperature on--Washington (State) Water temperature--Physiological effect--Washington (State) Water quality--Standards--Washington (State) Water quality management--Law and legislation--Washington (State) Water--Pollution--Law and legislation--Washington (State) Series: Publication; no. 00-10-070 Publication (Washington (State). Dept. of Ecology); no. 00-10-070. LC Classification: SH177.T45 E93 2002 Government Document No.: WA 574.5 Ec7eva s1 2002 wadocs

Evaluation and control of water pollution in Bhavani Basin [microform]: final report. Published/Created: Chennai: Centre for Water Resources, Anna University, 1998. Related Names: Anna University. Centre for Water Resources. Description: 76, [4] leaves: ill., maps; 29 cm. Summary: With reference to India. Notes: "Funded by Institute for water Studies, Water Resources Organisation (PWD)." Includes bibliographical references (leaves [77-80]. Additional Formats: Microfiche. New Delhi: Library of Congress Office; Washington, D.C.: Library of Congress Photoduplication Service, 2001. 2 microfiches. Master microform held by: DLC. Subject Keywords: Water Pollution Bhavani River Watershed (India) Evaluation. Water Pollution Bhavani River Watershed (India)a Control. LC Classification: Microfiche 2001/60246 (T)

Overseas Acquisitions No.: I-E-00-371651; 49

Evaluation of directive 76/464/EEC regarding list II substances on the quality of the most important surface waters in the community / European Commission. Published/Created: Luxembourg: Office for Official Publications of the European Communities; Lanham, MD: Bernan Associates [distributor], 1997. Related Names: European Commission. Description: 1 v. (various pagings): ill.; 30 cm. ISBN: 9282795888 Notes: "CR-03-96-224-EN-C"--P. [4] of cover. Summary also in Dutch and French. Subjects: Water quality management--European Union countries. Water--Pollution--Law and legislation--European Union countries. Cours d'eau--Aspect de l'environnement--Pays de l'Union europeí enne. ram Eau--Pollution--Droit--Pays de l'Union europeí enne. ram Eau--Qualiteí --Gestion--Pays de l'Union europeí enne. ram LC Classification: TD255 .E97 1997 Dewey Class No.: 341.7/6253/094 21 Language Code: eng dutfre

Evaluation of Ganga Action Plan. Published/Created: New Delhi: Ministry of Environment and Forests, Govt. of India, 1995. Related Names: Ganí‡gaí„ Pariyojanaí„ Nidesí aí„laya. India. Ministry of Environment and Forests. Description: 231 p.: ill., maps; 27 cm. Summary: Study undertaken at the behest of Ganí‡gaí„ Pariyojanaí„ Nidesí aí„laya. Subjects: Water--Pollution--Ganges River (India and Bangladesh) --Evaluation. LC Classification: TD304.G36 E84 1995 Overseas Acquisitions No.: I-E-2006-542777; 35-90 Reproduction/Stock No.: Library of Congress -- New Delhi Overseas Office Geographic Area Code: a-ii---

Evaluation of groundwater movement in the Frenchman Flat CAU using geochemical and isotopic analysis / prepared by R.L.

Hershey ... [et al.]; submitted to Nevada Site Office, National Nuclear Security Administration, U.S. Department of Energy, Las Vegas, Nevada. Published/Created: Las Vegas, Nev.: Desert Research Institute, 2005. Related Names: Hershey, R. L. (Ronald L.) University and Community College System of Nevada. Division of Hydrologic Sciences. Description: viii, 65 p.: col. ill., maps (some col.); 28 cm. Notes: "DOE/NV/13609-36". "March 2005." Includes bibliographical references (p. 61-65). Subjects: Groundwater flow--Nevada. Geochemistry--Nevada--Nevada Test Site. Radioactive pollution of water--Nevada Test Site. Groundwater--Pollution--Nevada--Nevada Test Site. Series: Publication; no. 45207 LC Classification: GB705.N3 E83 2005 Electronic File Information: Table of contents only http://www.loc.gov/catdir/toc/fy0713/2006 475534.html

Evaluation of Gulfwatch. Serial Key Title: Evaluation of Gulfwatch Abbreviated Title: Eval. Gulfwatch Published/Created: [Augusta, ME?]: Gulf of Maine Council on the Marine Environment, 1992- Related Names: Gulf of Maine Council on the Marine Environment. Description: v.: ill.; 28 cm. Vol. for 1992 published in 1994 and rev. in 1996. 1991- Current Frequency: Annual ISSN: 1093-4952 Notes: Title from cover. Latest issue consulted: 1992. SERBIB/SERLOC merged record Subjects: Environmental monitoring--Maine, Gulf of--Periodicals. Water quality biological assessment--Maine, Gulf of --Periodicals. Indicators (Biology)--Maine, Gulf of--Periodicals. Marine pollution--Environmental aspects--Maine, Gulf of --Periodicals. Marine pollution--Health aspects--Maine, Gulf of --Periodicals. LC Classification: QH541.15.M64 E945 Dewey Class No.: 577.7/345 21

Evaluation of livestock runoff as a source of water pollution in northern Utah / by Stephen T. Wieneke ... [et al.]. Published/Created: Logan, Utah: Utah Water Research Laboratory, College of Engineering, Utah State University, [1980] Related Names: Wieneke, Stephen T. Utah Water Research Laboratory. Description: xii, 167 p., [2] folded leaves of plates: ill., maps (some col.); 28 cm. Notes: "September 1980." Bibliography: p. 77-79. Subjects: Water--Pollution--Utah. Feedlot runoff--Utah. Series: Water quality series; UWRL/Q-80/02 Water quality series (Logan, Utah); UWRL/Q-80/02. LC Classification: TD424.35.U8 E93 1980 Dewey Class No.: 363.7/31 19 Geographic Area Code: n-us-ut

Evaluation of methods to minimize contamination hazards to wildlife using agricultural evaporation ponds in the San Joaquin Valley, California: final report / prepared for the California Department of Water Resources under contract no. STCA/DWR B57073 by David F. Bradford ... [et al.]. Published/Created: Los Angeles, CA: Environmental Science & Engineering Program, University of California, [1989] Related Names: Bradford, David F. California. Dept. of Water Resources. Description: 222 p.; 28 cm. Notes: "June 30, 1989." Includes bibliographical references (p. 201-211). Subjects: Waterfowl--Effect of water pollution on--California--San Joaquin Valley. Water birds--Effect of water pollution on--California--San Joaquin Valley. Sewage lagoons--California--San Joaquin Valley--Safety measures--Evaluation. Selenium in agriculture--Environmental aspects--California --San Joaquin Valley. Drainage--Environmental aspects--California--San Joaquin Valley. Series: ESE report; no. 89-64 LC Classification: QL696.A52 E93 1989 Dewey Class No.: 639.9/78 20 Geographic Area Code: n-us-ca

Evaluation of minimum data requirements for acute toxicity value extrapolation with aquatic organisms: final report / by Denny R. Buckler ... [et al.]. Published/Created: Gulf Breeze, FL (1 Sabine Island Drive, Gulf Breeze, 32561-5299): U.S. Environmental Protection Agency, National Health and Environmental Effects Research Laboratory, Gulf Ecology Division, [2003] Related Names: Buckler, Denny. National Health and Environmental Effects Research Laboratory (U.S.). Gulf Ecology Division. Description: vii, 165 p.; 28 cm. Notes: "July 2003." "EPA project no. DW-14-93900201-1." "EPA/600/R-03/104." Includes bibliographical references (p. 8). Subjects: Water quality bioassay. Acute toxicity testing. Aquatic organisms--Effect of water pollution on. LC Classification: QH90.57.B7 E93 2003 Dewey Class No.: 577.6/275 22

Evaluation of techniques for cost-benefit analysis of water pollution control programs and policies: report of the Administrator of the Environmental Protection Agency to the Congress of the United States in compliance with Public law 92-500, the Federal water pollution control act amendments of 1972. Published/Created: Washington: U.S. Govt. Print. Off., 1975. Related Names: United States. Environmental Protection Agency. Description: vii, 742 p.: ill.; 23 cm. Notes: Includes papers presented at a symposium conducted by the Agency in Sept. 1973. Bibliography: p. 69-70. Subjects: Water--Pollution--United States--Cost effectiveness. Series: Document - 93d Congress, 2d session, Senate; no. 93-132 United States. Congress. Senate. Document; no. 93-132. LC Classification: TD223 .E9 Dewey Class No.: 363.6 Geographic Area Code: n-us---

Evaluation of the economic costs and benefits of methods for reducing nutrient loads to the Gulf of Mexico: topic 6, report for the

integrated assessment on hypoxia in the Gulf of Mexico / Otto C. Doering ... [et al.]. Portion of Title: Report for the integrated assessment on hypoxia in the Gulf of Mexico Published/Created: Silver Spring, Md.: U.S. Dept. of Commerce, National Oceanic and Atmospheric administration, Coastal Ocean Program, [1999] Related Names: Doering, Otto C. (Otto Charles), 1940- Description: xiv, 115 p.: ill., map; 28 cm. Notes: "May 1999." Includes bibliographical references (p. 104-115). Subjects: Nutrient pollution of water--Economic aspects--Mexico, Gulf of. Water quality management--Economic aspects--Mississippi River Watershed. Environmental policy--Economic aspects--United States. Hypoxia (Water) Series: NOAA Coastal Ocean Program decision analysis series; no. 20 LC Classification: TD427.N87 E93 1999 Dewey Class No.: 363.739/42/0916364 21 Geographic Area Code: nm----- n-usm-- n-us---

Evaluation of treatment options to reduce water-borne selenium at coal mines in west-central Alberta. Published/Created: Roberts Creek, B.C.: Microbial Technologies, Inc., [2005-2006] Related Names: Microbial Technologies Inc. Alberta. Alberta Environment. Description: iii, 32, [5] p.: ill.; 28 cm. ISBN: 0778546047 (print) 9780778546047 (print) 0778546055 (online) 9780778546054 (online) Notes: "Report prepared for Alberta Environment, Water Research Users Group, Edmonton" -- t.p. Includes bibliographical references. Subjects: Selenium--Environmental aspects--Alberta. Coal mines and mining--Environmental aspects--Alberta. Birds--Effect of water pollution on--Alberta. Water--Pollution--Alberta. Series: Pub. No.; T/860 Publication (Alberta. Alberta Environment); no. T/860. Geographic Area Code: n-cn-ab

Factors affecting nutrient trends in major rivers of the Chesapeake Bay Watershed / by Lori A. Sprague ... [et al.]; prepared in cooperation with the Virginia Department of Environmental Quality ... [et al.]. Published/Created: Richmond, Va.: U.S. Dept. of the Interior, U.S. Geological Survey; Denver, CO: Branch of Information Services [distributor], 2000. Related Names: Sprague, Lori A. Virginia. Dept. of Environmental Quality. Geological Survey (U.S.) Description: vii, 98 p.: ill. (some col.), maps (some col.); 28 cm. Notes: Includes bibliographical references (p. 97-98). Subjects: Nutrient pollution of water--Chesapeake Bay Watershed. Agriculture--Environmental aspects--Chesapeake Bay Watershed. Streamflow--Chesapeake Bay Watershed. Series: Water-resources investigations report; 00-4218 LC Classification: GB701 .W375 no. 00-4218 TD427.N87 Dewey Class No.: 553.7/0973 s 363.738/4 21 Geographic Area Code: n-us-md n-us-va

Factors controlling elevated lead concentrations in water samples from aquifer systems in Florida / by Brian G. Katz ... [et al.]; prepared in cooperation with Florida Department of Environmental Protection. Published/Created: Tallahassee, Fla.: U.S. Dept. of the Interior, U.S. Geological Survey; Denver, CO: Branch of Information Services [distributor], 1999. Related Names: Katz, Brian G. Florida. Dept. of Environmental Protection. Geological Survey (U.S.) Description: iv, 22 p.: ill., maps; 28 cm. Notes: Shipping list no.: 99-0198-P. Includes bibliographical references (p. 20-22). Subjects: Lead--Environmental aspects--Florida. Groundwater--Pollution--Florida. Water--Sampling--Florida. Series: Water-resources investigations report; 99-4020 LC Classification: GB701 .W375 no. 99-4020 NAL Class No.: GB701.W375 no.99-4020 Dewey Class No.: 553.7/0973 s

363.738/4 21 Government Document No.: I 19.42/4:99-4020

Fate and movement of natural and synthetic compounds in the Calcasieu River, Louisana / edited by C.R. Demas and D.K. Demcheck. Published/Created: Denver, CO: For sale by the U.S. Geological Survey Earth Science Information Center, 1996. Projected Publication Date: 9602 Related Names: Demas, Charles R. Demcheck, Dennis K. Description: p. cm. Notes: Includes bibliographical references. Subjects: Water--Pollution--Louisiana--Calcasieu River. River sediments--Louisiana--Calcasieu River. Limnology--Louisiana--Calcasieu River. Series: U.S. Geological Survey water-supply paper; 2446 LC Classification: TD224.L8 F38 1996 Dewey Class No.: 551.48/3 20 Geographic Area Code: n-us-la

Fate of organic pesticides in the aquatic environment; a symposium sponsored by the Division of Pesticide Chemistry at the 161st meeting of the American Chemical Society, Los Angeles, Calif., March 29-31, 1971. Samuel D. Faust, symposium chairman. Published/Created: Washington, American Chemical Society, 1972. Related Names: Faust, Samuel Denton, 1929- American Chemical Society. Division of Pesticide Chemistry. Description: viii, 280 p. illus. 24 cm. ISBN: 084120151X Notes: Includes bibliographical references. Subjects: Pesticides--Environmental aspects--Congresses. Aquatic animals--Effect of water pollution on--Congresses. Series: Advances in chemistry series, 111 LC Classification: QD1 .A355 no. 111 QH545.P4 Dewey Class No.: 540/.8 s 574.5/263

Fate of pharmaceuticals in the environment and in water treatment systems / edited by Diana S. Aga. Published/Created: Boca Raton: Taylor & Francis, 2008. Projected Publication Date: 0711 Related Names: Aga, Diana S., 1967- Description: p. cm.

ISBN: 9781420052329 (alk. paper) Notes: Includes bibliographical references and index. Subjects: Drugs--Environmental aspects. Water--Pollution. Water--Purification. LC Classification: TD196.D78 F38 2008 Dewey Class No.: 628.5/2 22

Feasibility of new epidemiologic studies of low level arsenic / prepared by Allan H. Smith ... [et al.]; sponsored by AWWA Research Foundation. Published/Created: Denver, Colo: American Water Works Association, c1998. Related Names: Smith, Allan H. AWWA Research Foundation. Description: xvii, 142 p.; 28 cm. ISBN: 0898679613 Notes: Includes bibliographical references (p. 123-138). Subjects: Arsenic--Toxicology. Health risk assessment. Water--Pollution. LC Classification: RA1231.A7 F43 1998 Dewey Class No.: 615.9/25715 21

Fecal bacteria and general standard total maximum daily load development for Bluestone River / submitted ... by: New River-Highlands Resource Conservation and Development Area [and by] MapTech, Inc.; prepared for Virginia Department of Environmental Quality. Running Title: TMDL development, Bluestone River, VA Published/Created: [Richmond, Va.]: Dept. of Environmental Quality, [2004] Related Names: MapTech, Inc. (Blacksburg, Va.) Virginia. Dept. of Environmental Quality. Description: 1 v. (various pagings): col. ill., col. maps; 28 cm. Notes: Title from cover. "Submitted April 22, 2004, revised September 2004"--cover. Includes bibliographical references. "MapTech, Inc. of Blacksburg, Virginia supported this study as a subcontractor to New River-Highlands Resource Conservation and Development Area led by Gary Boring, through funding provided by Virginia's Department of Environmental Quality."--Acknowledgements, p. xxviii. c. 1-2, June 2005, dep., KFI Additional Formats:

Report also available in PDF format at Virginia DEQ web site. Subjects: Water quality biological assessment--Virginia--Bluestone River Watershed (Va. and W. Va.) Enterobacteriaceae--Bluestone River Watershed (Va. and W. Va.) Benthos--Bluestone River Watershed (Va. and W. Va.) Water--Pollution--Total maximum daily load--Bluestone River Watershed (Va. and W. Va.) LC Classification: TD224.V8 F41 2004 Dewey Class No.: 363.739/409755 22

Fecal bacteria and general standard total maximum daily load development for impaired streams in the Middle River and Upper South River watersheds, Augusta County, VA / submitted by MapTech, Inc.; prepared for Commonwealth of Virginia, Department of Conservation and Recreation, Division of Soil and Water. Running Title: TMDL development, Middle River, VA Published/Created: [Richmond, Va.]: Dept. of Environmental Quality, [2004] Related Names: MapTech, Inc. (Blacksburg, Va.) Virginia. Dept. of Conservation and Recreation. Virginia. Division of Soil and Water Conservation. Description: 1 v. (various pagings): col. ill., col. maps; 28 cm. Notes: Title from cover. "April 28, 2004"--cover. Includes bibliographical references. MapTech, Inc. of Blacksburg, Virginia, supported this study through funding provided by the Virginia Department of Conservation and Recreation contract #C199-02-735. c. 1-2, June 2005, dep., KFI Additional Formats: Report also available in PDF format at Virginia DEQ web site. Subjects: Water--Pollution--Total maximum daily load--Virginia --Middle River Watershed (Augusta County) Water--Pollution--Total maximum daily load--Virginia--Upper South River Watershed (Augusta County) Water quality biological assessment--Virginia--Middle River Watershed (Augusta County) Water quality biological assessment--Virginia--Upper South River

Watershed (Augusta County)
Enterobacteriaceae--Virginia--Middle
River Watershed (Augusta County)
Enterobacteriaceae--Virginia--Upper South
River Watershed (Augusta County)
Benthos--Virginia--Middle River
Watershed (Augusta County) Benthos--
Virginia--Upper South River Watershed
(Augusta County) LC Classification:
TD224.V8 F42 2004

Fecal coliform TMDL for Dodd Creek
Watershed, Virginia / submitted by
Virginia Department of Environmental
Quality [and] Virginia Dept. of
Conservation and Recreation; prepared by
the Louis Berger Group, Inc. Portion of
Title: Fecal coliform total maximum daily
load for Dodd Creek Watershed, Virginia
Published/Created: [Richmond, Va.]: Dept.
of Environmental Quality: Dept. of
Conservation and Recreation, [2002]
Related Names: Louis Berger Group.
Virginia. Dept. of Environmental Quality.
Virginia. Dept. of Conservation and
Recreation. Description: 1 v. (various
pagings): col. ill., col. maps; 28 cm. Notes:
Title from cover. "November 20, 2002"--
cover. "Draft report submitted July 2002"--
cover. Includes bibliographical references.
c. 1-2, December 2004, dep., KFI
Additional Formats: Report also available
in PDF format at Virginia DEQ web site.
Subjects: Enterobacteriaceae--Virginia--
Dodd Creek Watershed (Floyd County)
Water--Pollution--Total maximum daily
load--Virginia--Dodd Creek Watershed
(Floyd County) LC Classification:
TD224.V8 F437 2002

Fecal coliform TMDL for Mountain Run
Watershed, Culpeper County, Virginia /
submitted by Virginia Department of
Environmental Quality, Virginia
Department of Conservation and
Recreation; prepared by Gene Yagow,
Virginia Tech Department of Biological
Systems Engineering. Variant Title: Fecal

coliform total maximum daily load for
Mountain Run Watershed, Culpeper
County, Virginia Published/Created:
[Richmond, VA]: Virginia Dept. of
Environmental Quality: Virginia Dept. of
Conservation and Recreation, [2001]
Related Names: Virginia. Dept. of
Environmental Quality. Virginia. Dept. of
Conservation and Recreation. Virginia
Polytechnic Institute and State University.
Dept. of Biological Systems Engineering.
Description: xiv, 126, [2] p.: ill. (some
col.), maps (some col.); 28 cm. Notes: Title
from cover. "March 2001 Revised April
2001"--cover. Includes bibliographical
references. c. 1-2, January 2005, dep., KFI
Additional Formats: Report also available
in PDF format at Virginia DEQ web site.
Subjects: Water quality management--
Virginia--Mountain Run Watershed
(Culpeper County) Enterobacteriaceae--
Virginia--Mountain Run Watershed
(Culpeper County) Water--Pollution--Total
maximum daily load--Virginia --Mountain
Run Watershed (Culpeper County) LC
Classification: TD224.V8 F44 2001

Field manual for the investigation of fish kills /
edited by Fred P. Meyer and Lee A.
Barclay. Published/Created: [Washington,
D.C.]: U.S. Dept. of the Interior, Fish and
Wildlife Service: [For sale by the Supt. of
Docs., U.S. G.P.O.], 1990. Related Names:
Meyer, Fred P. Barclay, Lee A. U.S. Fish
and Wildlife Service. Description: vii, 120:
ill. (some col.); 26 cm. ISBN: 0160246865
Notes: Shipping list no.: 91-157-P. 024-
010-00685 Item 613-B Includes
bibliographical references. Subjects: Fish
kills--Handbooks, manuals, etc. Fish kills--
United States--Handbooks, manuals, etc.
Fishes--Diseases--Handbooks, manuals,
etc. Fishes--Effect of water pollution on--
Handbooks, manuals, etc. Series: Resource
publication; 177 Resource publication
(U.S. Fish and Wildlife Service); 177. LC
Classification: S914 .A3 no. 177 SH171
NAL Class No.: SH171.F54 Dewey Class

No.: 333.95/4/0973 s 639.3 20 Government
Document No.: I 49.66:177 Geographic
Area Code: n-us---

Field studies of radon in rocks, soils, and water
/ edited by Linda C.S. Gundersen, Richard
B. Wanty. Published/Created: Boca Raton,
Fla.: C.K. Smoley, c1993. Related Names:
Gundersen, L. C. S. Wanty, Richard B.
Description: xxi, 334 p.: ill., maps; 29 cm.
ISBN: 0873719557 Notes: "U.S.
Geological Survey research on the
geology, geophysics, and geochemistry of
radon in rocks, soils, and water." Includes
bibliographical references. Subjects:
Radon--Measurement. Radon--
Environmental aspects--Measurement.
Rocks--Analysis. Soils, Radioactive
substances in--Measurement. Radioactive
pollution of water--Measurement.
Groundwater--Analysis. LC Classification:
QC796.R6 F54 1993 Dewey Class No.:
628.5/35 20 Geographic Area Code: n-us--
- Electronic File Information: Publisher
description
http://www.loc.gov/catdir/enhancements/fy
0744/92038768-d .html

Field studies of radon in rocks, soils, and water
/ Linda C.S. Gundersen and Richard B.
Wanty, editors. Published/Created: Reston,
VA: U.S. Geological Survey; Washington,
D.C.: U.S. G.P.O., 1991. Related Names:
Gundersen, L. C. S. Wanty, Richard B.
Description: xxi, 334 p.: ill., maps; 28 cm.
Notes: Includes bibliographical references.
Subjects: Radon--Measurement. Radon--
Environmental aspects--Measurement.
Rocks--Analysis. Soils, Radioactive
substances in--Measurement. Radioactive
pollution of water--Measurement.
Groundwater--Analysis. Series: U.S.
Geological Survey bulletin; 1971 LC
Classification: QE75 .B9 no. 1971
QC796.R6 Dewey Class No.: 553/.97 20
Government Document No.: I 19.3:1971

Field testing of USEPA methods 1601 and
1602 for coliphage in groundwater /

prepared by Mohammad R. Karim ... [et
al.]; sponsored by Awwa Research
Foundation. Published/Created: Denver,
CO: Awwa Research Foundation and
American Water Works Association,
c2004. Related Names: Karim, Mohammad
R. AWWA Research Foundation.
Description: xxii, 104 p.: ill., map; 28 cm.
ISBN: 1583213481 Notes: Includes
bibliographical references (p. 97-101).
Subjects: Viral pollution of water--
Measurement. Bacteriophages--
Measurement. Groundwater--Analysis. LC
Classification: TD427.V55 F54 2004
Dewey Class No.: 628.1/61 22

Global freshwater quality: a first assessment /
edited by Michel Meybeck, Deborah V.
Chapman, Richard Helmer.
Published/Created: Oxford, UK;
Cambridge, Mass., USA: Published on
behalf of the World Health Organization
and the United Nations Environment
Programme by Blackwell Reference, 1990.
Description: x, 306 p.: ill.; 28 cm. ISBN:
0631173145:

Golob's oil pollution bulletin: the international
newsletter on oil pollution, prevention,
control, and cleanup from World
Information Systems and the Center for
Short-Lived Phenomena.
Published/Created: Cambridge, MA:
World Information Systems Description:
v.; 28 cm. Began in 1989; ceased in Jan.
1999. ISSN: 1051-6255

Governance of water-related conflicts in
agriculture: new directions in agri-
environmental and water policies in the EU
/ edited by Floor Brouwer, Ingo Heinz and
Thomas Zabel. Published/Created:
Dordrecht; Boston: Kluwer Academic
Publishers, c2003. Description: xiii, 222 p.:
ill.; 25 cm. ISBN: 1402015534 (hardback:
alk. paper)

Grays Harbor cooperatives water quality study:
1964 through 1966. Published/Created:

[Olympia, Washington Dept. of Fisheries] 1971. Description: vii, 114 p. illus. 28 cm.

Grazed pastures and surface water quality / Richard W. McDowell ... [et al.]. Published/Created: New York: Nova Science Publishers, c2007. Description: p. cm. Projected Publication Date: 0802 ISBN: 9781604560251 (hardcover: alk. paper)

Great Lakes binational toxics strategy ... progress report. Published/Created: Downsview, ON: Environment Canada, Environmental Protection Branch; Chicago, Ill.: U.S. Environmental Protection Agency, Great Lakes National Program Office Description: v.: col. ill.; 28 cm.

Great Lakes trends: into the new millennium / prepared by the Office of the Great Lakes, Michigan Department of Environmental Quality. Published/Created: [Lansing, Mich.]: The Office, [2000] Description: i leaf, 41 p.: ill.; 28 cm.

Great Lakes Water Quality Agreement of 1978: agreement, with annexes and terms of reference, between the United States and Canada signed at Ottawa November 22, 1978; and, Phosphorous Load Reduction Supplement signed October 7, 1983: as amended by protocol signed November 18, 1987 / consolidated by the International Joint Commission, United States and Canada. Edition Information: Rev. Published/Created: [Washington, D.C.]: The Commission, [1988] Description: ii, 130 p.; 21 cm.

Ground water models: scientific and regulatory applications / Water Science and Technology Board, Committee on Ground Water Modeling Assessment, Commission on Physical Sciences, Mathematics, and Resources, National Research Council. Published/Created: Washington, D.C.: National Academy Press, 1990.

Description: xv, 303 p.: ill.; 23 cm. ISBN: 0309039932

Ground water pollution potential of Clermont County, Ohio / by The Ohio Department of Natural Resources and the Center for Ground Water Management, Wright State University, Ronald G. Schmidt, project manager, James A. Wasserbauer, project assistant, Lori L. Wenz, project assistant: with a section on the general information about Clermont County by Mike Angle. Published/Created: [Columbus, Ohio]: Ohio Dept. of Natural Resources, Division of Water, Water Resources Section, [1996] Description: vi, 57 p.: ill., maps; 28 cm.

Ground water pollution potential of Warren County, Ohio / by the Center for Ground Water Management, Wright State University and the Ohio Department of Natural Resources, Division of Water, Ground Water Resources Section. Published/Created: Columbus, Ohio (1939 Fountain Square, Columbus 43224): Ohio Dept. of Natural Resources, Division of Water, Ground Water Resources Section, 1992. Description: vi, 73 p.: ill., maps; 28 cm.

Ground water quality / edited by C.H. Ward, W. Giger. P.L. McCarty. Published/Created: New York: Wiley, c1985. Description: xviii, 547 p.: ill.; 24 cm. ISBN: 0471815977

Ground water quality monitoring program / by Dale Trippler [and] Thomas P. Clark. Published/Created: [S.l.]: Minnesota Pollution Control Agency, Division of Solid and Hazardous Waste, Program Development and Facility Review Section, 1981- Description: v.: ill.; 28 cm. Vol. 3 (1980)- ISSN: 0738-1204

Ground water quantity and quality study, rural municipality of Reford. Published/Created: [Regina, Sask.]: Water Pollution Control Branch and Water Rights Branch,

Saskatchewan Dept. of the Environment, [1975] Description: vi, 71 p., 6 leaves of plates (some folded): ill.; 28 cm.

Ground water vulnerability assessment, Snake River Plain, Southern Idaho / produced through a cooperative effort by Michael Rupert ... [et al.]. Published/Created: [Boise]: Idaho Dept. of Health and Welfare, Division of Environmental Quality, [1991] Description: iii, 25 leaves: fold. maps; 28 cm.

Ground water vulnerability assessment: contamination potential under conditions of uncertainty / Committee on Techniques for Assessing Ground Water Vulnerability, Water Science and Technology Board, Commission on Geosciences, Environment, and Resources, National Research Council. Published/Created: Washington, D.C.: National Academy Press, 1993. Description: xiii, 204 p.: ill., maps; 24 cm. ISBN: 0309047994

Ground water vulnerability mapping / by Kang-Tsung Chang ... [et al.]; submitted to Division of Environmental Quality, Department of Health and Welfare. Published/Created: Moscow, Idaho: Idaho Water Resources Research Institute, University of Idaho, [1994] Description: vi, 121 p.: ill. (some col.); 28 cm.

Groundwater contamination and emergency response guide / by J.H. Guswa ... [et al.]. Published/Created: Park Ridge, N.J., U.S.A.: Noyes Publications, c1984. Description: xv, 490 p.: ill.; 25 cm. ISBN: 0815509995:

Groundwater contamination and reclamation: proceedings of a symposium held in Tucson, Arizona, August 14-15, 1985 / edited by Kenneth D. Schmidt; sponsored by American Water Resources Association. Published/Created: Bethesda, Md. (5410 Grosvenor La., Suite 220,

Bethesda 20814): The Association, c1985. Description: vi, 175 p.: ill.; 28 cm.

Groundwater contamination by organic pollutants: analysis and remediation / edited by Jagath J. Kaluarachchi. Published/Created: Reston, Va.: American Society of Civil Engineers, c2001. Description: x, 238 p.: ill.; 23 cm. ISBN: 0784405271

Groundwater contamination by toxic substances: a digest of reports: a report / prepared by the Environment and Natural Resources Division of the Congressional Research Service of the Library of Congress for the Committee on Environment and Public Works, U.S. senate. Published/Created: Washington: U.S. G.P.O., 1983 [i.e. 1984] Description: xxxv, 75 p.; 24 cm.

Groundwater contamination: optimal capture and containment / Steven M. Gorelick ...[et al.]. Published/Created: Boca Raton: Lewis Publishers, c1993. Description: xxviii, 385 p.: ill.; 24 cm. ISBN: 0873718720 (acid-free paper) Links: Publisher description http://www.loc.gov/catdir/enhancements/fy 0744/93023950-d.ht ml

Ground-water discharge and base-flow nitrate loads of nontidal streams, and their relation to a hydrogeomorphic classification of the Chesapeake Bay watershed, Middle Atlantic coast / by L. Joseph Bachman ... [et al.]. Published/Created: Baltimore, Md.: U.S. Dept. of the Interior, U.S. Geological Survey; Denver, CO: U.S. Geological Survey, Branch of Information Services [distributor], [1998] Description: iv, 71 p.: ill., maps (1 col.); 28 cm.

Groundwater monitoring survey: a survey / prepared by the staff of the Subcommittee on Oversight and Investigations of the Committee on Energy and Commerce, U.S. House of Representatives. Published/Created: Washington: U.S.

G.P.O., 1985. Description: vii, 200 p.: 1 form; 24 cm.

Groundwater pollution, aquifer recharge, and vulnerability / edited by N.S. Robins. Published/Created: London: Geological Society, 1998. Description: 224 p.: ill., maps; 26 cm. ISBN: 1897799985

Groundwater pollution: technology, economics, and management / edited by J. L. Wilson, R. L. Lenton, and J. Porras. Published/Created: Cambridge: Dept. of Civil Engineering, School of Engineering, Massachusetts Institute of Technology, 1976. Description: xiii, 311 p.: 17 ill.; 28 cm.

Ground-water quality of Texas: an overview of natural and man-affected conditions / compiled by Ground Water Protection Unit staff. Published/Created: Austin, Tex.: Texas Water Commission, [1989] Description: xiii, 197 p.: ill., maps (some col.); 28 cm.

Guest River total maximum daily load report: TMDL study for aquatic life use impairment / Virginia Department of Environmental Quality. Published/Created: Richmond, Va.: Dept. of Environmental Quality, [2003] Description: 1 v. (various pagings): ill. (some col.), maps (some col.); 28 cm. Links: Adobe Acrobat Reader required; viewed July 2005. http://www.deq.virginia.gov/tmdl/apptmdls/tenbigrvr/guestbc .pdf

Guidance document for aquatic effects assessment. Published/Created: Paris: Environment Directorate, Organisation for Economic Co-operation and Development, 1995. Description: 116 p.; 30 cm.

Guidance documents for natural resource damage assessment under the Oil Pollution Act of 1990 [electronic resource]. Published/Created: Silver Spring, MD: NOAA, Damage Assessment and

Restoration Program, [1996] Description: 1 CD-ROM; 4 3/4 in.

Guidance for development of total maximum daily loads. Published/Created: [Boise]: Surface Water Section, Idaho Division of Environmental Quality, [1999] Description: ii, 46 p.: ill., map; 28 cm.

Guide to operational procedures for the IGOSS Pilot Project on Marine Pollution (Petroleum) Monitoring / Intergovernmental Oceanographic Commission, World Meteorological Organization. Published/Created: [Paris]: Unesco, 1976, i.e. 1977. Description: 50 p.: ill.; 30 cm. ISBN: 923101501X (pbk.)

Guidelines for health related monitoring of coastal water quality: report of a group of experts jointly convened by WHO and UNEP, Rovinj, Yugoslavia, 23-25 February 1977. Published/Created: Copenhagen: World Health Organization, Regional Office for Europe, 1977. Description: 165 p.: ill.; 30 cm.

Guidelines for safe recreational water environments. Published/Created: Geneva: World Health Organization, 2003-2006. Description: 2 v.: ill.; 25 cm. ISBN: 9241545801 (v. 1) 9241546808 (v. 2)

Guidelines for the release of waste water from nuclear facilities with special reference to the public health significance of the proposed release of treated waste waters at Three Mile Island. Published/Created: Bethesda, Md.: National Council on Radiation Protection and Measurements, c1987. Description: v, 22 p.; 28 cm. ISBN: 091339288X (pbk.)

Guidelines of lake management / editors, S.E. JÃ‚rgensen and R.A. Vollenweider. Published/Created: [Otsu, Shiga, Japan]: International Lake Environment Committee: United Nations Environment

Programme, [1988]-<c1990 > Description: v. <1, 3 >: ill., maps; 24 cm.

Gulf of Mexico offshore operations monitoring experiment: final report: phase 1: sublethal responses to contaminant exposure / editor, Mahlon C. Kennicut II. Published/Created: New Orleans (1201 Elmwood Park Blvd., New Orleans 70123-2394): U.S. Dept. of the Interior, Minerals Management Service, Gulf of Mexico OCS Region, [1995] Description: 1 v. (various pagings): ill.; 28 cm.

Handbook for using a waste-reduction approach to meet aquatic toxicity limits / edited by Robert E. Holman and Jeri Gray. Published/Created: Raleigh, NC: Pollution Prevention Program of the North Carolina Dept. of Environment, Health, and Natural Resources, 1991. Description: 1 v. (various pagings): ill.; 28 cm.

Handbook of diagnostic procedures for petroleum-contaminated sites: (RESCOPP Project, EU813) / edited by Paul Lecomte and Claudio Mariotti. Published/Created: Chichester; New York: John Wiley, c1997. Description: xv, 192 p.: ill., maps; 24 cm. ISBN: 0471971081 (acid-free paper) Links: Publisher description http://www.loc.gov/catdir/description/wiley031/96039429.htm l Table of contents http://www.loc.gov/catdir/toc/wiley022/96039429.html

Handbook of groundwater remediation using permeable reactive barriers: applications to radionuclides, trace metals, and nutrients / edited by David L. Naftz ... [et al.]. Published/Created: San Diego, Calif.: Academic Press, c2002. Description: xxv, 539 p.: ill. (1 col.), maps; 24 cm. ISBN: 0125135637 (alk. paper) Links: Publisher description http://www.loc.gov/catdir/description/els031/2002101651.htm l Table of contents http://www.loc.gov/catdir/toc/els031/2002101651.html

Handbook of water resources and pollution control / edited by Harry W. Gehm and Jacob I. Bregman. Published/Created: New York: Van Nostrand Reinhold, c1976. Description: viii, 840 p.: ill.; 27 cm. ISBN: 0442210418

Hanford environmental oversight program: data summary report / Environmental Radiation Program. Published/Created: Olympia, WA: Environmental Radiation Section, Division of Radiation Protection, Dept. of Health Description: v.: ill., maps; 28 cm.

Hawaii's implementation plan for polluted runoff control. Published/Created: Honolulu, Hawaii: Hawaii Office of Planning, Coastal Zone Management Program; Hawaii Dept. of Health, Polluted Runoff Control Program, 2000. Description: 1 v. (various pagings): ill., maps; 28 cm.

Hazardous materials in the hydrologic environment: the role of research by the U. S. Geological Survey / Committee on U. S. Geological Survey Water Resources Research, Water Science and Technology Board, Commission on Geosciences, Environment, and Resources. Published/Created: Washington, D.C. National Academy Press. 1996. Description: x, 109 p.: ill.; 23 cm.

Hazardous substances in the European marine environment: trends in metals and persistent organic pollutants / Norman Green (task leader) ... [et al.]; EEA project manager, Anita Kuînitzer. Published/Created: Luxembourg: Office for Official Publications of the European Communities; Lanham, Md.: Bernan Associates [distributor], 2003. Description: vii, 75 p.: col. ill., col. maps.; 30 cm. ISBN: 9291676284

The ... annual clean water act corporate counsel retreat and information exchange: course materials / sponsored by the American Bar

Association Section of Natural Resources, Energy, and Environmental Law, Water Quality Committee, [and] Special Committee on Corporate Counsel. Published/Created: Chicago, Ill.: American Bar Association, Related Names: American Bar Association. Water Quality Committee. American Bar Association. Section of Natural Resources, Energy, and Enironmental Law. Special Committee on Corporate Counsel. Description: v.; 28 cm. Current Frequency: Annual Continues: Clean water act corporate counsel retreat and information exchange (DLC) 96640318 Notes: Description based on: 4th (May 2, 1996). SERBIB/SERLOC merged record Subjects: Water--Pollution--Law and legislation--United States. LC Classification: KF3790.Z9 C585 Dewey Class No.: 344.73/046343 21

The 1991 Gulf War: environmental assessments of IUCN and collaborators / A.R.G. Price ... [et al.]. Published/Created: Gland, Switzerland: IUCN in collaboration with WWF, IAEA, and IOC, 1994. Related Names: Price, Andrew, 1950- International Union for Conservation of Nature and Natural Resources. Description: xii, 48 p.: ill.; 30 cm. ISBN: 2831702054 Notes: Includes bibliographical references (p. [43]-45). Subjects: Persian Gulf War, 1991--Environmental aspects. Oil pollution of the sea--Persian Gulf Region. Oil pollution of the sea--Arabian Peninsula. Oil spills--Environmental aspects--Persian Gulf Region. Marine pollution--Persian Gulf Region. Fishes--Effect of water pollution on--Persian Gulf Region. Shrimp industry--Persian Gulf Region. Series: A marine conservation and development report LC Classification: GC1451 .A14 1994 Dewey Class No.: 363.73/82/0916535 20

The Agricultural industry and its effects on water quality: proceedings of a conference held at the University of Waikato,

Hamilton, New Zealand, 15-18 May 1979 / executive editor, S.H. Jenkins; sponsors, the New Zealand Committee for Water Pollution Research, the Royal Society of New Zealand, the Water Resources Council. Published/Created: Oxford; New York: Pergamon Press, 1979. Related Names: Jenkins, S. H. (Samuel Harry) New Zealand Committee for Water Pollution Research. Royal Society of New Zealand. Water Resources Council (N.Z.) Description: 727 p.: ill.; 25 cm. ISBN: 0080248896 (pbk.) Notes: Includes bibliographies and indexes. Subjects: Agricultural pollution--Congresses. Agricultural wastes--Congresses. Water--Pollution--Congresses. Series: Progress in water technology; v. 11, no. 6 0306-6746 LC Classification: TD428.A37 A35 1979 Dewey Class No.: 628.1/684 19 National Bibliography No.: GB***

The Assessment of sublethal effects of pollutants in the sea: a Royal Society discussion: held on 24 and 25 May 1978 / organized by H. A. Cole in collaboration with the Marine Pollution Subcommittee of the British National Committee on Oceanic Research. Published/Created: London: Royal Society; Great Neck, N.Y.: distributed by Scholium International, 1979. Related Names: Cole, Harry A., 1944- Royal Society (Great Britain) British National Committee on Oceanic Research. Marine Pollution Subcommittee. Description: viii, 235 p.: ill.; 31 cm. ISBN: 085403112X Notes: Distributor from label mounted on t.p. Pages also numbered 399-633. "First published in Philosophical transactions of the Royal Society of London, series B, volume 284 (no. 1015)" Includes bibliographies. Subjects: Marine animals--Effect of water pollution on. Marine plankton--Effect of water pollution on--Congresses. Marine pollution--Physiological effect--Congresses. LC Classification: QH545.W3 A85 Dewey

Class No.: 574.5/2636 National
Bibliography No.: GB***

The Brahmaputra Basin [microform]: basin
    sub-basin inventory of water pollution.
    Published/Created: New Delhi: Central
    Board for the Prevention and Control of
    Water Pollution, 1984- Description: v. <1-
    >: ill., maps; 28 cm. Summary: Study of
    water pollution in Assam. Incomplete
    Contents: pt. 1. The Dilli-Disang sub-
    basin. Notes: Master microform held by:
    DLC. Additional Formats: Microfiche.
    Washington, D.C.: Library of Congress
    Photoduplication Service, 1985- .
    microfiche <1- >; 11 x 15 cm. Series:
    Assessment and development study of
    river basin series; ADSORBS/11/1983-84
    LC Classification: Microfiche 85/60232
    (T) Overseas Acquisitions No.: I E 49661
    Geographic Area Code: a-ii---

The Brahmini-Baitarani Basin: basin sub-basin
    inventory of water pollution.
    Published/Created: New Delhi: Central
    Pollution Control Board, [1988?] Related
    Names: India. Central Pollution Control
    Board. Description: 114 p.: ill.; 28 cm.
    Notes: Brahmani-Baitarani river system
    drains the area lying between the
    Subarnarekha and Mahanadi basins in
    eastern India. Subjects: Water--Pollution--
    India--Brahmani River Watershed (Orissa)
    Water--Pollution--India--Baitarani River
    Watershed. Water quality management--
    India--Brahmani River Watershed (Orissa)
    Water quality management--India--
    Baitarani River Watershed. Series:
    Assessment and development study of
    river basin series; ADSORBS/19/1988-89
    LC Classification: TD304.B73 B73 1988
    Overseas Acquisitions No.: I E 60299
    Geographic Area Code: a-ii---

The Bridge over troubled water (Motion
    picture) Published/Created: [n.p.] Rob
    Mortarotti. Released by Eastman Kodak
    Co., 1970. Related Names: Mortarotti,
    Rob. [from old catalog] Eastman Kodak

Company. [from old catalog] Description:
    p. cm. Subjects: Water--Pollution--
    California--San Francisco Bay.

The Clean Water Act as amended by the Water
    Quality Act of 1987. Published/Created:
    Washington, D.C.: U.S. G.P.O.: For sale
    by the Supt. of Docs., Congressional Sales
    Office, U.S. G.P.O., 1988. Related Names:
    United States. Congress. Senate.
    Committee on Environment and Public
    Works. Related Titles: Clean Water Act.
    Description: iv, 214 p.; 24 cm. Notes: At
    head of title: 100th Congress, 2d session.
    Committee print. "Printed for the use of the
    Senate Committee on Environment and
    Public Works." Distributed to some
    depository libraries in microfiche.
    Shipping list no.: 88-195-P. "March 1988."
    Item 1045-A, 1045-B (microfiche)
    Includes bibliographical references.
    Subjects: Water--Pollution--Law and
    legislation--United States. Federal aid to
    water quality management--United States.
    Series: S. prt.; 100-91 LC Classification:
    KF3786.A55 C55 1988 NAL Class No.:
    KF3790.Z9C5 Dewey Class No.:
    344.73/046343/02632 347.3044634302632
    20 Government Document No.: Y 4.P
    96/10:S.prt.100-91 Geographic Area Code:
    n-us---

The Clean Water Act handbook / Mark A.
    Ryan, editor. Edition Information: 2nd ed.
    Published/Created: Chicago, Ill.: Section of
    Environment, Energy, and Resources,
    American Bar Association, c2003. Related
    Names: Ryan, Mark, 1957- American Bar
    Association. Section of Environment,
    Energy, and Resources. Related Titles:
    Clean Water Act handbook. Description:
    xxvi, 308 p.; 23 cm. ISBN: 1590312171
    Contents: Overview of the Clean Water
    Act / Theodore L. Garrett -- Water
    pollution control under the National
    Pollutant Discharge Elimination System /
    Karen M. McGaffey -- NPDES permit
    application and issuance procedures /

Randy Hill -- Publicly Owned Treatment Works (POTWs) / Alexandra Dapolito Dunn -- Pretreatment and indirect dischargers / Corinne A. Goldstein -- Wetlands:.Section 404 / Sylvia Quast, Steven T. Miano -- Oil and hazardous substance spills: Section 311 / David G. Dickman -- "Wet Weather" regulations / Randy Hill, David Allnutt -- Nonpoint source pollution control / Edward B. Witte, David P. Ross -- TMDLs: Section 303(d) / Laurie K. Beale, Karin Sheldon -- Enforcement: Section 309 / Beth S. Ginsberg, Jennifer E. Merrick -- Judicial review: Section 509 / Karen M. McGaffey. Notes: Rev. ed. of: The Clean Water Act handbook / Parthenia B. Evans, editor. Includes index. Subjects: United States. Federal Water Pollution Control Act. Water--Pollution--Law and legislation--United States. LC Classification: KF3790 .C545 2003 Dewey Class No.: 344.73/046343 21 Geographic Area Code: n-us--- Electronic File Information: Table of contents http://www.loc.gov/catdir/toc/ecip043/200 3010677.html

The Clean Water Act handbook / Parthenia B. Evans, editor. Published/Created: Chicago, Ill.: SONREEL, Section of Natural Resources, Energy and Environmental Law, American Bar Association, c1994. Related Names: Evans, Parthenia B. American Bar Association. Section of Natural Resources, Energy, and Environmental Law. Description: xxx, 282 p.; 23 cm. ISBN: 157073030X Notes: Includes bibliographical references and index. Subjects: United States. Federal Water Pollution Control Act. Water--Pollution--Law and legislation--United States. LC Classification: KF3790 .C545 1994 Dewey Class No.: 344.73/046343 347.30446343 20

The Clean water act showing changes made by the 1977 amendments. Published/Created:

Washington: U.S. Govt. Print. Off., 1977 [i.e. 1978] Related Names: United States. Congress. Senate. Committee on Environment and Public Works. United States. Law, statutes, etc. Federal water pollution control act. 1978. Description: iv, 125 p.; 24 cm. Notes: At head of title: 95th Congress, 1st session. Committee print. "Serial no. 95-12." "Printed for the use of the Senate Committee on Environment and Public Works." Subjects: Water--Pollution--Law and legislation--United States. LC Classification: KF3786.A55 C6 Dewey Class No.: 344/.73/0463 Geographic Area Code: n-us---

The Clean Water Act: 25th anniversary edition / Water Environment Federation. Cover Title: Clean Water Act, updated for 1997 Published/Created: Alexandria, VA (601 Wythe St., Alexandria 22314-1994): The Federation, 1997. Related Names: Water Environment Federation. Description: xxiii, 423 p.; 26 cm. Subjects: Water--Pollution--Law and legislation--United States. Water quality--United States. LC Classification: KF3787.122 .A2 1997 Dewey Class No.: 344.73/046343 21 Geographic Area Code: n-us---

The Clean Water Act: as amended by the Water Quality Act of 1987 / Michael A. Brown, chairman. Published/Created: New York, N.Y. (810 Seventh Ave., New York 10019): Practising Law Institute, 1987. Related Names: Brown, Michael A. (Michael Arthur), 1938- Practising Law Institute. Description: 376 p.: ill.; 22 cm. Notes: "Prepared for distribution at the The Clean Water Act as amended by the Water Quality Act of 1987 program, May-June 1987"--P. 5. "C4-4178." Subjects: Water--Pollution--Law and legislation--United States. Water quality--United States. Series: Litigation and administrative practice series Litigation course handbook series; no. 144 LC Classification: KF3790.Z9 C58 1987 Dewey Class No.:

344.73/046343 347.30446343 19
Geographic Area Code: n-us---

The Clean Water Act: new directions: a satellite
seminar / cosponsored by the Water
Environment Federation: and the American
Bar Association section of Natural
Resources, Energy, and Environmental
Law; and the Center for Continuing Legal
Education, in cooperation with U. S.
Environmental Protection Agency.
Published/Created: [Chicago, IL]: the
American Bar Association, 1996. Related
Names: American Bar Association. Section
on Natural Resources, Energy, and
Environmental Law. Water Environment
Federation. Center for Continuing Legal
Education (American Bar Association)
Description: [iii], 317 p.: ill., maps; 28 cm.
Notes: "A four hour ABA satelite seminar,
Broadcast live- to over 73 locations." --
Cover. "January 18, 1996." Panels discuss
watershed management, effluent trading,
NPDES permits, wet weather problems,
and enforcement issues. Subjects: United
States. Federal Water Pollution Control
Act. Water--Pollution--Law and
legislation--United States. Water quality
management--United States. LC
Classification: KF3790.Z9 C588 1996

The Clean Water Act: the next steps in new
directions: May 29, 1997, an ABA satellite
seminar / co-sponsored by the American
Bar Association, Section of Natural
Resources, Energy, and Environmental
Law ... [et al.]. Portion of Title: ABA
satellite seminar Published/Created:
[Chicago]: American Bar Association,
c1997. Related Names: American Bar
Association. Section of Natural Resources,
Energy, and Environmental Law.
Description: 218 p.; 28 cm. Subjects:
Water--Pollution--Law and legislation--
United States. LC Classification:
KF3790.Z9 C589 1997 Dewey Class No.:
344.73/046343 21 Geographic Area Code:
n-us---

The clean water and drinking water
infrastructure gap analysis / United States
Environmental Protection Agency, Office
of Water. Published/Created: [Washington
D.C.]: U.S. Environmental Protection
Agency, 2002. Related Names: United
States. Environmental Protection Agency.
Office of Water. Description: 50 p.: ill.; 28
cm. Notes: "September 2002." "EPA 816-
R-02-020." "2002--The Year of Clean
Water--Celebration & Recommitment"--
Cover. Includes bibliographical references.
Subjects: Water--Pollution--United States.
Water quality management--Economic
aspects--United States. Infrastructure
(Economics)--United States. LC
Classification: HC110.W32 G55 2002
Dewey Class No.: 333.91/22 21 Electronic
File Information: Table of contents
http://www.loc.gov/catdir/toc/fy036/20024
85419.html

The Condition of Illinois water resources /
Illinois Environmental Protection Agency,
Bureau of Water. Portion of Title: Illinois
water resources Published/Created:
Springfield, Ill.: Illinois Environmental
Protection Agency, Bureau of Water,
1995- Related Names: Illinois. Bureau of
Water. Illinois. Environmental Protection
Agency. Description: v.: ill., maps; 28 cm.
1972-1994- Current Frequency: Annual
Notes: "IEPA/BOW." Title from cover.
Subjects: Water quality--Illinois--
Periodicals. Water--Pollution--Illinois--
Periodicals. LC Classification: TD224.I3
C66

The Connecticut River ecological study (1965-
1973) revisited: ecology of the lower
Connecticut River 1973-2003 / edited by
Paul M. Jacobson ... [et al.].
Published/Created: Bethesda, Md.:
American Fisheries Society, 2004. Related
Names: Jacobson, Paul M. Description:
xxiv, 545 p.: ill., maps (some col.); 25 cm.
ISBN: 1888569662 (alk. paper) Notes:
Includes bibliographical references.

Subjects: Fishes--Effect of water pollution on--Connecticut River. Nuclear power plants--Environmental aspects--Connecticut River. Thermal pollution of rivers, lakes, etc.--Connecticut River. American shad. Series: American Fisheries Society monograph, 0362-1715;9 Monograph (American Fisheries Society); no. 9. LC Classification: SH177.T45 C662 2004 Dewey Class No.: 571.9/517/09746 22 Geographic Area Code: n-us-ct

The Connecticut River ecological study: the impact of a nuclear power plant / Daniel Merriman and Lyle M. Thorpe, editors. Published/Created: Washington: American Fisheries Society, 1976. Related Names: Merriman, Daniel. Thorpe, Lyle M. Description: xi, 252 p.: ill.; 26 cm. Notes: Includes bibliographies. Subjects: Connecticut Yankee Atomic Power Co. Fishes--Effect of water pollution on--Connecticut River. Nuclear power plants--Environmental aspects--Connecticut River. Thermal pollution of rivers, lakes, etc.--Connecticut River. American shad. Series: Monograph - American Fisheries Society; no. 1 Monograph (American Fisheries Society); no. 1. LC Classification: SH177.T45 C66 Dewey Class No.: 597/.5/04240974 Geographic Area Code: n-us-ct

The Control of oil pollution on the sea and inland waters: the effect of oil spills on the marine environment and methods of dealing with them / edited by J. Wardley-Smith. Published/Created: London: Graham and Trotman, 1976. Related Names: Wardley-Smith, J. Description: 251 p.: ill., maps; 24 cm. ISBN: 0860100219: Notes: Includes bibliographical references and index. Subjects: Oil spills. Oil pollution of water. LC Classification: TD427.P4 C66 Dewey Class No.: 363.6 National Bibliography No.: GB77-01573

The Cost of compliance with the proposed federal drinking water standards for radionuclides: final report / prepared by RCG/Hagler, Bailly, Inc. with assistance from Kennedy/Jenks Consultants, S.M. Stoller Corporation, Bruce Thompson. Published/Created: Denver, CO: Water Industry Technical Action Fund, American Water Works Association, c1992. Related Names: RCG/Hagler, Bailly, Inc. American Water Works Association. Water Industry Technical Action Panel. Related Titles: Cost of compliance--radionuclides. Description: 1 v. (various pagings): ill.; 28 cm. ISBN: 089867610X Notes: Spine title: Cost of compliance--radionuclides. "Prepared for American Water Works Association, Water Industry Technical Action Panel." "October 10, 1991." Includes bibliographical references. Subjects: Drinking water--Law and legislation--Compliance costs --United States. Radioactive pollution of water--United States--Costs. LC Classification: HC103 .C77 1992 Dewey Class No.: 628.1/685 20 Geographic Area Code: n-us---

The Costs of no wellhead protection in Maine: a study of the costs of cure vs. prevention / by Emery & Garrett Groundwater, Inc.; Peter Garrett, Project manager. Published/Created: [Augusta, Me.: Me. Bureau of Health], 1993. Related Names: Garrett, Peter. Emery & Garrett Groundwater. Drinking Water Program (Maine). Description: 2 v.: ill., maps; 28 cm. Notes: Cover title. "For Drinking Water Program, Bureau of Health, Maine Department of Human Services, Augusta, Maine." "November 1, 1993." Vol. 2 has special title: Appendices. Subjects: Groundwater--Pollution--Economic aspects--Maine. Water quality management--Economic aspects--Maine. Wellhead protection--Maine. LC Classification: HC107.M23 W323 1993

The decline of fisheries resources in New England: evaluating the impact of overfishing, contamination, and habitat degradation / edited by Robert Buchsbaum, Judith Pederson, William E. Robinson. Published/Created: Cambridge, Mass.: Massachusetts Institute of Technology, MIT Sea Grant College Program, c2005. Related Names: Buchsbaum, Robert. Pederson, Judith. Robinson, William E. Description: viii, 175 p.: ill., maps; 28 cm. Notes: Includes bibliographical references. Additional Formats: Also available online in PDF format. Subjects: Fishery management--New England. Fishery resources--New England. Fishery resources--Effect of water pollution on--New England. Fishery resources--Effect of human beings on--New England. Overfishing--New England. Series: MIT Sea Grant College Program publication; no. MITSG 05-5 MIT Sea Grant College Program report; no. MITSG 05-5. LC Classification: SH328 .D43 2005 Dewey Class No.: 333.95/61370974 22

The Delaware River drainage basin: report, recommended classifications and assignments of standards of quality and purity for designated waters of New York State. Published/Created: [Albany: Water Pollution Control Board, 1960] Related Names: New York (State). Water Pollution Control Board. Description: 305 p.: ill., maps, diagrs., tables; 28 cm. Subjects: Water--Pollution--Delaware River Watershed (N.Y.-Del. and N.J.) Water--Standards. Water resources development--New York (State) LC Classification: TD223 .N4

The Determination of organochlorine insecticides and polychlorinated biphenyls in sewages, sludges, muds, and fish, 1978. Organochlorine insecticides and polychlorinated biphenyls in water, an addition, 1984. Published/Created: London: H.M.S.O., 1985. Related Names:

Great Britain. Standing Committee of Analysts. Related Titles: Organochlorine insecticides and polychlorinated biphenyls in water. Determination of organochlorine insecticides and polychlorinated biphenyls in sewages, sludges, muds, and fish, 1978. Organochlorine insecticides and polychlorinated biphenyls in water, an addition, 1984. Description: 32 p.: ill.; 30 cm. ISBN: 0117517771 (pbk.): Notes: Cover title. Produced under the auspices of the Standing Committee of Analysts. Bibliography: p. 31. Subjects: Sewage--Analysis. Water chemistry. Insecticides--Analysis. Polychlorinated biphenyls--Analysis. Organochlorine compounds--Analysis. Fishes--Effect of water pollution on. Series: Methods for the examination of waters and associated materials LC Classification: TD735 .D48 1985

The Development of an automated biological monitoring system for water quality / John Cairns, Jr. ... [et al.]. Published/Created: Blacksburg: Virginia Water Resources Research Center, Virginia Polytechnic Institute and State University, 1973. Related Names: Cairns, John, 1923- Description: vi, 50 p.: ill.; 23 cm. Notes: Bibliography: p. 47-50. Subjects: Water--Pollution--Measurement. Indicators (Biology) Fishes--Effect of water pollution on. Series: Bulletin - Virginia Water Resources Research Center, Virginia Polytechnic Institute and State University; 59 Bulletin (Virginia Water Resources Research Center); 59. LC Classification: TD201 .V57 no. 59 TD423 Dewey Class No.: 551.4/8/08 s 628.1/68

The economics of a disaster: the Exxon Valdez oil spill / Bruce M. Owen ... [et al.]. Published/Created: Westport, Conn.: Quorum Books, 1995. Related Names: Owen, Bruce M. Description: xii, 200 p.: ill., map; 25 cm. ISBN: 0899309879 Notes: Includes bibliographical references (p. [195]-196) and index. Subjects: Exxon

Valdez (Ship) Water--Pollution--Economic aspects--Alaska--Prince William Sound. Oil spills--Economic aspects--Alaska--Prince William Sound. Tankers--Accidents. LC Classification: HC107.A47 W323 1995 Dewey Class No.: 338.3 20 Geographic Area Code: n-us-ak

The economics of clean water.
Published/Created: Washington, D.C.: U.S. Dept. of the Interior, Federal Water Pollution Control Administration: For sale by the Supt. of Docs., U.S. G.P.O., 1970- Related Names: United States. Federal Water Pollution Control Administration. United States. Environmental Protection Agency. Description: v.; 26 cm. Issued in parts. [3rd (1970)]- Current Frequency: Annual Continues: Cost of clean water and its economic impact (DLC)sn 87030151 (OCoLC)16850538 Merger of: United States. Environmental Protection Agency. Cost of clean air (1973) (DLC) 00236341 (OCoLC)5435224 United States. Environmental Protection Agency. Cost of clean air and clean water 0275-0384 (DLC) 80646706 (OCoLC)6045831 Cancelled/Invalid LCCN: sn 86020599 Notes: "Annual report of the administrator of the Environmental Protection Agency to the Congress of the United States." Vols. for issued in the Congressional Series as Senate documents. Vol. for 1970 issued by the Federal Water Pollution Control Administration; vol. for <1972>- issued by the U.S. Environmental Protection Agency. SERBIB/SERLOC merged record Continues: Cost of clean water and its economic impact. Merged with: United States. Environmental Protection Agency. Cost of clean air, to form: United States. Environmental Protection Agency. Cost of clean air and clean water. Subjects: Water--Pollution--United States--Periodicals. Pollution--Economic aspects--United States--Periodicals. LC Classification: HC110.P55 A24 Dewey Class No.: 333.9/1

The effect of turfgrass maintenance on surface-water quality in a suburban watershed, Inner Blue Grass, Kentucky / R. Michael Williams ... [et al.]. Published/Created: Lexington, Ky.: Kentucky Geological Survey, University of Kentucky, 2000. Related Names: Williams, R. Michael. Description: iv, 10 p.: ill., map; 28 cm. Notes: Includes bibliographical references (p. 8). Subjects: Turf management--Environmental aspects--Kentucky--Sinking Creek Region (Jessamine County) Water--Pollution--Kentucky--Sinking Creek Region (Jessamine County) Series: Report of investigations, 0075-5591; Series XII, 2000 Report of investigations (Kentucky Geological Survey); ser. XII, 5. LC Classification: SB433 .E37 2000 Dewey Class No.: 635.9/642/09769483 22

The Effects of stress and pollution on marine animals / by B.L. Bayne ... [et al.]. Published/Created: New York: Praeger, c1985. Related Names: Bayne, B. L. (Brian Leicester) Description: p. cm. ISBN: 0030570190 (alk. paper): Notes: Cataloging based on CIP information. Includes index. Bibliography: p. Subjects: Marine animals--Effect of water pollution on. Marine animals--Physiology. Stress (Physiology) LC Classification: QL121 .E34 1985 Dewey Class No.: 591.5/2636 19

The Effects of stress and pollution on marine animals / by B.L. Bayne ... [et al.]. Published/Created: New York: Praeger, c1985. Related Names: Bayne, B. L. (Brian Leicester) Description: p. cm. ISBN: 0030570190 (alk. paper): Notes: Cataloging based on CIP information. Includes index. Bibliography: p. Subjects: Marine animals--Effect of water pollution on. Marine animals--Physiology. Stress (Physiology) LC Classification: QL121 .E34 1985 Dewey Class No.: 591.5/2636 19

The Effects of uranium mill wastes on stream
biota [by] William F. Sigler [and others]
Published/Created: Logan, Utah
Agricultural Experiment Station, Utah
State University, 1966. Related Names:
Sigler, William F. Description: 76 p. illus.
(part col.) 23 cm. Notes: Cover title.
Bibliography: p. 41-45. Subjects: Uranium
mines and mining--Environmental aspects-
-Utah. Fishes--Effect of water pollution on.
Stream ecology--Utah. Radioactive waste
disposal in rivers, lakes, etc.--Utah. Series:
Utah Agricultural Experiment Station.
Bulletin 462 Bulletin (Utah Agricultural
Experiment Station); 462. LC
Classification: QH545.U7 E33 Dewey
Class No.: 574.5/26323 19 Geographic
Area Code: n-us-ut

The Effects of uranium mill wastes on stream
biota [by] William F. Sigler [and others]
Published/Created: Logan, Utah
Agricultural Experiment Station, Utah
State University, 1966. Related Names:
Sigler, William F. Description: 76 p. illus.
(part col.) 23 cm. Notes: Cover title.
Bibliography: p. 41-45. Subjects: Uranium
mines and mining--Environmental aspects-
-Utah. Fishes--Effect of water pollution on.
Stream ecology--Utah. Radioactive waste
disposal in rivers, lakes, etc.--Utah. Series:
Utah Agricultural Experiment Station.
Bulletin 462 Bulletin (Utah Agricultural
Experiment Station); 462. LC
Classification: QH545.U7 E33 Dewey
Class No.: 574.5/26323 19 Geographic
Area Code: n-us-ut

The Elizabeth River: an environmental
perspective / prepared by Virginia State
Water Control Board, Bureau of
Surveillance and Field Studies [and]
Chesapeake Bay Program.
Published/Created: [Richmond]: The
Bureau, [1983] Related Names: Virginia.
State Water Control Board. Bureau of
Surveillance and Field Studies.
Chesapeake Bay Program (U.S.)

Description: v, 65 leaves: ill.; 28 cm.
Notes: Bibliography: leaves 55-57.
Subjects: Water--Pollution--Environmental
aspects--Virginia --Elizabeth River
Watershed. Water quality bioassay--
Virginia--Elizabeth River Watershed.
Series: Basic data bulletin; 61 (Sept. 1983)
Basic data bulletin (Richmond, Va.); 61.
LC Classification: QH105.V8 E45 1983
Dewey Class No.: 628.1/68/09755 20
Geographic Area Code: n-us-va

The Elizabeth River: an environmental
perspective / prepared by Virginia State
Water Control Board, Bureau of
Surveillance and Field Studies [and]
Chesapeake Bay Program.
Published/Created: [Richmond]: The
Bureau, [1983] Related Names: Virginia.
State Water Control Board. Bureau of
Surveillance and Field Studies.
Chesapeake Bay Program (U.S.)
Description: v, 65 leaves: ill.; 28 cm.
Notes: Bibliography: leaves 55-57.
Subjects: Water--Pollution--Environmental
aspects--Virginia --Elizabeth River
Watershed. Water quality bioassay--
Virginia--Elizabeth River Watershed.
Series: Basic data bulletin; 61 (Sept. 1983)
Basic data bulletin (Richmond, Va.); 61.
LC Classification: QH105.V8 E45 1983
Dewey Class No.: 628.1/68/09755 20
Geographic Area Code: n-us-va

The Enduring Great Lakes / edited by John
Rousmaniere. Edition Information: 1st ed.
Published/Created: New York: Norton,
c1979. Related Names: Rousmaniere,
John. Related Titles: Natural history.
Description: x, 112 p.: ill. (some col.); 24
cm. ISBN: 0393011941: Notes: "A Natural
history book." "Adapted from an issue of
Natural history magazine." Bibliography:
p. 109-112. Subjects: Freshwater ecology--
Great Lakes (North America) Water--
Pollution--Great Lakes (North America)
Great Lakes (North America) LC
Classification: QH104.5.G7 E52 1979

Dewey Class No.: 574.5/2632 Geographic Area Code: nl-----

The Enduring Great Lakes / edited by John Rousmaniere. Edition Information: 1st ed. Published/Created: New York: Norton, c1979. Related Names: Rousmaniere, John. Related Titles: Natural history. Description: x, 112 p.: ill. (some col.); 24 cm. ISBN: 0393011941: Notes: "A Natural history book." "Adapted from an issue of Natural history magazine." Bibliography: p. 109-112. Subjects: Freshwater ecology--Great Lakes (North America) Water--Pollution--Great Lakes (North America) Great Lakes (North America) LC Classification: QH104.5.G7 E52 1979 Dewey Class No.: 574.5/2632 Geographic Area Code: nl-----

The environmental chemistry of aluminum / edited by Garrison Sposito. Edition Information: 2nd ed. Published/Created: Baca Raton, FL: Lewis Publishers, c1996. Related Names: Sposito, Garrison, 1939- Description: 464 p.: ill.; 25 cm. ISBN: 1566700302 (alk. paper) Notes: Includes bibliographical references and index. Subjects: Aluminum--Environmental aspects. Water--Pollution. Soil pollution. LC Classification: TD427.A45 E58 1995 Dewey Class No.: 628.5/2 20 Electronic File Information: Publisher description http://www.loc.gov/catdir/enhancements/fy 0731/95020082-d .html

The environmental chemistry of aluminum / edited by Garrison Sposito. Edition Information: 2nd ed. Published/Created: Baca Raton, FL: Lewis Publishers, c1996. Related Names: Sposito, Garrison, 1939- Description: 464 p.: ill.; 25 cm. ISBN: 1566700302 (alk. paper) LC Classification: TD427.A45 E58 1995 Dewey Class No.: 628.5/2 20 Electronic File Information: Publisher description http://www.loc.gov/catdir/enhancements/fy 0731/95020082-d .html

The Environmental crisis--opposing viewpoints / Neal Bernards, book editor. Published/Created: San Diego, CA: Greenhaven Press, c1991. Related Names: Bernards, Neal, 1963- Description: 288 p.: ill.; 23 cm. ISBN: 0899081754 (lib. bdg.) 0899081509 (pbk.) Summary: Presents opposing views on questions of environmental protection and damage resulting from air and water pollution, toxic wastes, pesticides, and the ever-growing tide of refuse. LC Classification: HC110.E5 E49835 1991 Dewey Class No.: 363.7 20 Geographic Area Code: n-us---

The Environmental crisis--opposing viewpoints / Neal Bernards, book editor. Published/Created: San Diego, CA: Greenhaven Press, c1991. Related Names: Bernards, Neal, 1963- Description: 288 p.: ill.; 23 cm. ISBN: 0899081754 (lib. bdg.) 0899081509 (pbk.) Summary: Presents opposing views on questions of environmental protection and damage resulting from air and water pollution, toxic wastes, pesticides, and the ever-growing tide of refuse. Notes: Includes bibliographical references and index. Subjects: Environmental policy--United States. Environmental protection. Series: Opposing viewpoints series Opposing viewpoints series (Unnumbered) LC Classification: HC110.E5 E49835 1991 Dewey Class No.: 363.7 20 Geographic Area Code: n-us---

The Great Lakes: causes of pollution [Filmstrip] Published/Created: [n.p.] Life, 1970. Made by Time-Life Photo Lab. Description: p. 73 fr. color. 35 mm.

The Great Lakes: results of pollution [Filmstrip] Published/Created: Life, 1970. Made by Time-Life Photo Lab. Description: p. 80 fr. color. 35 mm.

INDEX

M

N

O

P

pharmaceuticals, 4, 249
phosphate, 143
phosphorous, 4
phosphorus, 54, 55, 62, 63, 76, 142, 163
pigs, 71
pilot study, 154, 219
plankton, 257
planning, 30, 60, 66, 68, 70, 129, 131
plants, 5, 56, 91, 97, 98, 151, 152, 177, 194, 202, 223, 226, 239, 240
plastics, vii, 1, 5, 6, 7, 14, 19
PM, 134
Poland, 198
policy choice, xi, 90, 101
policymakers, xi, 89, 90
pollutants, ix, x, 2, 4, 5, 9, 16, 20, 51, 54, 55, 56, 57, 67, 69, 75, 76, 77, 78, 79, 80, 82, 84, 86, 149, 151, 152, 158, 163, 167, 176, 192, 197, 208, 254, 256, 257
pollution, vii, viii, xii, 1, 2, 3, 6, 7, 13, 14, 19, 20, 21, 22, 26, 38, 56, 57, 60, 62, 67, 68, 70, 76, 77, 82, 86, 93, 116, 119, 127, 128, 129, 130, 131, 132, 133, 135, 136, 137, 138, 139, 141, 142, 143, 144, 145, 146, 147, 148, 149, 150, 151, 152, 153, 154, 155, 156, 157, 158, 162, 163, 164, 165, 166, 167, 168, 169, 170, 171, 172, 173, 174, 175, 176, 177, 178, 179, 180, 181, 182, 183, 186, 188, 189, 190, 191, 192, 193, 194, 195, 196, 197, 198, 199, 200, 201, 202, 203, 204, 205, 206, 207, 208, 209, 210, 211, 212, 213, 214, 215, 216, 217, 218, 220, 222, 223, 226, 227, 231, 232, 233, 235, 236, 237, 238, 239, 240, 241, 243, 244, 245, 246, 247, 248, 249, 251, 252, 253, 255, 256, 257, 258, 259, 261, 262, 263, 264, 265
poor, 8, 9, 26, 95, 123, 233, 234
population, x, xi, 23, 26, 29, 89, 90, 91, 92, 94, 95, 96, 97, 98, 99, 100, 101, 102, 109, 113, 115, 120, 141, 189, 194
population growth, 98, 141
ports, 2, 4, 5, 6, 9, 13, 21
Potomac River, 162, 190
poultry, 53, 54, 56, 57, 59, 60, 61, 62, 64, 66, 72, 77, 79, 84, 86
power, 5, 202, 238, 239, 240, 261
power plants, 238, 239, 240, 261
precipitation, 79, 82, 83, 86, 87, 142, 155, 205
prediction, 139
preference, xi, 89
preservative, 103
president, 232, 233
President Clinton, 14
pressure, 39, 43

prevention, viii, xii, 2, 7, 21, 34, 35, 42, 70, 116, 119, 120, 123, 124, 132, 139, 190, 191, 199, 200, 219, 220, 224, 229, 245, 252, 261
prices, viii, 25, 39, 43
probability, 87
producers, 39, 40, 41, 43, 61, 67, 72, 84
production, viii, 26, 39, 40, 42, 43, 52, 53, 54, 57, 62, 63, 66, 67, 73, 78, 79, 82, 83, 113, 115, 117, 186, 200, 213, 214, 215, 216, 226
productivity, 220, 223
prognosis, 205
program, viii, xii, 8, 9, 12, 15, 23, 25, 26, 27, 29, 32, 34, 35, 41, 42, 45, 46, 56, 58, 60, 67, 68, 70, 72, 83, 84, 90, 91, 98, 99, 100, 101, 108, 119, 120, 121, 122, 123, 124, 125, 130, 141, 147, 153, 191, 207, 214, 242, 253, 256, 259
promote, vii, 1, 2, 4, 18
protocol, 13, 253
protocols, vii, 1, 2, 6, 19
protozoa, 55, 203
public administration, 198
public health, vii, 1, 4, 5, 11, 16, 30, 31, 32, 66, 67, 84, 97, 106, 113, 139, 255
public policy, 233
public service, 16
Public Works Committee, xi, 90, 92, 97, 100
Puerto Rico, 102
purification, 128, 129, 236, 245

Q

quality assurance, 179, 196
quality research, 165

R

radiation, 141, 238, 239, 240
Radiation, 180, 201, 208, 238, 255, 256
radio, 230, 231, 232, 233
radionuclides, 180, 199, 208, 256, 261
radon, 201, 207, 252
rain, 82
rainfall, 56, 58, 62, 78, 82, 87
range, ix, 4, 9, 10, 33, 35, 39, 41, 46, 51, 55, 76, 107, 114
rangeland, 54, 86
reasoning, 83
reception, 5
recognition, 85
reconcile, 79
reconstruction, 208
recovery, 35, 120, 142

S

T